Generalized Lie Theory in Mathematics, Physics and Beyond

Sergei Silvestrov · Eugen Paal
Viktor Abramov · Alexander Stolin
Editors

Generalized Lie Theory in Mathematics, Physics and Beyond

Springer

Sergei Silvestrov (Editor in Chief)
Centre for Mathematical Sciences
Division of Mathematics
Lund Institute of Technology
Lund University Box 118
221 00 Lund
Sweden
Sergei.Silvestrov@math.lth.se

Viktor Abramov
Institute of Pure Mathematics
University of Tartu
J. Liivi 2
50409 Tartu
Estonia
Viktor.Abramov@ut.ee

Eugen Paal
Department of Mathematics
Tallinn University of Technology
Ehitajate tee 5
19086 Tallinn Estonia
eugen.paal@ttu.ee

Alexander Stolin
Mathematical Sciences
Chalmers University of Technology
and Göteborg University
Department of Mathematical Sciences
412 96 Göteborg
Sweden
astolin@math.chalmers.se

ISBN: 978-3-540-85331-2 e-ISBN: 978-3-540-85332-9

Library of Congress Control Number: 2008933563

Math.Subject Classification (2000): 17-06, 16-06, 81-06, 17A-XX, 17B-XX, 81R-XX

Cover design: WMX Design GmbH, Heidelberg

Printed on acid-free paper

springer.com

Preface

The aim of this book is to extend the understanding of the fundamental role of generalizations of Lie and related non-commutative and non-associative structures in Mathematics and Physics.

This is a thematic volume devoted to the interplay between several rapidly expanding research fields in contemporary Mathematics and Physics, such as generalizations of the main structures of Lie theory aimed at quantization and discrete and non-commutative extensions of differential calculus and geometry, non-associative structures, actions of groups and semi-groups, non-commutative dynamics, non-commutative geometry and applications in Physics and beyond.

The specific fields covered by this volume include:

- Applications of Lie, non-associative and non-commutative associative structures to generalizations of classical and quantum mechanics and non-linear integrable systems, operadic and group theoretical methods;
- Generalizations and quasi-deformations of Lie algebras such as color and super Lie algebras, quasi-Lie algebras, Hom–Lie algebras, infinite-dimensional Lie algebras of vector fields associated to Riemann surfaces, quasi-Lie algebras of Witt type and their central extensions and deformations important for integrable systems, for conformal field theory and for string theory;
- Non-commutative deformation theory, moduli spaces and interplay with non-commutative geometry, algebraic geometry and commutative algebra, q-deformed differential calculi and extensions of homological methods and structures;
- Crossed product algebras and actions of groups and semi-groups, graded rings and algebras, quantum algebras, twisted generalizations of coalgebras and Hopf algebra structures such as Hom-coalgebras, Hom-Hopf algebras, and super Hopf algebras and their applications to bosonisation, parastatistics, parabosonic and parafermionic algebras, orthoalgebas and root systems in quantum mechanics;
- Commutative subalgebras in non-commutative algebras and their connections to algebraic curves, important for the extension of algebra–geometric methods to discrete versions of non-linear equations and representation theory, Lie and generalized Lie methods for differential equations and geometry.

The volume will stimulate further advances in the topics represented as well as in related directions and applications.

This book will be a source of inspiration for a broad spectrum of researchers and for research students, as the contributions lie at the intersection of the research directions and interests of several large research communities and research groups on modern Mathematics and Physics.

This volume consists of 5 parts comprising 25 chapters, which were contributed by 32 researchers from 12 different countries. All contributions in the volume have been refereed.

Lund, Sweden
June 2008

Sergei Silvestrov
Eugen Paal
Viktor Abramov
Alexander Stolin

Acknowledgements

This volume is an outcome of Baltic-Nordic and international cooperation initiatives by Centre for Mathematical Sciences at Lund University, Department of Mathematics at Chalmers University of Technology and Göteborg University in Sweden and the Departments of Mathematics at Tallinn University of Technology and University of Tartu in Estonia. This cooperation, originally aimed at strengthening contacts between Estonia and Sweden in Mathematics and Physics, has rapidly expanded to Baltic-Nordic and European level resulting by now in several major jointly organized international research events and projects.

Contributions in the volume were reported at the international workshop of the Baltic-Nordic "Algebra, Geometry and Mathematical Physics" Network, that took place at the Centre for Mathematical Sciences in Lund University, Lund, Sweden on October 12–14, 2006. This conference was co-organized by Lund University (Sergei Silvestrov, Chair), Tallinn University of Technology (Eugen Paal), University of Tartu (Viktor Abramov), Chalmers University of Technology and Göteborg University (Alexander Stolin). This workshop was the first such major international conference that was organized in Sweden in the framework of this cooperation. More than fifty researchers from different countries took part and gave talks. This volume was created as a result of this successful thematically focused research event.

In Lund University, the conference and preparation of this volume was organized and supported as an international initiative associated to activities and projects of Noncommutative Geometry and Noncommutative Analysis Seminar lead by Sergei Silvestrov at the Center of Mathematical Sciences. The administrative and logistic support and facilities for the conference provided by the Centre for Mathematical Sciences at Lund University is gratefully acknowledged. This conference and the efforts in preparation of this volume in Lund were also supported from the Swedish side by The Crafoord Foundation, The Swedish Foundation of International Cooperation in Research and Higher Education (STINT), The Royal Swedish Academy of Sciences and The Swedish Research Council.

The co-organizers of the conference Viktor Abramov and Eugen Paal gratefully acknowledge financial support of their research and participation in preparation of this volume and in the work of the AGMF Conference in Lund rendered by the

Estonian Science Foundation under the research grants ETF6206, ETF5634 and ETF6912. Excellent Web support for the conference and for the AGMF network as well has been provided by Astralgo Science in Tallinn.

We are also grateful to Johan Öinert from the Centre for Mathematical Sciences, Lund University for his invaluable technical assistance during the preparation of the volume for publication.

Contents

Contributors

Viktor Abramov Institute of Mathematics, University of Tartu, Liivi 2, Tartu, Estonia, viktor.abramov@ut.ee

Chris Athorne Department of Mathematics, University of Glasgow, Glasgow G12 8QW, Scotland, UK, c.athorne@maths.gla.ac.uk

Čestmír Burdík Department of Mathematics, Faculty of Nuclear Sciences and Physical Engineering, Czech Technical University, Trojanova 13, 120 00 Prague 2, Czech Republic, burdik@kmalpha.fjfi.cvut.cz

G. D'Este Dipartimento di Matematica, Università di Milano, Via Saldini 50, 20133 Milan, Italy, gabriella.deste@mat.unimi.it

C. Daskaloyannis Department of Mathematics, Aristotle University of Thessaloniki, Thessaloniki 54124, Greece, daskalo@math.auth.gr

Marcel de Jeu Mathematical Institute, Leiden University, P.O. Box 9512, 2300 RA Leiden, The Netherlands, mdejeu@math.leidenuniv.nl

Aristophanes Dimakis Department of Financial and Management Engineering, University of the Aegean, 31 Fostini Str., 82100 Chios, Greece, dimakis@aegean.gr

Eivind Eriksen Oslo University College, P.O. Box 4, St Olavs Plass, 0130 Oslo, Norway, eeriksen@hio.no

Helge Glöckner Institut für Mathematik, Universität Paderborn, Warburger Str. 100, 33098 Paderborn, Germany, glockner@math.uni-paderborn.de

Tatjana Gramushnjak Institute of Mathematics and Natural Sciences, Tallinn University, Narva road 25, 10120 Tallinn, Estonia, tatjana@tlu.ee

Trond Stølen Gustavsen BI Norwegian School of Management, 0442 Oslo, Norway, trond.s.gustavsen@bi.no

Lars Hellström Sand 216, 881 91 Sollefteå, Sweden, Lars.Hellstrom@residenset.net

Hilja L. Huru Department of Mathematics and Statistics, University of Tromsø, 9037 Tromsø, Norway, hilja.huru@matnat.uit.no

K. Kanakoglou Department of Physics, Aristotle University of Thessaloniki, Thessaloniki 54124, Greece, kanakoglou@hotmail.com

Pavel Kolesnikov Sobolev Institute of Mathematics, 630090 Novosibirsk, Russia, pavelsk@math.nsc.ru

Daniel Larsson Department of Mathematics, Uppsala University, Box 480, 751 06 Uppsala, Sweden, daniel.larsson@math.uu.se

Abdenacer Makhlouf Laboratoire de Mathématiques, Informatique et Application, Université de Haute Alsace, 4 rue des Frères Lumière, 68093 Mulhouse, France, Abdenacer.Makhlouf@uha.fr

Folkert Müller-Hoissen Max-Planck-Institute for Dynamics and Self-Organization, Bunsenstrasse 10, 37073 Göttingen, Germany, folkert.mueller-hoissen@ds.mpg.de

Ondřej Navrátil Department of Mathematics, Faculty of Transportation Sciences, Czech Technical University, Na Florenci 25, 110 00 Prague, Czech Republic, navratil@fd.cvut.cz

Johan Öinert Centre for Mathematical Sciences, Lund Institute of Technology, Lund University, Box 118, 22100 Lund, Sweden, Johan.Oinert@math.lth.se

Eugen Paal Department of Mathematics, Tallinn University of Technology, Ehitajate tee 5, 19086 Tallinn, Estonia, eugen.paal@ttu.ee

Peeter Puusemp Department of Mathematics, Tallinn University of Technology, Ehitajate tee 5, 19086 Tallinn, Estonia, puusemp@staff.ttu.ee

Maido Rahula Institute of Mathematics, University of Tartu, J. Liivi Str. 2-613, 50409 Tartu, Estonia, maido.rahula@ut.ee

Vitali Retšnoi Institute of Mathematics, University of Tartu, J. Liivi Str. 2-613, 50409 Tartu, Estonia, vitali@ut.ee

Lionel Richard School of Mathematics of the University of Edinburgh and Maxwell Institute for Mathematical Sciences, JCMB – King's Buildings, EH9 3JZ Edinburgh, UK, lionel.richard@ed.ac.uk

Artur E. Ruuge Faculteit Ingenieurswetenschappen, Vrije Universiteit Brussel (VUB), Pleinlaan 2, 1050 Brussel, Belgium, artur.ruuge@ua.ac.be

Martin Schlichenmaier Institute of Mathematics, University of Luxembourg, Campus Limpertsberg, 162 A, Avenue de la Faiencerie, 1511 Luxembourg, Grand-Duchy of Luxembourg, Martin.Schlichenmaier@uni.lu

Gunnar Sigurdsson Department of Theoretical Physics,
School of Engineering Sciences, Royal Institute of Technology,
AlbaNova University Center, 106 91 Stockholm, Sweden, gunnsi@kth.se

Sergei D. Silvestrov Centre for Mathematical Sciences,
Lund University, Box 118, 22100 Lund, Sweden, Sergei.Silvestrov@math.lth.se

Christian Svensson Mathematical Institute, Leiden University, P.O. Box 9512,
2300 RA Leiden, The Netherlands
and
Centre for Mathematical Sciences, Lund University, Box 118, SE-22100 Lund,
Sweden, chriss@math.leidenuniv.nl

Fred Van Oystaeyen Department Wiskunde-Informatica,
Universiteit Antwerpen (UA), Middelheimlaan 1, 2020 Antwerp, Belgium,
fred.vanoystaeyen@ua.ac.be

Jüri Virkepu Department of Mathematics, Tallinn University of Technology,
Ehitajate tee 5, 19086 Tallinn, Estonia, jvirkepu@staff.ttu.ee

Part I
Non-Associative and Non-Commutative Structures for Physics

Chapter 1
Moufang Transformations and Noether Currents

Eugen Paal

Abstract The Noether currents generated by continuous Moufang tranformations are constructed and their equal-time commutators are found. The corresponding charge algebra turns out to be a birepresentation of the tangent Mal'ltsev algebra of an analytic Moufang loop.

1.1 Introduction

The Noether currents generated by the Lie transformation groups are well known and widely exploited in modern field theory and theory of elementary particles. Nevertheless, it may happen that group theoretical formalism of physics is too rigid and one has to extend it beyond the Lie groups and algebras. From this point of view it is interesting to elaborate an extension of the group theoretical methods based on Moufang loops as a minimal nonassociative generalization of the group concept. In particular, the Mal'tsev algebra structure of the quantum chiral gauge theory was established in [1,2].

In this paper, the Noether currents generated by continuous Moufang transformations (Sect. 2) are constructed. The method is based on the generalized Lie–Cartan theorem (Sect. 3). It turns out that the resulting charge algebra is a birepresentation of the tangent Mal'ltsev algebra of an analytic Moufang loop (Sect. 4). Throughout the paper $\mathrm{i} \doteq \sqrt{-1}$.

E. Paal
Department of Mathematics, Tallinn University of Technology, Ehitajate tee 5,
19086 Tallinn, Estonia
e-mail: eugen.paal@ttu.ee

S. Silvestrov et al. (eds.), *Generalized Lie Theory in Mathematics, Physics and Beyond,*

1.2 Moufang Loops and Mal'tsev Algebras

A Moufang loop [3, 4] is a quasigroup G with the unit element $e \in G$ and the Moufang identity

$$(ag)(ha) = a(gh)a, \qquad a,g,h \in G\,.$$

Here the multiplication is denoted by juxtaposition. In general, the multiplication need not be associative: $gh \cdot a \neq g \cdot ha$. Inverse element g^{-1} of g is defined by

$$gg^{-1} = g^{-1}g = e\,.$$

A Moufang loop G is said [5] to be *analytic* if G is also a real analytic manifold and main operations – multiplication and inversion map $g \mapsto g^{-1}$ – are analytic mappings.

As in case of the Lie groups, structure constants c^i_{jk} of an analytic Moufang loop are defined by

$$c^i_{jk} \doteq \left.\frac{\partial^2 (ghg^{-1}h^{-1})^i}{\partial g^j \partial h^k}\right|_{g=h=e} = -c^i_{kj}, \qquad i,j,k = 1,\ldots,r \doteq \dim G\,.$$

Let $T_e(G)$ be the tangent space of G at the unit element $e \in G$. For all $x,y \in T_e(G)$, their (tangent) product $[x,y] \in T_e(G)$ is defined in component form by

$$[x,y]^i \doteq c^i_{jk} x^j y^k = -[y,x]^i, \qquad i = 1,\ldots,r\,.$$

The tangent space $T_e(G)$ being equipped with such an anti-commutative multiplication is called the *tangent algebra* of the analytic Moufang loop G. We shall use notation $\Gamma \doteq \{T_e(G),[\cdot,\cdot]\}$ for the tangent algebra of G.

The tangent algebra of G need not be a Lie algebra. There may exist such a triple $x,y,z \in T_e(G)$ that does not satisfy the Jacobi identity:

$$J(x,y,z) \doteq [x,[y,z]] + [y,[z,x]] + [z,[x,y]] \neq 0\,.$$

Instead, for all $x,y,z \in T_e(G)$ one has a more general *Mal'tsev identity* [5]

$$[J(x,y,z),x] = J(x,y,[x,z])\,.$$

Anti-commutative algebras with this identity are called the *Mal'tsev algebras*. Thus every Lie algebra is a Mal'tsev algebra as well.

1.3 Birepresentations

Consider a pair (S,T) of the maps $g \mapsto S_g$, $g \mapsto T_g$ of a Moufang loop G into GL_n. The pair (S,T) is called a (linear) *birepresentation* of G if the following conditions hold:

- $S_e = T_e = id$,
- $T_g S_g S_h = S_{gh} T_g$,
- $S_g T_g T_h = T_{hg} S_g$.

The birepresentation (S,T) is called *associative*, if the following simultaneous relations hold:

$$S_g S_h = S_{gh}, \quad T_g T_h = T_{hg}, \quad S_g T_h = T_h S_g \qquad \forall g,h \in G.$$

In general, one can consider nonassociative birepresentations even for groups.

The *generators* of a differentiable birepresentation (S,T) of an analytic Moufang loop G are defined as follows:

$$S_j \doteq \left.\frac{\partial S_g}{\partial g^j}\right|_{g=e}, \quad T_j \doteq \left.\frac{\partial T_g}{\partial g^j}\right|_{g=e}, \qquad j = 1,\dots,r.$$

Theorem 1.1 (generalized Lie–Cartan theorem [6, 7]). *The generators of a differentiable birepresentation of an analytic Moufang loop satisfy the commutation relations*

$$\begin{aligned}
[S_j,S_k] &= 2Y_{jk} + \frac{1}{3}c^p_{jk}S_p + \frac{2}{3}c^p_{jk}T_p, \\
[S_j,T_k] &= -Y_{jk} + \frac{1}{3}c^p_{jk}S_p - \frac{1}{3}c^p_{jk}T_p, \\
[T_j,T_k] &= 2Y_{jk} - \frac{2}{3}c^p_{jk}S_p - \frac{1}{3}c^p_{jk}T_p,
\end{aligned}$$

where the Yamaguti operators Y_{jk} obey the relations

$$Y_{jk} + Y_{kj} = 0, \qquad c^p_{jk}Y_{pl} + c^p_{kl}Y_{pj} + c^p_{lj}Y_{pk} = 0,$$

satisfy the reductivity relations

$$\begin{aligned}
\left[Y_{jk},S_n\right] &= d^p_{jkn}S_p, \\
\left[Y_{jk},T_n\right] &= d^p_{jkn}T_p
\end{aligned}$$

and commutation relations

$$[Y_{jk},Y_{ln}] = d^p_{jkl}Y_{pn} + d^p_{jkn}Y_{lp}$$

with the Yamaguti constants

$$6d^p_{jkl} \doteq c^p_{js}c^s_{kl} - c^p_{ks}c^s_{jl} + c^p_{sl}c^s_{jk}.$$

Dimension of this Lie algebra does not exceed $2r + r(r-1)/2$. The Jacobi identities are guaranteed by the defining identities of the Lie [8] and general Lie [9, 10] *triple systems* associated with the tangent Mal'tsev algebra $T_e(G)$ of G.

These commutation relations are known from the theory of alternative algebras [11] as well. This is due to the fact that commutator algebras of the *alternative* algebras turn out to be the Mal'tsev algebras. In a sense, one can also say that the differential of a birepresentation (S,T) of the analytic Moufang loop is a *birepresentation of its tangent Mal'tsev algebra* Γ.

1.4 Moufang–Noether Currents and ETC

Let us now introduce conventional canonical notations. The coordinates of a space–time point x are denoted by x^α $(\alpha = 0,1,\dots,d-1)$, where $x^0 = t$ is the time coordinate and x^i are the spatial coordinates denoted concisely as $\mathbf{x} \doteq (x^1,\dots,x^{d-1})$. The Lagrange density $\mathscr{L}(\psi,\partial\psi)$ is supposed to depend on a system of independent (bosonic or fermionic) fields $\psi^A(x)$ $(A = 1,\dots,n)$ and their derivatives $\partial_\alpha\psi^A \doteq \psi^A_{,\alpha}$. The canonical d-momenta are denoted by

$$p^\alpha_A \doteq \frac{\partial\mathscr{L}}{\partial\psi^A_{,\alpha}}\,.$$

The Moufang–Noether currents are defined in vector (matrix) notations as follows:

$$s^\alpha_j \doteq p^\alpha S_j\psi\,, \qquad t^\alpha_j \doteq p^\alpha T_j\psi$$

and the corresponding Moufang–Noether charges are defined as spatial integrals by

$$\sigma_j(t) \doteq -\mathrm{i}\int s^0_j(x)\mathrm{d}\mathbf{x}\,, \qquad \tau_j(t) \doteq -\mathrm{i}\int t^0_j(x)\mathrm{d}\mathbf{x}\,.$$

By following the canonical prescription assume that the following equal-time commutators (or anti-commutators when u^A are fermionic fields) hold:

$$\begin{aligned}
\left[p^0_A(\mathbf{x},t),\psi^B(\mathbf{y},t)\right] &= -\mathrm{i}\delta^B_A\delta(\mathbf{x}-\mathbf{y})\,,\\
\left[\psi^A(\mathbf{x},t),\psi^B(\mathbf{y},t)\right] &= 0\,,\\
\left[p^0_A(\mathbf{x},t),p^0_B(\mathbf{x},t)\right] &= 0\,.
\end{aligned}$$

As a matter of fact, these equal-time commutators (ETC) do not depend on the associativity property of either G nor (S,T). Nonassociativity hides itself in the structure constants of G and in the commutators $[S_j,T_k]$. Due to this, computation of the ETC of the Noether–Moufang charge densities can be carried out in standard way and nonassociativity reveals itself only in the final step when the commutators $[S_j,T_k]$ are required.

First recall that in the associative case the Noether charge densities obey the ETC

$$\begin{aligned}
\left[s^0_j(\mathbf{x},t),s^0_k(\mathbf{y},t)\right] &= \mathrm{i}c^p_{jk}s^0_p(\mathbf{x},t)\delta(\mathbf{x}-\mathbf{y})\,,\\
\left[t^0_j(\mathbf{x},t),t^0_k(\mathbf{y},t)\right] &= -\mathrm{i}c^p_{jk}t^0_p(\mathbf{x},t)\delta(\mathbf{x}-\mathbf{y})\,,\\
\left[s^0_j(\mathbf{x},t),t^0_k(\mathbf{y},t)\right] &= 0\,.
\end{aligned}$$

It turns out that for non-associative Moufang transformations these ETC are violated minimally. The Moufang–Noether charge density algebra reads

$$\begin{aligned}\left[s^0_j(\mathbf{x},t), s^0_k(\mathbf{y},t)\right] &= \mathrm{i}c^p_{jk}s^0_p(\mathbf{x},t)\delta(\mathbf{x}-\mathbf{y}) - 2\left[s^0_j(\mathbf{x},t), t^0_k(\mathbf{y},t)\right] ,\\ \left[t^0_j(\mathbf{x},t), t^0_k(\mathbf{y},t)\right] &= -\mathrm{i}c^p_{jk}t^0_p(\mathbf{x},t)\delta(\mathbf{x}-\mathbf{y}) - 2\left[s^0_j(\mathbf{x},t), t^0_k(\mathbf{y},t)\right] .\end{aligned}$$

The ETCs

$$[s^0_j(\mathbf{x},t), t^0_k(\mathbf{y},t)] = [t^0_j(\mathbf{y},t), s^0_k(\mathbf{x},t)]$$

represent an associator of an analytic Moufang loop and so may be called the associator as well. Associators of a Moufang loop are not arbitrary but have to fulfil certain constraints [6], the *generalized Lie and Maurer–Cartan equations*. In the present situation the constraints can conveniently be listed by closing the above ETC, which in fact means construction of a Lie algebra generated by the Moufang–Noether charge densities.

Start by rewriting the Moufang–Noether algebra as follows:

$$\left[s^0_j(\mathbf{x},t), s^0_k(\mathbf{y},t)\right] = \mathrm{i}\left[2Y^0_{jk}(x) + \frac{1}{3}c^p_{jk}s^0_p(x) + \frac{2}{3}c^p_{jk}t^0_p(x)\right]\delta(\mathbf{x}-\mathbf{y}) , \quad (1.1)$$

$$\left[s^0_j(\mathbf{x},t), t^0_k(\mathbf{y},t)\right] = \mathrm{i}\left[-Y^0_{jk}(x) + \frac{1}{3}c^p_{jk}s^0_p(x) - \frac{1}{3}c^p_{jk}t^0_p(x)\right]\delta(\mathbf{x}-\mathbf{y}) , \quad (1.2)$$

$$\left[t^0_j(\mathbf{x},t), s^0_k(\mathbf{y},t)\right] = \mathrm{i}\left[2Y^0_{jk}(x) - \frac{2}{3}c^p_{jk}s^0_p(x) - \frac{1}{3}c^p_{jk}t^0_p(x)\right]\delta(\mathbf{x}-\mathbf{y}) . \quad (1.3)$$

Here (1.2) can be seen as a definition of the *Yamagutian* $Y^0_{jk}(x)$. The Yamagutian is thus a recapitulation of the associator. It can be shown that

$$Y^0_{jk}(x) + Y^0_{kj}(x) = 0 , \quad (1.4)$$

$$c^p_{jk}Y^0_{pl}(x) + c^p_{kl}Y^0_{pj}(x) + c^p_{lj}Y^0_{pk}(x) = 0 . \quad (1.5)$$

The constraints (1.4) trivially descend from the anti-commutativity of the commutator bracketing, but the proof of (1.5) needs certain effort. Further, it turns out that the following *reductivity* conditions hold:

$$\left[Y^0_{jk}(\mathbf{x},t), s^0_n(\mathbf{y},t)\right] = \mathrm{i}d^p_{jkn}s^0_p(\mathbf{x},t)\delta(\mathbf{x}-\mathbf{y}) , \quad (1.6)$$

$$\left[Y^0_{jk}(\mathbf{x},t), s^0_n(\mathbf{y},t)\right] = \mathrm{i}d^p_{jkn}s^0_p(\mathbf{x},t)\delta(\mathbf{x}-\mathbf{y}) . \quad (1.7)$$

Finally, by using the reductivity conditions, one can check that the Yamagutian obeys the Lie algebra

$$[Y^0_{jk}(\mathbf{x},t), Y^0_{ln}(\mathbf{y},t)] = \mathrm{i}\left[d^p_{jkl}Y^0_{pn}(\mathbf{x},t) + d^p_{jkn}Y^0_{lp}(\mathbf{x},t)\right]\delta(\mathbf{x}-\mathbf{y}) . \quad (1.8)$$

When integrating the above ETC (1.1)–(1.8) one can finally obtain the

Theorem 1.2 (Moufang–Noether charge algebra). *The Moufang–Noether charge algebra* (σ, τ) *is a birepresentation of the Mal'tsev algebra* Γ.

Acknowledgements The research was in part supported by the Estonian Science Foundation, Grant 6912.

References

1. Jo, S.-G.: Commutator in anomalous non-abelian chiral gauge theory. Phys. Lett. B **163**, 353–359 (1985)
2. Niemi, A.J., Semenoff, G.W.: Quantum holonomy and the chiral gauge anomaly. Phys. Rev. Lett. B **55**, 927–930 (1985)
3. Moufang, R.: Zur Struktur von Alternativkörpern. Math. Ann. **B110**, 416–430 (1935)
4. Pflugfelder, H.: *Quasigroups and Loops: Introduction.* (Heldermann, Berlin, 1990)
5. Mal'tsev, A.I.: Analytical loops. Matem. Sbornik. **36**, 569–576 (1955, in Russian)
6. Paal, E.: Continuous Moufang transformations. Acta Appl. Math. **50**, 77–91 (1998)
7. Paal, E.: Moufang loops and integrability of generalized Lie equations. Czech. J. Phys. **54**, 1375–1379 (2004)
8. Loos, O.: Über eine Beziehung zwischen Malcev-Algebren und Lie-Tripelsystemen. Pacific J. Math. **18**, 553–562 (1966)
9. Yamaguti, K.: Note on Malcev algebras. Kumamoto J. Sci. **A5**, 203–207 (1962)
10. Yamaguti, K.: On the theory of Malcev algebras. Kumamoto J. Sci. **A6**, 9–45 (1963)
11. Schafer, R.D.: *An Introduction to Nonassociative algebras*. (Academic, New York, 1966)

Chapter 2
Weakly Nonassociative Algebras, Riccati and KP Hierarchies

Aristophanes Dimakis and Folkert Müller-Hoissen

Abstract It has recently been observed that certain nonassociative algebras (called 'weakly nonassociative', WNA) determine, via a universal hierarchy of ordinary differential equations, solutions of the KP hierarchy with dependent variable in an associative subalgebra (the middle nucleus). We recall central results and consider a class of WNA algebras for which the hierarchy of ODEs reduces to a matrix Riccati hierarchy, which can be easily solved. The resulting solutions of a matrix KP hierarchy determine, under a 'rank one condition', solutions of the scalar KP hierarchy. We extend these results to the discrete KP hierarchy. Moreover, we build a bridge from the WNA framework to the Gelfand–Dickey–Sato formulation of the KP hierarchy.

2.1 Introduction

The Kadomtsev–Petviashvili (KP) equation is an extension of the famous Korteweg–deVries (KdV) equation to $2+1$ dimensions. It first appeared in a stability analysis of KdV solitons [41, 43]. In particular, it describes nonlinear fluid surface waves in a certain approximation and explains to some extent the formation of network patterns formed by line wave segments on a water surface [41]. It is 'integrable' in several respects, in particular in the sense of the inverse scattering method. Various remarkable properties have been discovered that allow to access (subsets of) its solutions in different ways, see in particular [15,47,48]. Apart from its direct relevance

A. Dimakis
Department of Financial and Management Engineering, University of the Aegean, 31 Fostini Str., 82100 Chios, Greece
e-mail: dimakis@aegean.gr

F. Müller-Hoissen
Max-Planck-Institute for Dynamics and Self-Organization, Bunsenstrasse 10, 37073 Göttingen, Germany
e-mail: folkert.mueller-hoissen@ds.mpg.de

S. Silvestrov et al. (eds.), *Generalized Lie Theory in Mathematics, Physics and Beyond,*

in physics, the KP equation and its hierarchy (see [15, 44], for example) is deeply related to the theory of Riemann surfaces (Riemann–Schottky problem, see [9] for a review). Some time ago, this stimulated discussions concerning the role of KP in string theory (see [4,30,49,51], for example). Later the Gelfand–Dickey hierarchies, of which the KdV hierarchy is the simplest and which are reductions of the KP hierarchy, made their appearance in matrix models, first in a model of two-dimensional quantum gravity (see [50,52] and references therein). This led to important developments in algebraic geometry (see [69], for example). Of course, what we mentioned here by far does not exhaust what is known about KP and there is probably even much more in the world of mathematics and physics linked to the KP equation and its descendants that still waits to be uncovered.

In fact, an apparently completely different appearance of the KP hierarchy has been observed in [20]. On a freely generated 'weakly nonassociative' (WNA) algebra (see Sect. 2.2) there is a family of commuting derivations[1] that satisfy identities which are in correspondence with the equations of the KP hierarchy (with dependent variable in a noncommutative associative subalgebra). As a consequence, there is a hierarchy of ordinary differential equations (ODEs) on this WNA algebra that implies the KP hierarchy. More generally, this holds for *any* WNA algebra. In this way WNA algebras determine classes of solutions of the KP hierarchy.

In Sect. 2.2 we recall central results of [20] and present a new result in Proposition 2.1. Section 2.3 applies the WNA approach to derive a matrix Riccati[2] hierarchy, the solutions of which are solutions of the corresponding matrix KP hierarchy (which under certain conditions determines solutions of the scalar KP hierarchy). In Sect. 2.4 we extend these results to the discrete KP hierarchy [3, 14, 19, 26, 33]. Furthermore, in Sect. 2.5 we show how the Gelfand–Dickey–Sato formulation [15] of the KP hierarchy (with dependent variable in any associative algebra) emerges in the WNA framework. Section 2.6 contains some conclusions.

2.2 Nonassociativity and KP

In [20] we called an algebra $(\mathbb{A}, \circ)$ (over a commutative ring) *weakly nonassociative (WNA)* if

$$(a, b \circ c, d) = 0 \qquad \forall a,b,c,d \in \mathbb{A}\,, \tag{2.1}$$

where $(a,b,c) := (a \circ b) \circ c - a \circ (b \circ c)$ is the associator in $\mathbb{A}$. The *middle nucleus* of $\mathbb{A}$ (see e.g. [54]),

$$\mathbb{A}' := \{ b \in \mathbb{A} \mid (a,b,c) = 0 \ \forall a,c \in \mathbb{A} \}\,, \tag{2.2}$$

[1] Families of commuting derivations on certain algebras also appeared in [39, 40], for example. In fact, the ideas underlying the work in [20] grew out of our work in [17] which has some algebraic overlap with [40].

[2] Besides their appearance in control and systems theory, matrix Riccati equations (see [1, 55, 70], for example) frequently showed up in the context of integrable systems, see in particular [12, 13, 22–25, 27, 31, 32, 35, 36, 62–64, 67, 71].

is an *associative* subalgebra and a two-sided ideal. We fix $f \in \mathbb{A}$, $f \notin \mathbb{A}'$, and define $a \circ_1 b := a \circ b$,

$$a \circ_{n+1} b := a \circ (f \circ_n b) - (a \circ f) \circ_n b\,, \qquad n = 1,2,\ldots\,. \tag{2.3}$$

As a consequence of (2.1), these products only depend on the equivalence class $[f]$ of f in $\mathbb{A}/\mathbb{A}'$. The subalgebra $\mathbb{A}(f)$, generated by f in the WNA algebra $\mathbb{A}$, is called δ*-compatible* if, for each $n \in \mathbb{N}$,

$$\delta_n(f) := f \circ_n f \tag{2.4}$$

extends to a *derivation* of $\mathbb{A}(f)$. In the following we recall some results from [20].

Theorem 2.1. *Let $\mathbb{A}(f)$ be δ-compatible. The derivations δ_n commute on $\mathbb{A}(f)$ and satisfy identities that are in correspondence via $\delta_n \mapsto \partial_{t_n}$ (the partial derivative operator with respect to a variable t_n) with the equations of the potential Kadomtsev–Petviashvili (pKP) hierarchy with dependent variable in $\mathbb{A}'$.*

This is a central observation in [20] with the following immediate consequence.

Theorem 2.2. *Let $\mathbb{A}$ be any WNA algebra over the ring of complex functions of independent variables $t_1, t_2, \ldots$. If $f \in \mathbb{A}$ solves the hierarchy of ODEs*[3]

$$f_{t_n} := \partial_{t_n}(f) = f \circ_n f\,, \qquad n = 1,2,\ldots\,, \tag{2.5}$$

then $-f_{t_1}$ lies in $\mathbb{A}'$ and solves the KP hierarchy with dependent variable in $\mathbb{A}'$.

Corollary 2.1. *If there is a constant $\nu \in \mathbb{A}$, $\nu \notin \mathbb{A}'$, with $[\nu] = [f] \in \mathbb{A}/\mathbb{A}'$, then, under the assumptions of Theorem 2.2,*

$$\phi := \nu - f \in \mathbb{A}' \tag{2.6}$$

solves the potential KP (pKP) hierarchy[4]

$$\sum_{i,j,k=1}^{3} \varepsilon_{ijk} \left(\lambda_i^{-1}(\phi_{[\lambda_i]} - \phi) + \phi \circ \phi_{[\lambda_i]}\right)_{[\lambda_k]} = 0\,, \tag{2.7}$$

where ε_{ijk} is totally antisymmetric with $\varepsilon_{123} = 1$, λ_i, $i = 1,2,3$, are indeterminates, and $\phi_{\pm[\lambda]}(\mathbf{t}) := \phi(\mathbf{t} \pm [\lambda])$, where $\mathbf{t} = (t_1, t_2, \ldots)$ and $[\lambda] := (\lambda, \lambda^2/2, \lambda^3/3, \ldots)$.

Remark 1. If $C \in \mathbb{A}'$ is constant, then $f = \nu' - (\phi + C)$ with constant $\nu' := \nu + C$ satisfying $[\nu'] = [\nu] = [f]$. Hence, with ϕ also $\phi + C$ is a solution of the pKP hierarchy. This can also be checked directly using (2.7), of course.

The next result will be used in Sect. 2.4.

---

[3] f has to be differentiable, of course, which requires a corresponding (e.g. Banach space) structure on $\mathbb{A}$. The flows given by (2.5) indeed commute [20]. Furthermore, (2.5) implies δ-compatibility of the algebra $\mathbb{A}(f)$ generated by f in $\mathbb{A}$ over $\mathbb{C}$ [20].

[4] This functional representation of the potential KP hierarchy appeared in [7, 8]. See also [20, 22] for equivalent formulae.

Proposition 2.1. *Suppose f and f' solve (2.5) in a WNA algebra $\mathbb{A}$, and $[f] = [f']$. Then*

$$f' \circ f = \alpha\,(f' - f) \tag{2.8}$$

is preserved under the flows (2.5), for all $\alpha \in \mathbb{C}$.

Proof.

$$\begin{aligned}(f' \circ f)_{t_n} &= f'_{t_n} \circ f + f' \circ f_{t_n} = (f' \circ_n f') \circ f + f' \circ (f \circ_n f) \\ &= (f' \circ_n f') \circ f - f' \circ_n (f' \circ f) + \alpha\, f' \circ_n (f' - f) \\ &\quad + f' \circ (f \circ_n f) - (f' \circ f) \circ_n f + \alpha\,(f' - f) \circ_n f \\ &= -f' \circ_{n+1} f + f' \circ_{n+1} f + \alpha\,(f' \circ_n f' - f \circ_n f) = \alpha\,(f' - f)_{t_n} \,.\end{aligned}$$

In the third step we have added terms that vanish as a consequence of (2.8). Then we used (3.11) in [20] (together with the fact that the products $\circ_n$ only depend on the equivalence class $[f] = [f'] \in \mathbb{A}/\mathbb{A}'$), and also (2.3), to combine pairs of terms into products of one degree higher. □

Remark 2. In functional form, (2.5) can be expressed (e.g. with the help of results in [20]) as

$$\lambda^{-1}(f - f_{-[\lambda]}) - f_{-[\lambda]} \circ f = 0 \,. \tag{2.9}$$

Setting $f' = f_{-[\lambda]}$ (which also solves (2.9) if f solves it), this takes the form (2.8) with $\alpha = -\lambda^{-1}$.

In order to apply the above results, we need examples of WNA algebras . For our purposes, it is sufficient to recall from [20] that any WNA algebra with $\dim(\mathbb{A}/\mathbb{A}') = 1$ is isomorphic to one determined by the following data:

(1) An associative algebra $\mathscr{A}$ (e.g. any matrix algebra)
(2) A fixed element $g \in \mathscr{A}$
(3) Linear maps $\mathscr{L}, \mathscr{R} : \mathscr{A} \to \mathscr{A}$ such that

$$[\mathscr{L}, \mathscr{R}] = 0\,, \qquad \mathscr{L}(a \circ b) = \mathscr{L}(a) \circ b\,, \qquad \mathscr{R}(a \circ b) = a \circ \mathscr{R}(b)\,. \tag{2.10}$$

Augmenting $\mathscr{A}$ with an element f such that

$$f \circ f := g\,, \qquad f \circ a := \mathscr{L}(a)\,, \qquad a \circ f := \mathscr{R}(a)\,, \tag{2.11}$$

leads to a WNA algebra $\mathbb{A}$ with $\mathbb{A}' = \mathscr{A}$, provided that the following condition holds,

$$\exists a, b \in \mathscr{A} : \mathscr{R}(a) \circ b \neq a \circ \mathscr{L}(b)\,. \tag{2.12}$$

This guarantees that the augmented algebra is *not* associative. Particular examples of $\mathscr{L}$ and $\mathscr{R}$ are given by multiplication from left, respectively right, by fixed elements of $\mathscr{A}$ (see also the next section).

2.3 A Class of WNA Algebras and a Matrix Riccati Hierarchy

Let $\mathcal{M}(M,N)$ be the vector space of complex $M \times N$ matrices, depending smoothly on independent real variables $t_1, t_2, \ldots$, and let S, L, R, Q be constant matrices of dimensions $M \times N$, $M \times M$, $N \times N$ and $N \times M$, respectively. Augmenting with a constant element ν and setting[5]

$$\nu \circ \nu = -S\,, \qquad \nu \circ A = LA\,, \qquad A \circ \nu = -AR\,, \qquad A \circ B = AQB\,, \quad (2.13)$$

for all $A, B \in \mathcal{M}(M,N)$, we obtain a WNA algebra $(\mathbb{A}, \circ)$. The condition (2.12) requires

$$RQ \neq QL\,. \quad (2.14)$$

For the products $\circ_n$, $n > 1$, we have the following result.

Proposition 2.2.

$$\nu \circ_n \nu = -S_n\,, \quad \nu \circ_n A = L_n A\,, \quad A \circ_n \nu = -AR_n\,, \quad A \circ_n B = AQ_n B\,, \quad (2.15)$$

where

$$\begin{pmatrix} R_n & Q_n \\ S_n & L_n \end{pmatrix} = H^n \qquad \textit{with} \quad H := \begin{pmatrix} R & Q \\ S & L \end{pmatrix}\,. \quad (2.16)$$

Proof. Using the definition (2.3), one proves by induction that

$$S_{n+1} = LS_n + SR_n\,,\ L_{n+1} = LL_n + SQ_n\,,\ R_{n+1} = QS_n + RR_n\,,\ Q_{n+1} = QL_n + RQ_n\,,$$

for $n = 1, 2, \ldots$, where $S_1 = S$, $L_1 = L$, $R_1 = R$, $Q_1 = Q$. This can be written as

$$\begin{pmatrix} R_{n+1} & Q_{n+1} \\ S_{n+1} & L_{n+1} \end{pmatrix} = H \begin{pmatrix} R_n & Q_n \\ S_n & L_n \end{pmatrix}\,,$$

which implies (2.16). □

Using (2.6) and (2.15) in (2.5), leads to the matrix Riccati equations[6]

$$\phi_{t_n} = S_n + L_n \phi - \phi R_n - \phi Q_n \phi\,, \qquad n = 1, 2, \ldots\,. \quad (2.17)$$

Solutions of (2.17) are obtained in a well-known way (see [1, 63], for example) via

$$\phi = YX^{-1} \quad (2.18)$$

---

[5] Using (2.6), in terms of f this yields relations of the form (2.11).

[6] The corresponding functional form is $\lambda^{-1}(\phi_{[\lambda]} - \phi) + \phi Q \phi_{[\lambda]} = S + L\phi_{[\lambda]} - \phi R$, which is easily seen to imply (2.7), see also [21]. The appendix provides a FORM program [37, 66] which independently verifies that any solution of (2.17), reduced to $n = 1, 2, 3$, indeed solves the matrix pKP equation in $(\mathcal{M}(M,N), \circ)$.

from the linear system

$$Z_{t_n} = H^n Z, \qquad Z = \begin{pmatrix} X \\ Y \end{pmatrix} \tag{2.19}$$

with an $N \times N$ matrix X and an $M \times N$ matrix Y, provided X is invertible. This system is solved by

$$Z(\mathbf{t}) = e^{\xi(H)} Z_0 \qquad \text{where} \quad \xi(H) := \sum_{n \geq 1} t_n H^n \,. \tag{2.20}$$

If Q has rank 1, then

$$\varphi := \mathrm{tr}(Q\phi) \tag{2.21}$$

defines a homomorphism from $(\mathcal{M}(M,N), \circ)$ into the scalars (with the ordinary product of functions). Hence, if ϕ solves the pKP hierarchy in $(\mathcal{M}(M,N), \circ)$, then φ solves the scalar pKP hierarchy.[7] More generally, if $Q = VU^T$ with V, U of dimensions $N \times r$, respectively $M \times r$, then $U^T \phi V$ solves the $r \times r$-matrix pKP hierarchy.

The general linear group $GL(N+M, \mathbb{C})$ acts on the space of all $(N+M) \times (N+M)$ matrices H by similarity transformations. In a given orbit this allows to choose for H some 'normal form', for which we can evaluate (2.20) and then elaborate the effect of $GL(N+M, \mathbb{C})$ transformations (see also Remark 3 below) on the corresponding solution of the pKP hierarchy, with the respective Q given by the normal form of H. By a similarity transformation we can always achieve that $Q = 0$ and the problem of solving the pKP hierarchy (with some non-zero Q) can thus in principle be reduced to solving its linear part. Alternatively, we can always achieve that $S = 0$ and the next two examples take this route.

Example 1. If $S = 0$, we can in general not achieve that also $Q = 0$. In fact, the matrices

$$H = \begin{pmatrix} R & Q \\ 0 & L \end{pmatrix} \quad \text{and} \quad H_0 := \begin{pmatrix} R & 0 \\ 0 & L \end{pmatrix} \tag{2.22}$$

are similar (i.e. related by a similarity transformation) if and only if the matrix equation $Q = RK - KL$ has an $N \times M$ matrix solution K [5, 28, 53, 56, 60], and then

$$H = \mathscr{T} H_0 \mathscr{T}^{-1} \,, \qquad \mathscr{T} = \begin{pmatrix} I_N & -K \\ 0 & I_M \end{pmatrix} \,. \tag{2.23}$$

It follows that

[7] For related results and other perspectives on the rank one condition, see [29] and the references cited there. The idea to look for (simple) solutions of matrix and more generally operator versions of an 'integrable' equation, and to generate from it (complicated) solutions of the scalar equation by use of a suitable map, already appeared in [46] (see also [2, 6, 10, 11, 34, 38, 58, 59]).

$$H^n = \mathscr{T} H_0^n \mathscr{T}^{-1} = \begin{pmatrix} R^n & R^n K - K L^n \\ 0 & L^n \end{pmatrix} \tag{2.24}$$

and thus

$$e^{\xi(H)} = \begin{pmatrix} e^{\xi(R)} & e^{\xi(R)} K - K e^{\xi(L)} \\ 0 & e^{\xi(L)} \end{pmatrix} . \tag{2.25}$$

If (2.14) holds, we obtain the following solution of the matrix pKP hierarchy in $(\mathscr{M}(M,N),\circ)$,

$$\phi = e^{\xi(L)} \phi_0 \, (I_N + K\phi_0 - e^{-\xi(R)} K e^{\xi(L)} \phi_0)^{-1} e^{-\xi(R)} , \tag{2.26}$$

where $\phi_0 = Y_0 X_0^{-1}$. This in turn leads to

$$\begin{aligned} \varphi &= \mathrm{tr}\Big(e^{-\xi(R)} (RK - KL) e^{\xi(L)} \phi_0 \, (I_N + K\phi_0 - e^{-\xi(R)} K e^{\xi(L)} \phi_0)^{-1} \Big) \\ &= \mathrm{tr}\Big(\log(I_N + K\phi_0 - e^{-\xi(R)} K e^{\xi(L)} \phi_0) \Big)_{t_1} \\ &= (\log \tau)_{t_1} , \qquad \tau := \det(I_N + K\phi_0 - e^{-\xi(R)} K e^{\xi(L)} \phi_0) . \end{aligned} \tag{2.27}$$

If $\mathrm{rank}(Q) = 1$, then φ solves the scalar pKP hierarchy. Besides (2.14) and this rank condition, further conditions will have to be imposed on the (otherwise arbitrary) matrices R, K, L and ϕ_0 to achieve that φ is a *real* and *regular* solution. See [18], and references cited there, for classes of solutions obtained from an equivalent formula or restrictions of it. This includes multi-solitons and soliton resonances (KP-II), and lump solutions (passing to KP-I via $t_{2n} \mapsto \mathrm{i} t_{2n}$ and performing suitable limits of parameters).

Example 2. If $M = N$ and $S = 0$, let us consider

$$H = \mathscr{T} H_0 \mathscr{T}^{-1} , \qquad H_0 = \begin{pmatrix} L & I \\ 0 & L \end{pmatrix} , \qquad \mathscr{T} = \begin{pmatrix} I & -K \\ 0 & I \end{pmatrix} , \tag{2.28}$$

with $I = I_N$ and a constant $N \times N$ matrix K. As a consequence,

$$Q = I + [L, K] . \tag{2.29}$$

We note that H_0 is *not* similar to $\mathrm{diag}(L, L)$ [60]. Now we obtain

$$H^n = \mathscr{T} H_0^n \mathscr{T}^{-1} = \begin{pmatrix} L^n & n L^{n-1} + [L^n, K] \\ 0 & L^n \end{pmatrix} \tag{2.30}$$

and furthermore

$$e^{\xi(H)} = \begin{pmatrix} e^{\xi(L)} & \sum_{n \geq 1} n t_n L^{n-1} e^{\xi(L)} + [e^{\xi(L)}, K] \\ 0 & e^{\xi(L)} \end{pmatrix} . \tag{2.31}$$

If $[L, [L, K]] \neq 0$ (which is condition (2.14)), we obtain the solution

$$\phi = e^{\xi(L)} \phi_0 \, (I + K\phi_0 + F)^{-1} e^{-\xi(L)} \tag{2.32}$$

of the matrix pKP hierarchy in $(\mathcal{M}(N,N), \circ)$, where

$$F := \Big(\sum_{n \geq 1} n t_n L^{n-1} - e^{-\xi(L)} K e^{\xi(L)} \Big) \phi_0 \,. \tag{2.33}$$

Furthermore, using $F_{t_1} = e^{-\xi(L)} (I + [L,K]) e^{\xi(L)} \phi_0$, we find

$$\varphi = \mathrm{tr}((I + [L,K])\phi) = \mathrm{tr}(F_{t_1} (I + K\phi_0 + F)^{-1}) = (\mathrm{tr} \log(I + K\phi_0 + F))_{t_1} \tag{2.34}$$

and thus

$$\varphi = (\log \tau)_{t_1} \,, \quad \tau := \det \Big(I + K\phi_0 + (\sum_{n \geq 1} n t_n L^{n-1} - e^{-\xi(L)} K e^{\xi(L)}) \phi_0 \Big) . \tag{2.35}$$

If $\mathrm{rank}(I + [L,K]) = 1$ (see also [29,61,68] for appearances of this condition), then φ solves the scalar pKP hierarchy. Assuming that ϕ_0 is invertible, we can rewrite τ as follows,

$$\tau = \det \Big(e^{\xi(L)} (\phi_0^{-1} + K) e^{-\xi(L)} + \sum_{n \geq 1} n t_n L^{n-1} - K \Big) \tag{2.36}$$

(dropping a factor $\det(\phi_0)$). This simplifies considerably if we set $\phi_0^{-1} = -K$.[8] Choosing moreover

$$L_{ij} = -(q_i - q_j)^{-1} \quad i \neq j \,, \quad L_{ii} = -p_i \,, \quad K = \mathrm{diag}(q_1, \ldots, q_N) \,, \tag{2.37}$$

(2.36) reproduces a polynomial (in any finite number of the t_n) tau function associated with Calogero–Moser systems [29,61,68]. Alternatively, we may choose

$$L = \mathrm{diag}(q_1, \ldots, q_N) \,, \quad K_{ij} = (q_i - q_j)^{-1} \quad i \neq j \,, \quad K_{ii} = p_i \,. \tag{2.38}$$

The corresponding solutions of the KP-I hierarchy ($t_{2n} \mapsto \mathrm{i} t_{2n}$) include the rational soliton solutions ('lumps') originally obtained in [45]. In particular, $N = 2$ and $q_2 = -\bar{q}_1$, $p_2 = \bar{p}_1$ (where the bar means complex conjugation), yields the single lump solution given by

$$\tau = |p_1 + \xi'(q_1)|^2 + \frac{1}{4\Re(q_1)^2} \quad \text{where } \xi'(q) := \sum_{n \geq 1} n t_n q^{n-1} \Big|_{\{t_{2k} \mapsto \mathrm{i} t_{2k}, k=1,2,\ldots\}} \tag{2.39}$$

Example 3. Let $M = N$ and $L = S\pi_-$, $R = \pi_+ S$, $Q = \pi_+ S \pi_-$, with constant $N \times N$ matrices π_+, π_- subject to $\pi_+ + \pi_- = I$. The matrix H can then be written as

[8] Note that in this case $\phi = (\sum_{n \geq 1} n t_n L^{n-1} - K)^{-1}$, which is rational in any finite number of the variables t_n.

$$H = \begin{pmatrix} \pi_+ \\ I \end{pmatrix} S \begin{pmatrix} I & \pi_- \end{pmatrix} , \tag{2.40}$$

which lets us easily calculate

$$H^n = \begin{pmatrix} \pi_+ S^n & \pi_+ S^n \pi_- \\ S^n & S^n \pi_- \end{pmatrix} . \tag{2.41}$$

As a consequence, we obtain

$$\phi = (-C_+ + e^{\xi(S)} C_-)(\pi_- C_+ + \pi_+ e^{\xi(S)} C_-)^{-1} , \tag{2.42}$$

where $C_\pm := I \mp \pi_\pm \phi_0$. This solves the matrix pKP hierarchy in $\mathcal{M}(M,N)$ with the product $A \circ B = A\pi_+ S\pi_- B$ if (2.14) holds, which is $\pi_+ S(\pi_+ - \pi_-)S\pi_- \neq 0$. If furthermore $\mathrm{rank}(\pi_+ S \pi_-) = 1$, then

$$\varphi = \mathrm{tr}(Q\phi) = -\mathrm{tr}(\pi_+ S) + (\log \tau)_{t_1} , \qquad \tau = \det(\pi_- C_+ + \pi_+ e^{\xi(S)} C_-) \tag{2.43}$$

solves the scalar pKP hierarchy (see also [21]). We will meet the basic structure underlying this example again in Sect. 2.5.

Remark 3. A $GL(N+M,\mathbb{C})$ matrix

$$\mathcal{T} = \begin{pmatrix} A & B \\ C & D \end{pmatrix} \tag{2.44}$$

can be decomposed as follows,

$$\mathcal{T} = \begin{pmatrix} I_N & BD^{-1} \\ 0 & I_M \end{pmatrix} \begin{pmatrix} S_D & 0 \\ 0 & D \end{pmatrix} \begin{pmatrix} I_N & 0 \\ D^{-1}C & I_M \end{pmatrix} , \tag{2.45}$$

if D and its Schur complement $S_D = A - BD^{-1}C$ are both invertible. Let us see what effect the three parts of $\mathcal{T}$ induce on ϕ when acting on Z.

(1) Writing $P = D^{-1}C$, the first transformation leads to $\phi \mapsto \phi + P$, a shift by the constant matrix P.
(2) The second transformation amounts to $\phi \mapsto D\phi S_D^{-1}$ (where ϕ is now the result of the previous transformation).
(3) Writing $K = -BD^{-1}$, the last transformation is $\phi \mapsto \phi\,(I_N - K\phi)^{-1}$.

2.4 WNA Algebras and Solutions of the Discrete KP Hierarchy

The potential discrete KP (pDKP) hierarchy in an associative algebra $(\mathcal{A}, \circ)$ can be expressed in functional form as follows,[9]

[9] This functional representation of the pDKP hierarchy is equivalent to (3.32) in [19].

$$\Omega(\lambda)^+ - \Omega(\lambda)_{-[\mu]} = \Omega(\mu)^+ - \Omega(\mu)_{-[\lambda]}\,, \tag{2.46}$$

where λ, μ are indeterminates,

$$\Omega(\lambda) := \lambda^{-1}(\phi - \phi_{-[\lambda]}) - (\phi^+ - \phi_{-[\lambda]}) \circ \phi\,, \tag{2.47}$$

and $\phi = (\phi_k)_{k\in\mathbb{Z}}$, $\phi_k^+ := \phi_{k+1}$. The pDKP hierarchy implies that each component ϕ_k, $k \in \mathbb{Z}$, satisfies the pKP hierarchy and its remaining content is a special pKP Bäcklund transformation (BT) acting between neighbouring sites on the linear lattice labeled by k [3, 19]. This suggests a way to extend the method of Sect. 2.3 to construct exact solutions of the pDKP hierarchy. What is needed is a suitable extension of (2.5) that accounts for the BT and this is offered by Proposition 2.1.

Theorem 2.3. *Let $\mathbb{A}$ be a WNA algebra with a constant element $\nu \in \mathbb{A}$, $\nu \notin \mathbb{A}'$. Any solution*

$$f = (\nu - \phi_k)_{k\in\mathbb{Z}}\,, \tag{2.48}$$

of the hierarchy (2.5) together with the compatible constraint[10]

$$f^+ \circ f = 0 \tag{2.49}$$

yields a solution $\phi = (\phi_k)_{k\in\mathbb{Z}}$ of the pDKP hierarchy in $\mathbb{A}'$.

Proof. Since $[f^+] = [f]$, the compatibility follows by setting $f' = f^+$ and $\alpha = 0$ in Proposition 2.1. Using $f_{t_1} = f \circ f$, we rewrite (2.9) as

$$\lambda^{-1}(f - f_{-[\lambda]}) + (f - f_{-[\lambda]}) \circ f - f_{t_1} = 0\,.$$

Inserting $f = \nu - \phi$, this takes the form

$$\lambda^{-1}(\phi - \phi_{-[\lambda]}) - \phi_{t_1} - (\phi - \phi_{-[\lambda]}) \circ \phi = \theta - \theta_{-[\lambda]}$$

with $\theta := -\phi \circ \nu$. Next we use (2.49) and $f_{t_1} = f \circ f$ to obtain $(f^+ - f) \circ f + f_{t_1} = 0$, which is

$$\phi_{t_1} - (\phi^+ - \phi) \circ \phi = \theta^+ - \theta\,.$$

Together with the previous equation, this leads to

$$\lambda^{-1}(\phi - \phi_{-[\lambda]}) - (\phi^+ - \phi_{-[\lambda]}) \circ \phi = \theta^+ - \theta_{-[\lambda]}$$

(which is actually equivalent to the last two equations), so that

$$\Omega(\lambda) = \theta^+ - \theta_{-[\lambda]}\,.$$

---

[10] Note that (2.49) implies $f^{n+} \circ_n f = 0$, where $f_k^{n+} := f_{k+n}$. This follows by induction from $f^{(n+1)+} \circ_{n+1} f = f^{(n+1)+} \circ (f^{n+} \circ_n f) - (f^{(n+1)+} \circ f^{n+}) \circ_n f = f^{(n+1)+} \circ (f^{n+} \circ_n f) - (f^+ \circ f)^{n+} \circ_n f$, where we used (2.3) and $[f^{n+}] = [f]$ in the first step.

This is easily seen to solve (2.46). □

Let us choose the WNA algebra of Sect. 2.3.[11] Evaluation of (2.5) leads to the matrix Riccati hierarchy (2.17), and (2.49) with $f^+ = \nu + C - \phi^+$ becomes

$$S + CR + (L + CQ)\phi - \phi^+ R - \phi^+ Q\phi = 0\,, \tag{2.50}$$

which can be rewritten as

$$\phi^+ = (S + L\phi)(R + Q\phi)^{-1} + C = Y^+ (X^+)^{-1} \tag{2.51}$$

(assuming that the inverse matrices exist), where X^+, Y^+ are the components of

$$Z^+ = THZ = THe^{\xi(H)} Z^{(0)}\,, \tag{2.52}$$

with Z, H, T taken from Sect. 2.3. Deviating from the notation of Sect. 2.3, we write $Z^{(0)}$ for the constant vector, since Z_0 should now denote the component of Z at the lattice site 0. In order that (2.52) defines a pDKP solution on the whole lattice, we need H invertible. Since the matrix C, and thus also T, may depend on the lattice site k, solutions of (2.46) are determined by

$$Z_k = T_k H T_{k-1} H \cdots T_1 H Z_0,\; Z_{-k} = (T_{-k}H)^{-1} (T_{-k+1}H)^{-1} \cdots (T_{-1}H)^{-1} Z_0, \tag{2.53}$$

where $k \in \mathbb{N}$. This corresponds to a sequence of transformations applied to the matrix pKP solution ϕ_0 determined by Z_0, which generate new pKP solutions (cf. [3]). ϕ_1 is then given by (2.51) in terms of ϕ_0, and

$$\begin{aligned} \phi_2 = {} & [LS + SR + LC_1 R + (L^2 + SQ + LC_1 Q)\phi_0] \\ & \times [R^2 + QS + QC_1 R + (QL + RQ + QC_1 Q)\phi_0]^{-1} + C_2 \end{aligned} \tag{2.54}$$

shows that the action of the T_k becomes considerably more involved for $k > 1$. In the special case $T_k = I_{N+M}$ (so that $C_k = 0$), we have

$$Z_k = e^{\xi(H)} (H^k Z_0^{(0)}) \qquad k \in \mathbb{Z}\,. \tag{2.55}$$

If $X_k^{(0)}, Y_k^{(0)}$ are the components of the vector $H^k Z_0^{(0)}$, the lattice component ϕ_k of the pDKP solution determined in this way is therefore just given by the pKP solution of Sect. 2.3 with initial data (at $\mathbf{t} = 0$)

$$\phi_k^{(0)} = Y_k^{(0)} (X_k^{(0)})^{-1} = L^k \phi_0^{(0)} \, [R^k + (R^k K - K L^k)\phi_0^{(0)}]^{-1}\,. \tag{2.56}$$

With the restrictions of Example 1 in Sect. 2.3, assuming that L and R are invertible (so that H is invertible), the corresponding solution of the matrix pDKP hierarchy (in the matrix algebra with product $A \circ B = A(RK - KL)B$) is

[11] Since there is only a single element ν, matrices L, R, S do not depend on the discrete variable k.

$$\phi_k = e^{\xi(L)} L^k \phi_0^{(0)} [R^k (I_N + K\phi_0^{(0)}) - e^{-\xi(R)} K e^{\xi(L)} L^k \phi_0^{(0)}]^{-1} e^{-\xi(R)}, \, k \in \mathbb{Z}, \quad (2.57)$$

which leads to

$$\varphi_k = (\log \tau_k)_{t_1} \quad \text{with } \tau_k = \det\Big(R^k (I_N + K\phi_0^{(0)}) - e^{-\xi(R)} K e^{\xi(L)} L^k \phi_0^{(0)}\Big). \quad (2.58)$$

If $Q = RK - KL$ has rank 1, this is a solution of the scalar pDKP hierarchy.[12] As a special case, let us choose $M = N$, $L = \mathrm{diag}(p_1, \ldots, p_N)$, $R = \mathrm{diag}(q_1, \ldots, q_N)$, and K with entries $K_{ij} = (q_i - p_j)^{-1}$.[13] Then Q has rank 1 and we obtain N-soliton tau functions of the scalar discrete KP hierarchy. These can also be obtained via the Birkhoff decomposition method using appropriate initial data as in [57,65].

With the assumptions made in Example 2 of Sect. 2.3, setting $\phi_0^{(0)} = -K^{-1}$, assuming that K and L are invertible, and choosing for T_k the identity, we find the matrix pDKP solution

$$\phi_k = \Big(\sum_{n\geq 1} n t_n L^{n-1} + kL^{-1} - K\Big)^{-1}, \qquad k \in \mathbb{Z}. \quad (2.59)$$

If $\mathrm{rank}(I_N + [L,K]) = 1$, this leads to the following solution of the scalar pDKP hierarchy,

$$\varphi_k = (\log \tau_k)_{t_1} \qquad \text{with} \quad \tau_k = \det\Big(\sum_{n\geq 1} n t_n L^{n-1} + kL^{-1} - K\Big). \quad (2.60)$$

In Example 3 of Sect. 2.3, H is not invertible, so that (2.52) does not determine a pDKP solution.

2.5 From WNA to Gelfand–Dickey–Sato

Let $\mathfrak{R}$ be the complex algebra of pseudo-differential operators [15]

$$\mathcal{V} = \sum_{i \ll \infty} v_i \partial^i, \quad (2.61)$$

with coefficients $v_i \in \mathfrak{A}$, where $\mathfrak{A}$ is the complex differential algebra of polynomials in (in general noncommuting) symbols $u_n^{(m)}$, $m = 0,1,2,\ldots$, $n = 2,3,\ldots$, where $\partial(u_n^{(m)}) = u_n^{(m+1)}$ and $\partial(vw) = \partial(v)w + v\partial(w)$ for $v,w \in \mathfrak{A}$. We demand that $u_n^{(m)}$, $n = 2,3,\ldots$, $m = 0,1,2,\ldots$, are algebraically independent in $\mathfrak{A}$, and we introduce the following linear operators on $\mathfrak{R}$,

$$S(\mathcal{V}) := \mathfrak{L}\mathcal{V}, \qquad \pi_+(\mathcal{V}) := \mathcal{V}_{\geq 0}, \qquad \pi_-(\mathcal{V}) := \mathcal{V}_{<0} := \mathcal{V} - \mathcal{V}_{\geq 0}, \quad (2.62)$$

[12] Recall that $\varphi = \mathrm{tr}(Q\phi)$ (cf. 2.21) determines a homomorphism if Q has rank 1. As a consequence, if ϕ solves the matrix pDKP hierarchy (2.46), then φ solves the scalar pDKP hierarchy.

[13] The condition (2.14) requires $q_i \neq p_j$ for all $i,j = 1,\ldots,N$.

where $\mathscr{V}_{\geq 0}$ is the projection of a pseudo-differential operator $\mathscr{V}$ to its differential operator part, and

$$\mathfrak{L} = \partial + u_2 \partial^{-1} + u_3 \partial^{-2} + \cdots . \tag{2.63}$$

Let I denote the identity of $\mathfrak{R}$ (which we identify with the identity in $\mathfrak{A}$), and let $\mathscr{O}$ be the subspace of linear operators on $\mathfrak{R}$ spanned by S and elements of the form $S\pi_\pm S\pi_\pm \cdots \pi_\pm S$ (with any combination of signs). $\mathscr{O}$ becomes an algebra with the product given by

$$A \circ B := A\pi_+ S\pi_- B . \tag{2.64}$$

$(\mathscr{O}, \circ)$ is then generated by the elements $(S\pi_-)^m S(\pi_+ S)^n$, $m,n = 0,1,\ldots$. Let us furthermore introduce $\mathscr{A} := \{v \in \mathfrak{A} : v = \mathrm{res}(A(I)),\, A \in \mathscr{O}\}$, where res takes the residue (the coefficient of ∂^{-1}) of a pseudo-differential operator. This is a subalgebra of $\mathfrak{A}$, since for $A,B \in \mathscr{O}$ we have

$$\mathrm{res}(A(I))\,\mathrm{res}(B(I)) = \mathrm{res}(A\pi_+ S\pi_- B(I)) , \tag{2.65}$$

so that the product of elements of $\mathscr{A}$ is again in $\mathscr{A}$. As a consequence of this relation (read from right to left), $\mathscr{A}$ is generated by the elements $\mathrm{res}((S\pi_-)^m S(\pi_+ S)^n(I))$, $m,n = 0,1,\ldots$. Based on the following preparations, we will argue that $\mathscr{A}$ and $(\mathscr{O}, \circ)$ are actually isomorphic algebras.

Lemma 2.1. *For all* $\mathscr{V} \in \mathfrak{R}$,

$$\mathrm{res}((S\pi_-)^m \mathscr{V}) = \mathrm{res}(\mathscr{D}_m \mathscr{V}) , \qquad m = 0,1,\ldots , \tag{2.66}$$

where $\mathscr{D}_0 = I$ *and* $\{\mathscr{D}_m\}_{m=1}^\infty$ *are the differential operators recursively determined by* $\mathscr{D}_m = (\mathscr{D}_{m-1}\mathfrak{L})_{\geq 0}$.

Proof. We do the calculation for $m = 2$. This is easily generalized to arbitrary $m \in \mathbb{N}$.

$$\begin{aligned}\mathrm{res}((S\pi_-)^2 \mathscr{V}) &= \mathrm{res}(\mathfrak{L}(\mathfrak{L}\mathscr{V}_{<0})_{<0}) = \mathrm{res}(\mathfrak{L}_{\geq 0}\mathfrak{L}\mathscr{V}_{<0}) = \mathrm{res}((\mathfrak{L}_{\geq 0}\mathfrak{L})_{\geq 0}\mathscr{V}) \\ &= \mathrm{res}(\mathscr{D}_2 \mathscr{V}) .\end{aligned}$$

□

Proposition 2.3. *For* $m,n = 0,1,\ldots$,

$$\mathrm{res}((S\pi_-)^m S(\pi_+ S)^n(I)) = \sum_{k=0}^m \binom{m}{k} u^{(k)}_{m+n+2-k} + \textit{terms nonlinear in } u^{(j)}_k . \tag{2.67}$$

Proof. According to the preceding lemma, we have

$$\mathrm{res}((S\pi_-)^m S(\pi_+ S)^n(I)) = \mathrm{res}(\mathscr{D}_m S(\pi_+ S)^n(I)) .$$

Next we note that $\mathscr{D}_m = \partial^m + D_m$, $(\pi_+ S)^n(I) = \partial^n + D'_n$ with differential operators D_m, D'_n (of degree smaller than m, respectively n) such that each of its summands contains factors from $\{u_k^{(j)}\}$ (so their coefficients are non-constant polynomials in the $u_k^{(j)}$). It follows that

$$\begin{aligned}\mathrm{res}((S\pi_-)^m S(\pi_+ S)^n(I)) &= \mathrm{res}((\partial^m + D_m)\mathfrak{L}_{<0}(\partial^n + D'_n)) \\ &= \mathrm{res}(\partial^m \mathfrak{L}_{<0}\partial^n) + \text{terms nonlinear in } u_k^{(j)} \,.\end{aligned}$$

It remains to evaluate

$$\begin{aligned}\mathrm{res}(\partial^m \mathfrak{L}_{<0}\partial^n) &= \sum_{j=1}^{\infty} \mathrm{res}(\partial^m u_{1+j}\partial^{n-j}) = \sum_{j=1}^{\infty} \mathrm{res}\Big(\sum_{k=0}^{m} \binom{m}{k} u_{1+j}^{(k)} \partial^{m+n-j-k}\Big) \\ &= \sum_{k=0}^{m} \binom{m}{k} u_{m+n+2-k}^{(k)} \,.\end{aligned}$$

□

According to the last proposition, the linear term with the highest derivative[14] in the residue of $(S\pi_-)^m S(\pi_+ S)^n(I)$ is given by $u_{n+2}^{(m)}$. We conclude that the monomials $(S\pi_-)^m S(\pi_+ S)^n$, $m,n = 0,1,\ldots$, are algebraically independent in $(\mathscr{O},\circ)$, since any algebraic relation among them would induce a corresponding algebraic relation in the set of $u_n^{(m)}$, but we assumed the $u_n^{(m)}$ to be algebraically independent. Together with (2.65), this implies that $\mathscr{A}$ and $(\mathscr{O},\circ)$ are isomorphic algebras.

The last result allows us to introduce a WNA structure directly on $\mathscr{A}$ as follows.[15] Augmenting $\mathscr{A}$ with f such that, for $\mathscr{V},\mathscr{W} \in \mathscr{O}(I)$,

$$\begin{aligned} f \circ f &:= -\mathrm{res}(\mathfrak{L}) \,, & f \circ \mathrm{res}(\mathscr{V}) &:= \mathrm{res}(\mathfrak{L}\mathscr{V}_{<0}) \,, \\ \mathrm{res}(\mathscr{V}) \circ f &:= -\mathrm{res}(\mathscr{V}_{<0}\mathfrak{L}) \,, & \mathrm{res}(\mathscr{V}) \circ \mathrm{res}(\mathscr{W}) &:= \mathrm{res}(\mathscr{V})\,\mathrm{res}(\mathscr{W}) \,, \end{aligned} \tag{2.68}$$

indeed defines a WNA algebra $\mathbb{A} = \mathbb{A}(f)$. The relations (2.68) are well-defined since $\mathrm{res}(A(I))$ uniquely determines $A \in \mathscr{O}$. By induction we obtain

$$\begin{aligned} f \circ_n f &= -\mathrm{res}(\mathfrak{L}^n) \,, & f \circ_n \mathrm{res}(\mathscr{V}) &= \mathrm{res}(\mathfrak{L}^n \mathscr{V}_{<0}) \,, \\ \mathrm{res}(\mathscr{V}) \circ_n f &= -\mathrm{res}(\mathscr{V}_{<0}\mathfrak{L}^n) \,, & \mathrm{res}(\mathscr{V}) \circ_n \mathrm{res}(\mathscr{W}) &= \mathrm{res}(\mathscr{V}_{<0}\mathfrak{L}^n \mathscr{W}_{<0}) \,. \end{aligned} \tag{2.69}$$

Let the u_n now depend on variables $t_1, t_2, \ldots$, and set $\partial = \partial_{t_1}$. The hierarchy (2.5) of ODEs,

$$f_{t_n} = f \circ_n f = -\mathrm{res}(\mathfrak{L}^n) \,, \qquad n = 1,2,\ldots \,, \tag{2.70}$$

by use of the WNA structure implies

[14] If $m = 0$, the linear term is simply u_{n+2} and thus again 'the linear term with the highest derivative'.

[15] Note that the corresponding WNA structure for $(\mathscr{O},\circ)$ resembles that of Example 3 in Sect. 2.3.

$$\begin{aligned}\partial_{t_n}(\mathrm{res}(\mathfrak{L}^m)) &= -\partial_{t_n}(f \circ_m f) = -f_{t_n} \circ_m f - f \circ_m f_{t_n} \\ &= -(f \circ_n f) \circ_m f - f \circ_m (f \circ_n f) = \mathrm{res}\Big(\mathfrak{L}^m (\mathfrak{L}^n)_{<0} - (\mathfrak{L}^n)_{\geq 0} \mathfrak{L}^m\Big) \\ &= \mathrm{res}\Big([(\mathfrak{L}^n)_{\geq 0}, \mathfrak{L}^m]\Big) . \end{aligned} \tag{2.71}$$

Since also $\partial_{t_n}(\mathrm{res}(\mathfrak{L}^m)) = \mathrm{res}([(\mathfrak{L}^m)_{\geq 0}, \mathfrak{L}^n]) = \partial_{t_m}(\mathrm{res}(\mathfrak{L}^n))$, we conclude that if we extend $\mathscr{A}$ to $\tilde{\mathscr{A}}$ by adjoining an element $\phi = \partial^{-1}(u_2)$, then

$$\phi_{t_n} = \mathrm{res}(\mathfrak{L}^n) , \qquad n = 1, 2, \ldots . \tag{2.72}$$

It follows that $v := f + \phi$ satisfies $\partial_{t_n}(v) = 0$, $n = 1, 2, \ldots$, and is therefore constant. Equation (2.72) determines all the u_k in terms of the derivatives of ϕ (see [16], for example). From (2.72) with $n = 2, 3$, and (2.71) with $m = n = 2$, we recover the pKP equation

$$(4\phi_{t_3} - \phi_{t_1 t_1 t_1} - 6\,{\phi_{t_1}}^2)_{t_1} - 3\,\phi_{t_2 t_2} + 6\,[\phi_{t_1}, \phi_{t_2}] = 0 , \tag{2.73}$$

in accordance with the general theory. More generally, (2.71) determines the whole pKP hierarchy . They are the residues of

$$\partial_{t_n}(\mathfrak{L}^m) = [(\mathfrak{L}^n)_{\geq 0}, \mathfrak{L}^m] , \qquad m, n = 1, 2, \ldots . \tag{2.74}$$

This is equivalent to the Gelfand–Dickey–Sato system $\partial_{t_n}(\mathfrak{L}) = [(\mathfrak{L}^n)_{\geq 0}, \mathfrak{L}]$, $n = 1, 2, \ldots$, which is a well-known formulation of the KP hierarchy (see [15], for example).

We have thus shown how the Gelfand–Dickey–Sato formulation of the KP hierarchy can be recovered in the WNA framework. In fact, for the particular WNA algebra chosen above, the hierarchy (2.5) of ODEs is equivalent to the Gelfand–Dickey–Sato formulation of the KP hierarchy.

2.6 Conclusions

In this work we extended our previous results [18, 20] on the relation between weakly nonassociative (WNA) algebras and solutions of KP hierarchies to discrete KP hierarchies. We also provided further examples of solutions of matrix KP hierarchies and corresponding solutions of the scalar KP hierarchy. In particular we recovered a well-known tau function related to Calogero–Moser systems in this way (Example 2 in Sect. 2.3). Furthermore, we established a connection with the Gelfand–Dickey–Sato formulation of the KP hierarchy. As a byproduct, in Sect. 2.5 we obtained a new realization of the *free* WNA algebra generated by a single element, which also has a realization in terms of quasi-symmetric functions [20]. There is more, however, we have to understand in the WNA framework. In particular this concerns the multi-component KP hierarchy (see [42] and references therein) and its reductions, which include the Davey–Stewartson, two-dimensional Toda lattice and N-wave hierarchies. Our hope is that also in these cases the WNA approach leads in a quick way to relevant classes of exact solutions.

Appendix: From Riccati to KP with FORM

The following FORM program [37, 66] verifies that any solution of the first three equations of the Riccati hierarchy (2.17) solves the pKP equation in an algebra with product $A \circ B = AQB$.

```
Functions phi,phix,phiy,phit,L,Q,R,S,dx,dy,dt;
Symbol n;
Local pKP=dx*(4*phit-6*phix*Q*phix-dx^2*phix)
-3*dy*phiy+6*(phix*Q*phiy-phiy*Q*phix); *pKP equation
repeat;
id phix=S+L*phi-phi*R-phi*Q*phi;        *Riccati system
id phiy=S(2)+L(2)*phi-phi*R(2)-phi*Q(2)*phi;
id phit=S(3)+L(3)*phi-phi*R(3)-phi*Q(3)*phi;
id dx*phi= phix+phi*dx;             *differentiation rule
id dy*phi= phiy+phi*dy; id dt*phi=phit+phi*dt;
id dx?{dx,dy,dt}*L?{L,Q,R,S}=L*dx;  *constant L,Q,R,S
* recursion relations (see proof of proposition 2.2):
id L(n?{2,3})=L*L(n-1)+S*Q(n-1);
id R(n?{2,3})=Q*S(n-1)+R*R(n-1);
id S(n?{2,3})=L*S(n-1)+S*R(n-1);
id Q(n?{2,3})=Q*L(n-1)+R*Q(n-1);
id L?{L,Q,R,S}(1)=L; endrepeat; id dx?{dx,dy,dt}=0;
print pKP;             *should return zero
.end
```

This program provides an elementary and quick way toward the classes of exact solutions of the KP equation given in the examples in Sect. 2.3.

References

1. Abou-Kandil, H., Freiling, G., Ionescu, V., Jank, G.: Matrix Riccati Equations in Control and Systems Theory. Systems and Control: Foundations and Applications. Birkhäuser, Basel (2003)
2. Aden, H., Carl, B.: On realizations of solutions of the KdV equation by determinants on operator ideals. J. Math. Phys. **37**, 1833–1857 (1996)
3. Adler, M., van Moerbeke, P.: Vertex operator solutions of the discrete KP-hierarchy. Commun. Math. Phys. **203**, 185–210 (1999)
4. Alvarez-Gaumé, L., Gomez, C., Reina, C.: New methods in string theory. In: L. Alvarez-Gaumé, M.B. Green, M.T. Grisaru, R. Iengo, E. Sezgin (eds.) Superstrings 1987, pp. 341–422. World Scientific, Teaneck, NJ (1987)

5. Bhatia, R., Rosenthal, P.: How and why to solve the operator equation $AX - XB = Y$. Bull. London Math. Soc. **29**, 1–21 (1997)
6. Blohm, H.: Solution of nonlinear equations by trace methods. Nonlinearity **13**, 1925–1964 (2000)
7. Bogdanov, L.: Analytic-Bilinear Approach to Integrable Hierarchies, *Mathematics and its Applications*, vol. 493. Kluwer, Dordrecht (1999)
8. Bogdanov, L., Konopelchenko, B.: Analytic-bilinear approach to integrable hierarchies. II. Multicomponent KP and 2D Toda lattice hierarchies. J. Math. Phys. **39**, 4701–4728 (1998)
9. Buchstaber, V., Krichever, I.: Integrable equations, addition theorems, and the Riemann–Schottky problem. Russ. Math. Surveys **61**(1), 19–78 (2006)
10. Carl, B., Schiebold, C.: Nonlinear equations in soliton physics and operator ideals. Nonlinearity **12**, 333–364 (1999)
11. Carl, B., Schiebold, C.: Ein direkter Ansatz zur Untersuchung von Solitonengleichungen. Jber. Dt. Math.-Verein. **102**, 102–148 (2000)
12. Chau, L.L.: Chiral fields, self-dual Yang-Mills fields as integrable systems, and the role of the Kac-Moody algebra. In: K.B. Wolf (ed.) Nonlinear Phenomena, *Lecture Notes in Physics*, vol. 189, pp. 110–127. Springer, Berlin (1983)
13. Common, A., Roberts, D.: Solutions of the Riccati equation and their relation to the Toda lattice. J. Phys. A: Math. Gen. **19**, 1889–1898 (1986)
14. Dickey, L.: Modified KP and discrete KP. Lett. Math. Phys. **48**, 277–289 (1999)
15. Dickey, L.: Soliton Equations and Hamiltonian Systems. World Scientific, Singapore (2003)
16. Dimakis, A., Müller-Hoissen, F.: Explorations of the extended ncKP hierarchy. J. Phys. A: Math. Gen. **37**, 10,899–10,930 (2004)
17. Dimakis, A., Müller-Hoissen, F.: An algebraic scheme associated with the noncommutative KP hierarchy and some of its extensions. J. Phys. A: Math. Gen. **38**, 5453–5505 (2005)
18. Dimakis, A., Müller-Hoissen, F.: From nonassociativity to solutions of the KP hierarchy. Czech. J. Phys. **56**, 1123–1130 (2006)
19. Dimakis, A., Müller-Hoissen, F.: Functional representations of integrable hierarchies. J. Phys. A: Math. Gen. **39**, 9169–9186 (2006)
20. Dimakis, A., Müller-Hoissen, F.: Nonassociativity and integrable hierarchies. nlin.SI/0601001 (2006)
21. Dimakis, A., Müller-Hoissen, F.: Burgers and KP hierarchies: A functional representation approach. Theor. Math. Phys. **152**, 933–947 (2007)
22. Dorfmeister, J., Neher, E., Szmigielski, J.: Automorphisms of Banach manifolds associated with the KP-equation. Quart. J. Math. Oxford **40**, 161–195 (1989)
23. Faibusovich, L.: Generalized Toda flows, Riccati equations on the Grassmanian, and the QR-algorithm. Funct. Anal. Appl. **21**(2), 166–168 (1987)
24. Falqui, G., Magri, F., Pedroni, M.: Bihamiltonian geometry, Darboux coverings, and linearization of the KP hierarchy. Commun. Math. Phys. **197**, 303–324 (1998)
25. Falqui, G., Magri, F., Pedroni, M., Zubelli, J.: An elementary approach to the polynomial τ-functions of the KP hierarchy. Theor. Math. Phys. **122**, 17–28 (2000)
26. Felipe, R., Ongay, F.: Algebraic aspects of the discrete KP hierarchy. Linear Algebra and its Applications **338**, 1–18 (2001)
27. Ferreira, L., Gomes, J., Razumov, A., Saveliev, M., Zimerman, A.: Riccati-type equations, generalized WZNW equations, and multidimensional Toda systems. Commun. Math. Phys. **203**, 649–666 (1999)
28. Flanders, H., Wimmer, H.: On the matrix equations $AX - XB = C$ and $AX - YB = C$. SIAM J. Appl. Math. **32**, 707–710 (1977)
29. Gekhtman, M., Kasman, A.: On KP generators and the geometry of the HBDE. J. Geom. Phys. **56**, 282–309 (2006)
30. Gilbert, G.: The Kadomtsev-Petviashvili equations and fundamental string theory. Commun. Math. Phys. **117**, 331–348 (1988)
31. Grundlach, A., Levi, D.: Higher order Riccati equations as Bäcklund transformations. preprint CRM-2469 (1997)

32. Haak, G.: Negative flows of the potential KP-hierarchy. Trans. Am. Math. Soc. **348**, 375–390 (1996)
33. Haine, L., Iliev, P.: Commuting rings of difference operators and an adelic flag manifold. Int. Math. Res. Notices (6), 281–323 (2000)
34. Han, W., Li, Y.: Remarks on the solutions of the Kadomtsev–Petviashvili equation. Phys. Lett. A **283**, 185–194 (2001)
35. Harnad, J., Saint-Aubin, Y., Shnider, S.: The soliton correlation matrix and the reduction problem for integrable systems. Commun. Math. Phys. **93**, 33–56 (1984)
36. Hazewinkel, M.: Ricatti and soliton equations. In: K. Ito, T. Hida (eds.) Gaussian Random Fields: The Third Nagoya Levy Seminar, *Series on Probability and Statistics*, vol. 1, pp. 187–196. World Scientific, Singapore (1991)
37. Heck, A.: FORM for Pedestrians. NIKHEF, Amsterdam (2000)
38. Huang, S.Z.: An operator method for finding exact solutions to vector Korteweg–deVries equations. J. Math. Phys. **44**, 1357–1388 (2003)
39. Ihara, K.: Derivations and automorphisms on the algebra of non-commutative power series. Math. J. Okayama Univ. **47**, 55–63 (2005)
40. Ihara, K., Kaneko, M., Zagier, D.: Derivation and double shuffle relations for multiple zeta values. Compositio Math. **142**, 307–338 (2006)
41. Infeld, E., Rowlands, G.: Nonlinear Waves, Solitons and Chaos. Cambridge University Press, Cambridge (2000)
42. Kac, V., van der Leur, J.: The *n*-component KP hierarchy and representation theory. J. Math. Phys. **44**, 3245–3293 (2003)
43. Kadomtsev, B., Petviashvili, V.: On the stability of solitary waves in a weakly dispersing medium. Sov. Phys. Doklady **15**, 539–541 (1970)
44. Kupershmidt, B.: KP or mKP, *Mathematical Surveys and Monographs*, vol. 78. American Math. Soc., Providence (2000)
45. Manakov, S., Zakharov, V., Bordag, L., Its, A., Matveev, V.: Two-dimensional solitons of the Kadomtsev–Petviashvili equation and their interaction. Phys. Lett. A **63**, 205–206 (1977)
46. Marchenko, V.: Nonlinear Equations and Operator Algebras. Mathematics and Its Applications. Reidel, Dordrecht (1988)
47. Matveev, V., Salle, M.: Darboux Transformations and Solitons. *Springer Series in Nonlinear Dynamics*. Springer, Berlin (1991)
48. Miwa, T., Jimbo, M., Date, E.: Solitons: Differential Equations, Symmetries and Infinite Dimensional Algebras. Cambridge University Press, Cambridge (2000)
49. Morozov, A.: String theory: what is it? Sov. Phys. Usp. **35**, 671–714 (1992)
50. Morozov, A.: Integrability and matrix models. Phys. Uspekhi **37**, 1–55 (1994)
51. Mulase, M.: KP equations, strings, and the Schottky problem. In: M. Kashiwara, T. Kawai (eds.) Algebraic Analysis, vol. II, pp. 473–492. Academic, Boston (1988)
52. Mulase, M.: Matrix integrals and integrable systems. In: K.F. et. al. (ed.) Topology, Geometry and Field Theory, pp. 111–127. World Scientific, Singapore (1994)
53. Olshevsky, V.: Similarity of block diagonal and block triangular matrices. Integr. Equat. Oper. Th. **15**, 853–863 (1992)
54. Pflugfelder, H.: Quasigroups and Loops, *Sigma Series in Pure Mathematics*, vol. 7. Heldermann, Berlin (1990)
55. Reid, W.: Riccati Differential Equations. Academic, New York (1972)
56. Roth, W.: The equations $AX - YB = C$ and $AX - XB = C$ in matrices. Proc. Am. Math. Soc. **3**, 392–396 (1952)
57. Sakakibara, M.: Factorization methods for noncommutative KP and Toda hierarchies. J. Phys. A: Math. Gen. **37**, L599–L604 (2004)
58. Schiebold, C.: Solitons of the sine-Gordon equation coming in clusters. Rev. Mat. Complutense **15**, 265–325 (2002)
59. Schiebold, C.: From the non-Abelian to the scalar two-dimensional Toda lattice. Glasgow Math. J. **47A**, 177–189 (2005)
60. Schweinsberg, A.: The operator equation $AX - XB = C$ with normal A and B. Pac. J. Math. **102** (1982)

61. Shiota, T.: Calogero-Moser hierarchy and KP hierarchy. J. Math. Phys. **35**, 5844–5849 (1994)
62. Takasaki, K.: A new approach to the self-dual Yang-Mills equations II. Saitama Math. J. **3**, 11–40 (1985)
63. Takasaki, K.: Geometry of universal Grassmann manifold from algebraic point of view. Rev. Math. Phys. **1**, 1–46 (1989)
64. Turbiner, A., Winternitz, P.: Solutions of nonlinear differential and difference equations with superposition formulas. Lett. Math. Phys. **50**, 189–201 (1999)
65. Ueno, K., Takasaki, K.: Toda lattice hierarchy. In: K. Okamoto (ed.) Group Representations and Systems of Differential Equations, *Advanced Studies in Pure Mathematics*, vol. 4, pp. 1–95. North-Holland, Amsterdam (1984)
66. Vermaseren, J.: FORM Reference Manual. NIKHEF, Amsterdam (2002)
67. Wadati, M., Sanuki, H., Konno, K.: Relationships among inverse method, Bäcklund transformation and an infinite number of conservation laws. Prog. Theor. Phys. **53**, 419–436 (1975)
68. Wilson, G.: Collisions of Calogero-Moser particles and an adelic Grassmannian. Invent. math. **133**, 1–41 (1998)
69. Witten, E.: Algebraic geometry associated with matrix models of two-dimensional gravity. In: L. Goldberg, A. Phillips (eds.) Topological methods in modern mathematics (Stony Brook, NY, 1991), pp. 235–269. Publish or Perish, Houston, TX (1993)
70. Zakhar-Itkin, M.: The matrix Riccati differential equation and the semi-group of linear fractional transformations. Russ. Math. Surv. **78**, 89–131 (1973)
71. Zelikin, M.: Geometry of operator cross-ratio. Sbornik: Math. **197**, 37–51 (2006)

Chapter 3
Applications of Transvectants

Chris Athorne

Abstract We discuss the role of the transvectant, a device dating from the early history of representation theory, in the theory of Padé approximants and hyperelliptic $\wp$-functions.

3.1 Introduction

This paper address a tetrahedron of ideas, the vertices being labelled: *Hirota derivative*, *Padé approximant*, *Transvectant* and *hyperelliptic function*. With one exception the edges (relations between these vertices of this tetrahedron) will be described.

Section 3.2 deals with the historical notion of transvectants and illustrates their geometrical significance as invariants of the Lie algebra $\mathfrak{sl}_2(\mathbb{C})$.

Transvectants suggest the idea of the generalised Hirota map, the original of which appears in soliton theory and the theory of integrable systems. These generalisations are viewed as intertwining operators between (possibly infinite dimensional) $\mathfrak{sl}_2(\mathbb{C})$-modules.

In Sect. 3.4 we deal with infinite dimensional modules which arise from rational functions. These are connected with the theory of Padé approximants. The value of the current approach is that it presents a clean, Lie theoretic treatment of material which is calculationally complex. It is also unexpected.

Section 3.5 deals with the theory of genus two hyperelliptic functions, generalising to the sextic curve results familiar from the theory of the Weierstrass $\wp$-function on the cubic. The transvectant/Hirota theory provides a powerful tool for dealing with the theory in a unified, algebraic way which appears to be a practical alternative to the classical analytic approach.

C. Athorne
Department of Mathematics, University of Glasgow, Glasgow, G12 8QW, Scotland
e-mail: c.athorne@maths.gla.ac.uk

S. Silvestrov et al. (eds.), *Generalized Lie Theory in Mathematics, Physics and Beyond,*

We do not here describe the relation between Padé approximants and hyperelliptic functions.

3.2 Transvectants

Consider the quadratic bilinear form with coefficients in $\mathbb{C}$,

$$a(x,y) = a_2y^2 + 2a_1xy + a_0x^2 \tag{3.1}$$

and the triple of differential operators defined by [10]

$$\begin{aligned} e &= y\partial_x - a_0\partial_{a_1} - 2a_1\partial_{a_2} && (3.2)\\ f &= x\partial_y - 2a_1\partial_{a_0} - a_2\partial_{a_1} && (3.3)\\ h &= y\partial_y - x\partial_x + 2a_0\partial_{a_0} - 2a_2\partial_{a_2}. && (3.4)\end{aligned}$$

Then e, f and h form a three dimensional representation of $\mathfrak{sl}_2$ and the triple a_0, a_1, a_2 a three dimensional $\mathfrak{sl}_2$-module. Further we have the *covariance* of the quadratic form:

$$\begin{aligned} e(a(x,y)) &= 0 && (3.5)\\ f(a(x,y)) &= 0 && (3.6)\\ h(a(x,y)) &= 0. && (3.7)\end{aligned}$$

The transvectant [13] is essentially a way of building higher dimensional modules out of such simple ones.

Suppose

$$b(x,y) = b_2y^2 + 2b_1xy + b_0x^2 \tag{3.8}$$

is another quadratic form covariant under a similar e, f and h action, then define the first and second transvectants as:

$$\begin{aligned} (a,b)^{(1)} &= \partial_x a\partial_y b - \partial_x a\partial_x b && (3.9)\\ &= 4((a_1b_2 - a_2b_1)y^2 + (a_0b_2 - a_2b_0)xy + (a_0b_1 - a_1b_0)x^2) && (3.10)\\ (a,b)^{(2)} &= \partial_x^2 a\partial_y^2 b - 2\partial_{xy}^2 a\partial_{xy}^2 b + \partial_y^2 a\partial_x^2 b && (3.11)\\ &= 4(a_0b_2 - 2a_1b_1 + a_2b_0) && (3.12)\end{aligned}$$

Of these $(a,b)^{(1)}$ is again a covariant quadratic form whose coefficients are a basis for the antisymmetric three dimensional module sitting inside the tensor product of $\{a_0, a_1, a_2\}$ and $\{b_0, b_1, b_2\}$; $(a,b)^{(2)}$, on the other hand, is independent of x and y, a one dimensional module (invariant) of $\mathfrak{sl}_2$

The covariance of these objects translates into geometry. In particular the vanishing of transvectants is geometrically significant.

Thus $(a,a)^{(2)}$ is the discriminant of the form $a(x,y)$ and $((a,b)^{(1)},(a,b)^{(1)}))^{(2)}$ is the resultant of the forms $a(x,y)$ and $b(x,y)$, whose vanishing implies a common factor. The vanishing of $(a,b)^{(2)}$ on the other hand is equivalent to the statement that the roots α_1,α_2 and β_1,β_2 of the corresponding inhomogeneous quadratics are harmonic [6]:

$$\frac{(\alpha_1-\beta_1)(\alpha_2-\beta_2)}{(\alpha_1-\beta_2)(\alpha_2-\beta_1)}=-1. \tag{3.13}$$

3.3 Hirota

Now consider a formal Laurent series

$$a^{[n]}(x,y)=\sum_{i+j=n} a_{ij}x^iy^j \tag{3.14}$$

where $i,j,n\in\mathbb{Z}$. The form $a^{[n]}$ is covariant under the $\mathfrak{sl}_2$ action,

$$e=y\partial_x-\sum(i+1)a_{i+1\,j-1}\partial_{a_{ij}} \tag{3.15}$$
$$f=x\partial_y-\sum(j+1)a_{i-1\,j+1}\partial_{a_{ij}} \tag{3.16}$$
$$h=y\partial_y-x\partial_x+\sum(i-j)\partial_{a_{ij}}. \tag{3.17}$$

The "monomials" x^iy^j are a basis for an infinite dimensional module $V^{[n]}=\{x^iy^j|i+j=n\}$ and the a_{ij} for a dual module $A^{[n]}=\{a_{ij}|i+j=n\}$. There is a submodule, $V_0^{[n]}=<x^iy^j|i+j=n,i\geq 0>$. Then $(V_0^{[n]})^{\perp}$ is a submodule of $A^{[n]}$ and we can think of the covariant forms analytic at $x=0$ as having coefficients in $F_0^{[n]}=A^{[n]}/(V_0^{[n]})^{\perp}$.

The covariant form $a^{[n]}$, or an analytic from, is then to be thought of as $a^{[n]}=<a,z>$ for $a\in A^{[n]}$ and $z\in V^{[n]}$ or, correspondingly, $a\in F_0^{[n]}$ and $z\in V_0^{[n]}$. Covariance is exactly the definition of the dual action:

$$e<a,z>=<e(a),z>+<a,e(z)>=0, \tag{3.18}$$

etc.

Hirota maps are most simply defined as maps between tensor products of modules:

$$E:V^{[n]}\otimes V^{[m]}\rightarrow V^{[n+1]}\otimes V^{[m+1]} \tag{3.19}$$
$$\alpha\otimes\beta\mapsto x\alpha\otimes y\beta-y\alpha\otimes x\beta \tag{3.20}$$
$$F:V^{[n]}\otimes V^{[m]}\rightarrow V^{[n-1]}\otimes V^{[m-1]} \tag{3.21}$$
$$\alpha\otimes\beta\mapsto \partial_x\alpha\otimes\partial_y\beta-\partial_y\alpha\otimes\partial_x\beta \tag{3.22}$$

Using the pairing of $A^{[n]}\otimes A^{[m]}$ and $V^{[n]}\otimes V^{[m]}$,

$$< a\otimes b, z\otimes w >=< a,z >< b,w > \tag{3.23}$$

we can define the dual action of E and F on the modules of coefficients, $A^{[n]}$, by requiring the covariance condition,

$$E < a\otimes b, z\otimes w >=< E(a\otimes b), z > + < a, E(z\otimes w) >= 0, \tag{3.24}$$

etc. The action restricts to the analytic submodules so we have:

$$\begin{aligned} E : F_0^{[n]}\otimes F_0^{[m]} &\to F_0^{[n-1]}\otimes F_0^{[m-1]} \\ a_{ij}^{[n]}\otimes b_{kl}^{[m]} &\mapsto -a_{i-1\,j}^{[n]}\otimes b_{kl-1}^{[m]} + a_{ij-1}^{[n]}\otimes b_{k-1\,l}^{[m]} \\ F : F_0^{[n]}\otimes F_0^{[m]} &\to F_0^{[n-1]}\otimes F_0^{[m-1]} \\ a_{ij}^{[n]}\otimes b_{kl}^{[m]} &\mapsto -(i+1)(l+1)a_{i+1\,j}^{[n]}\otimes b_{kl+1}^{[m]} + (j+1)(k+1)a_{ij+1}^{[n]}\otimes b_{k+1\,l}^{[m]}. \end{aligned} \tag{3.25}$$

These Hirota maps have the following important properties [2, 14]:

- E, F and $H=[E,F]$ are intertwining operators, commuting with the e, f and h action on order two tensor products.
- E, F and H themselves a representation of $\mathfrak{sl}_2$, making the space $\{F_0^{[n]}\otimes F_0^{[m]} | n,m\in\mathbb{Z}\}$ into an $\mathfrak{sl}_2$-module.
- The decomposition of tensor products into direct sums of modules (plethysm) can be recursively generated using E or F.
- F generalises the transvectant construction. A symmetrization over tensor products is involved so that the second operator, E, is trivialised.
- The generalizations to higher order tensor products and to higher rank $\mathfrak{sl}_n$ are straightforward.

Finally, a word on history and nomenclature. In soliton theory in the 1970s, Hirota [11] introduced an operator called the Hirota derivative:

$$D(f(\xi)\cdot g(\xi)) = \lim_{\eta\to 0}\frac{d}{d\eta} f(\xi+\eta)g(\xi-\eta) = (f_\xi g - f g_\xi)(\xi) \tag{3.26}$$

D is not, of course, a derivative, but it does satisfy the condition [9]

$$D(hf\cdot hg) = h^2 D(f\cdot g) \tag{3.27}$$

for arbitrary functions $h(\xi)$. More generally it commutes with an SL_2 action [2]. Starting with the F operator above it is easy to see, if we restrict attention to the module $F_0^{[0]}$ and write $y = x\xi$, that after symmetrization over the tensor product we obtain

$$Symm(F(f\otimes g)) = D(f\cdot g). \tag{3.28}$$

In soliton theory use of the formalism of the Hirota derivative allows the expression of integrable partial differential equations in "bilinear" form, in the sense alluded to in (3.27), an observation belonging to the Sato theory of the KdV and KP equations and their descendents [12].

3.4 Padé

In an attempt to motivate the observations of the following section consider the following Ricatti differential equation:

$$\frac{d\psi}{d\xi} = -\psi^2 - \xi\psi + m \tag{3.29}$$

m being a constant integer. The whole class of Ricatti equations is permuted under Möbius transformations, the group PSL_2,

$$\psi \mapsto \frac{a(\xi)\psi + b(\xi)}{c(\xi)\psi + d(\xi)}$$

and the particular equation (3.29) for solution $\psi = \psi_0$ is moved to

$$\frac{d\psi_1}{d\xi} = -\psi_1^2 - \xi\psi_1 + m - 1 \tag{3.30}$$

by the map

$$\psi_0 = \frac{m}{\xi + \psi_1}. \tag{3.31}$$

If m is a positive integer we can by a sequence of such maps reduce the equation to the case $m = 0$ and express the solution $\psi = \psi_0$ as the continued fraction,

$$\psi = \frac{m}{\xi +} \frac{m-1}{\xi +} \frac{m-2}{\xi +} \cdots \frac{1}{\xi}. \tag{3.32}$$

But continued fractions are a particular case of Padé approximants and so we anticipate a connection between the latter and the algebra $\mathfrak{sl}_2$.

Padé approximants are rational functions approximating analytic functions and as such they generalize the polynomial approximation obtained by truncating a Taylor series. Let $f^{[\sigma]}(x,y)$ be a function analytic in x and homogeneous of degree σ in x and y. It is associated with the module $F_0^{[\sigma]}$ of Sect. 3.4. Let $P^{[n,m]}(x,y)$ and $Q^{[n,m]}(x,y)$ be homogeneous polynomials in x and y of degrees n and m respectively (thus corresponding to finite dimensional modules of dimensions $n+1$ and $m+1$ respectively), satisfying,

$$P^{[n,m]} - Q^{[n,m]} f^{[\sigma]} = R^{[n,m]} = O(x^{n+m+1} y^{-m-1}). \tag{3.33}$$

These linear equations allow one to solve for the coefficients of P and Q (up to an overall factor) in terms of the first $n+m+1$ coefficients of f. Because we can interpret (3.33) as a relation between modules, all such relations are covariant and decompose into irreducible modules whose basis elements are permuted by e, f and h. This must hold no less for the classical recursion relations [8],

$$P^{[n,m]}Q^{[n',m']} - P^{[n',m']}Q^{[n,m]} = R^{[n,m]}Q^{[n',m']} - R^{[n',m']}Q^{[n,m]} \tag{3.34}$$

where $n-m = n'-m' = \sigma$

$$\alpha_{n+2,m+2}P^{[n+2,m+2]} + \alpha_{n+1,m+1}P^{[n+1,m+1]} + \alpha_{n,m}P^{[n,m]} = 0 \tag{3.35}$$

where $n-m=\sigma$.

The key element from the point of view of representation theory is the highest weight element

$$\Delta^{[n,m]} = \begin{vmatrix} f^{[\sigma]}_{n+1,-m-1} & f^{[\sigma]}_{n,-m} & f^{[\sigma]}_{n-1,-m+1} & \cdots & 0 \\ f^{[\sigma]}_{n+2,-m-2} & f^{[\sigma]}_{n+1,-m-1} & f^{[\sigma]}_{n,-m} & \cdots & 0 \\ \vdots & & & & \vdots \\ f^{[\sigma]}_{n+m+1,-2m-1} & f^{[\sigma]}_{n+m,-2m} & f^{[\sigma]}_{n+m-1,-2m+1} & \cdots & f^{[\sigma]}_{n+1,-m-1} \end{vmatrix} \tag{3.36}$$

This element generates the module associated with the analytic remainder function, $R^{[\sigma]}$, which can be regarded as a module of relations that should be satisfied were $f^{[\sigma]}$ to be an analytic expansion of a rational function. The element can also be identified, using the Casimir operator, as belonging to the symmetrization of the $m+1$-fold tensor product of $F_0^{[\sigma]}$ with itself.

Further, $\Delta^{[n,m]}$ is a higher order transvectant generated by application of Hirota maps to $\otimes^{m+1} f^{[\sigma]}_{0,-n+m}$.

Again, if we interpret the coefficients $f^{[\sigma]}_{i,i-\sigma}$ as, up to numerical factors, the Taylor coefficients of an analytic function, then the vanishing of $\Delta^{[n,m]}$ is a multilinear Hirota, ordinary differential equation whose general solution is the rational function $P(\xi)/Q(\xi)$.

Finally, there are recursion relations amongst the $\Delta^{[n,m]}$, most simply a four point relation on the lattice, $\mathbb{N}\times\mathbb{N}$, having the shape,

$$\Delta^{[n+1,m+1]}\Delta^{[n,m-1]} = \Delta^{[n,m]} \wedge \Delta^{[n+1,m]} \tag{3.37}$$

where $\wedge$ again denotes a transvectant-like operation [1].

3.5 Hyperelliptic

We concentrate in this section on hyperelliptic functions on Jacobians of genus two hyperelliptic curves [3]. Note that the roles of x and y in this section are not as in earlier sections.

The generic hyperelliptic, sextic curve is [5]:

$$y^2 = g(x) = g_6 x^6 + 6g_5 x^5 + 15g_4 x^4 + 20g_3 x^3 + 15g_2 x^2 + 6g_1 x + 6g_0. \tag{3.38}$$

This relation is covariant under an $\mathfrak{sl}_2$ action given by

$$e = \partial_x - \sum_{i=0}^{6} (6-i) g_{i+1} \partial_{g_i} \tag{3.39}$$

$$f = -x^2 \partial_x - 3xy\partial_y - \sum_{i=0}^{6} i g_{i-1} \partial_{g_i}. \tag{3.40}$$

in the sense that

$$e(y^2 - g(x)) = 0 \tag{3.41}$$

$$f(y^2 - g(x)) = -6x(y^2 - g(x)), \tag{3.42}$$

that is the polynomial ideal generated by the relation is $\mathfrak{sl}_2$-invariant.

The Jacobian variety of the curve is a two complex dimensional torus with local coordinates u_1 and u_2 obtained by integrating the following holomorphic 1-forms along paths:

$$du_1 = \frac{dx_1}{y_1} + \frac{dx_2}{y_2} \tag{3.43}$$

$$du_2 = \frac{x_1 dx_1}{y_1} + \frac{x_2 dx_2}{y_2} \tag{3.44}$$

where (x_1, y_1) and (x_2, y_2) are a pair of points on the curve.

du_1 and du_2 are a basis for holomorphic 1-forms as is easily seen by expanding in a local parameter. Further they are a two dimensional representation of $\mathfrak{sl}_2$:

$$du_2 \xrightarrow{e} du_1 \xrightarrow{e} 0 \tag{3.45}$$

$$0 \xleftarrow{f} du_2 \xleftarrow{f} du_1. \tag{3.46}$$

The genus two $\wp$-function is defined on the Jacobi variety by the algebro-differential relations,

$$\frac{1}{x_1 - x_2}(\wp_{,11} + (x_1 + x_2)\wp_{,12} + x_1 x_2 \wp_{,22}) - \frac{1}{(x_1 - x_2)^3}(\tilde{F}(x_1, x_2) - y_1 y_2) = 0 \tag{3.47}$$

where $\wp_{,ij} = -\partial_{u_i}\partial_{u_j} \ln \sigma(u_1, u_2)$, the σ-function playing the role of a potential. The function $\tilde{F}(x_1, x_2)$ is a polarization of $g(x)$ modified in such a way as to retain covariance.

The derivatives of the $\wp$-function of all orders, have to satisfy certain algebraic relations. Traditionally these are derived by first transforming a branch point of the curve to infinity, using a Möbius transformation, then expanding the relation (3.47) in a Taylor expansion in the local parameter of the place at infinity, [7]. One has then to undo the transformation to return the branch point to its finite, generic location. The simplest relations thus obtained are the fourth order, linear relations [4]:

$$-\frac{1}{3}\wp_{,2222}+2\wp_{,22}^2=g_2g_6-4g_3g_5+3g_4^2+g_4\wp_{,22}-2g_5\wp_{,12}+g_6\wp_{,11}$$

$$-\frac{1}{3}\wp_{,1222}+2\wp_{,22}\wp_{,12}=\frac{1}{2}(g_1g_6-3g_2g_5+2g_3g_4)+g_3\wp_{,22}-2g_4\wp_{,12}+g_5\wp_{,11}$$

$$-\frac{1}{3}\wp_{,1122}+\frac{2}{3}\wp_{,22}\wp_{,11}+\frac{4}{3}\wp_{,12}^2=\frac{1}{6}(g_0g_6-9g_2g_4+8g_3^2)$$
$$+g_2\wp_{,22}-2g_3\wp_{,12}+g_4\wp_{,11}$$

$$-\frac{1}{3}\wp_{,1112}+2\wp_{,11}\wp_{,12}=\frac{1}{2}(g_0g_5-3g_1g_4+2g_2g_3)+g_1\wp_{,22}-2g_2\wp_{,12}+g_3\wp_{,11}$$

$$-\frac{1}{3}\wp_{,1111}+2\wp_{,11}^2=g_0g_4-4g_1g_3+3g_2^2+g_0\wp_{,22}-2g_1\wp_{,12}+g_2\wp_{,11}$$

One of the consequences of the covariance of the algebraic relations is that identities for the $\wp$-function must group together into modules. It is easily checked, for instance, that the above five relations are related by applications of e and f and form a five dimensional module.

Further, one can rewrite the identities in transvectant form. The last, for instance is,

$$\frac{1}{6}D_1^4(\sigma\cdot\sigma)=(g_0,g_0)^{(4)}\sigma^2-(g_0,D_1^2(\sigma\cdot\sigma))^{(2)} \tag{3.48}$$

where D_1 is the Hirota derivative with respect to u_1 and the tranvectants in the g's are an appropriate choice of coefficient from exactly the sort of operation introduced in Sect. 3.3.

These Baker equations for the $\wp$ function are themselves compatibility relations for differential operators (a Lax pair) defined on the Jacobian. This allows one, in particular, to construct genus two solutions of the Boussinesq hierarchy of integrable partial differential equations.

The representation theory gives one a way of deriving and classifying all such relations and will hopefully provide a machine that can, for arbitrary genus, replace the somewhat ad hoc methods of singularity expansions with a systematic algorithm.

I would like to thank the University of Lund, Sweden, and my home institution, Glasgow University, for supporting my visit to the AGMF workshop in Lund.

References

1. Athorne, C.: A novel approach to the theory of Padé approximants. J. Nonlinear Math. Phys. **12**, 15–27 (2005)
2. Athorne, C.: Algebraic Hirota maps. In Faddeev, L., Van Moerbeke, P., Lambert, F. (Eds.) Bilinear Integrable Systems: from Classical to Quantum, Continuous to Discrete, pp. 17–33. Springer, Dordrecht (2006)

3. Athorne, C., Eilbeck, J.C., Enolskii, V.Z.: Identities for the classical genus two $\wp$-function. J. Geom. Phys. **48**, 354–368 (2003)
4. Baker, H.F.: Multiply Periodic Functions. Cambridge University Press, Cambridge (1907)
5. Baker, H.F.: Abelian Functions. Cambridge University Press, Cambridge (1987)
6. Coxeter, H.S.M.: Projective Geometry. Springer, New York (1994)
7. Eilbeck, J.C., Enolskii, V.Z., Leykin, D.V.: On the Kleinian construction of abelian functions of canonical algebraic curves. In D. Levi, O. Ragnisco (Eds.) SIDE III – Symmetries and Integrability of Difference Equations, pp. 121–138. CRM Proceedings and Lecture Notes, vol. 25. Am. Math. Soc., Providence, RI (2000)
8. Frobenius, F.G., Stickelberger, L.: Zur Theorie der elliptischen Functionen. J. Reine Angew. Math. **83**, 175–179 (1877)
9. Grammaticos, B., Ramani, A., Hietarinta, J.: Multilinear operators: the natural extension of Hirota's bilinear formalism. Phys. Lett. A **190**, 65–70 (1994)
10. Hilbert, D.: Lectures on Algebraic Invariant Theory. Cambridge University Press, Cambridge (1993)
11. Hirota, Ryogo: Direct method of finding exact solutions of nonlinear evolution equations. In R.M. Miura (Ed.) Bäcklund Transformations, the Inverse Scattering Method, Solitons, and their Applications, pp. 40–68. Lecture Notes in Mathematics, vol. 515, Springer, Berlin (1976)
12. Miwa, T., Jimbo, M., Date, E.: Solitons: Differential Equations, Symmetries and Infinite-Dimensional Algebras. Cambridge Tracts in Mathematics, vol. 135. Cambridge University Press, Cambridge (2000)
13. Olver, Peter J.: Classical invariant theory. London Mathematical Society Student Texts, vol. 44. Cambridge University Press, Cambridge (1999)
14. Sanders, Jan A.: Multilinear Hirota operators, modular forms and the Heisenberg algebra. In International Conference on Differential Equations, pp. 812–823. World Scientific, River Edge, NJ (2000)

Chapter 4
Automorphisms of Finite Orthoalgebras, Exceptional Root Systems and Quantum Mechanics

Artur E. Ruuge and Fred Van Oystaeyen

Abstract An orthoalgebra is a partial abelian monoid whose structure captures some properties of the direct sum operation of the subspaces of a Hilbert space. Given a physical system (quantum or classical), the collection of all its binary observables (properties) may be viewed as an orthoalgebra. In the quantum case, in contrast to the classical, the orthoalgebra cannot have a "bivaluation" (a morphism ending in a two-element orthoalgebra). An interesting combinatorial problem is to construct finite orthoalgebras not admitting bivaluations. In this paper we discuss the construction of a family such examples closely related to the irreducible root systems of exceptional type.

4.1 Introduction

In this paper we discuss some features of a concrete family of examples of finite orthoalgebras. The concept of an orthoalgebra is a little more general than a well known concept of a *lattice*. The basic idea is that instead of a *globally* defined binary operation $\vee$, one considers just a *partially* defined binary operation $\oplus$ with similar properties.

Let S be a set equipped with a relation $R \subset S \times S$. Let $\oplus : R \to S$, $(x,y) \mapsto x \oplus y$ be a map. Let $\mathbf{0}$ and $\mathbf{1}$ be two distinct elements of S. The collection of these data is called an *orthoalgebra* if for any $x, y, z \in S$ the following properties are satisfied:

A.E. Ruuge
Faculteit Ingenieurswetenschappen, Vrije Universiteit Brussel (VUB), Pleinlaan 2, 1050 Brussels, Belgium
e-mail: artur.ruuge@ua.ac.be

F.V. Oystaeyen
Dept. Wiskunde-Informatica, Universiteit Antwerpen (UA), Middelheimlaan 1, 2020 Antwerp, Belgium
e-mail: fred.vanoystaeyen@ua.ac.be

S. Silvestrov et al. (eds.), *Generalized Lie Theory in Mathematics, Physics and Beyond,*

(1) if $x \oplus y$ is defined, then $y \oplus x$ is defined and $y \oplus x = x \oplus y$; (2) if $(x \oplus y) \oplus z$ is defined, then $x \oplus (y \oplus z)$ is defined and $x \oplus (y \oplus z) = (x \oplus y) \oplus z$; (3) $x \oplus \mathbf{0}$ is defined and $x \oplus \mathbf{0} = x$; (4) there exists a unique $x^* \in S$ such that $x^* \oplus x = \mathbf{1}$; (5) if $x \oplus x$ is defined, then $x = \mathbf{0}$. A morphism between two orthoalgebras is defined in a natural way as a map between the underlying sets which respects $\oplus$, $\mathbf{0}$ and $\mathbf{1}$.

A prototype example of an orthoalgebra is the *Hilbert space orthoalgebra* (which is in fact a lattice) consisting of all closed linear manifolds of a Hilbert space $\mathscr{H}$: the partial operation $(x,y) \mapsto x \oplus y$ is just the direct sum of two orthogonal subspaces x and y, the $\mathbf{0}$ element is the trivial subspace and the $\mathbf{1}$ element is the space $\mathscr{H}$ itself. Denote this orthoalgebra by $L(\mathscr{H})$. At the same time, there exist orthoalgebras which are not embeddable into a Hilbert space orthoalgebra.

If we have an abstract physical system X described by non-relativistic quantum mechanics, then its zero-one observables (also termed *properties*) may be identified with the underlying set of a Hilbert space orthoalgebra $L(\mathscr{H})$. At the same time, if one has an abstract physical system Y which is described by classical mechanics, then its properties form an orthoalgebra as well: one has a measurable space $(\Omega, \mathscr{F})$ termed the *phase space*, the properties are identified with the elements of $\mathscr{F}$, which is taken as the ground subset for the orthoalgebra, the domain of definition of $\oplus$ is the set of all disjoint pairs (U,V), $U \oplus V = U \cup V$, $\mathbf{0}$ is the empty subset of Ω and $\mathbf{1}$ is the set Ω itself.

Imagine the following situation. Suppose that we have a physical system Z, but we *do not know* a priori if it's classical or quantum. What is necessary to do to answer this question by observing the system? This motivates the following mathematical problem: if we have an orthoalgebra defined in set-theoretic terms, how could one see the fact that it corresponds to a quantum but not to a classical system?

Given an orthoalgebra $(S,R,\oplus,\mathbf{0},\mathbf{1})$, one may consider all decompositions of the unit element $\mathbf{1} = \oplus_i x_i$. If this orthoalgebra corresponds to a measurable space $(\Omega, \mathscr{F})$, $S = \mathscr{F}$, then there exists a function $v : S \longrightarrow \{0,1\}$ such that for any decomposition $\mathbf{1} = \oplus_i x_i$ there exists a unique i_0 such that $v(x_{i_0}) = 1$. It suffices to take any $\omega \in \Omega$ and then put $v(x) = 1$ if $x \ni \omega$ and 0 – otherwise. Being put in other terms, the existence of such a function v is equivalent to the existence of a morphism from the considered orthoalgebra to a two-element orthoalgebra. One terms such a morphism a *bivaluation*.

If we now look at the Hilbert space orthoalgebra $L(\mathscr{H})$, assuming that $\mathscr{H}$ is complex and $\dim \mathscr{H} > 2$, then functional analysis tells us that a $L(\mathscr{H})$ *does not* admit a bivaluation (invoke the Gleason's theorem). Therefore we have a way to decide (in principle) if our system is quantum or not by looking at the existence of bivaluations. In case the dimension of $\mathscr{H}$ is finite, then the situation is more interesting. It turns out, that it suffices to consider just a *finite* number of rays in $\mathscr{H}$ in order to establish the non-existence of a bivaluation (i.e. assuming that a bivaluation exists and then restricting it to these rays leads to a contradiction). For the case where $\dim \mathscr{H} = 3$, this fact has been understood by Kochen and Specker [1].

Since there exist orthoalgebras having just a *finite* number of elements, it is interesting to consider the following problems:

- How to construct a finite orthoalgebra which does not admit a bivaluation?
- Is it possible to embed such an orthoalgebra into a Hilbert space orthoalgebra?

In the following sections we will construct examples of orthoalgebras of these kind and discuss their relation to the exceptional irreducible root systems.

4.2 Saturated Configurations

Let $\mathscr{H}$ be a complex or real Hilbert space of finite dimension d. Let $A = \{\rho_1, \rho_2, \dots, \rho_N\} \subset \mathbb{P}(\mathscr{H})$ be a finite collection of rays. Then, if we look at all tuples of mutually orthogonal elements of A, the maximal size of the tuple we can get is the dimension of space d. For $k = 1, 2, \dots, d$, denote the set of all k-tuples of mutually orthogonal elements of A by $\mathscr{P}_\perp^{(k)}(A)$. We say that A is *saturated* if for all k and for all $B \in \mathscr{P}_\perp^{(k)}(A)$ there exists $M \in \mathscr{P}_\perp^{(d)}(A)$ such that $M \supset B$.

If A is saturated, then the collection of all subspaces span$U$, $U \in \mathscr{P}_\perp^{(k)}(A)$, $k = 1, 2, \dots, d$, adjoined with the trivial subspace of $\mathscr{H}$, forms a finite orthoalgebra.

Where can one find saturated collections of rays? It turns out that good examples come the irreducible root systems of exceptional type. Recall that these are G_2, F_4, E_6, E_7 and E_8. Their root systems are given by collections of vectors in the vector spaces of dimensions $2, 4, 6, 7$ and 8, respectively; the cardinalities of the root systems are $12, 48, 72, 126$ and 240, respectively. Consider the rays represented by the vectors of the root systems. This yields the collections of rays which have cardinalities $6, 24, 36, 63$ and 120, respectively.

We have the following fact (it can be verified on a personal computer in a straightforward way):

Proposition 4.1. *(1) The collections of rays stemming from the root systems G_2, F_4, E_7 and E_8 are saturated;*

(2) The collection of rays stemming from E_6 is not saturated, but has the following property: for all $k = 1, 2, 3, 4$, and for all $B \in \mathscr{P}_\perp^{(k)}(A)$ there exists $N \in \mathscr{P}_\perp^{(4)}(A)$ such that $N \supset B$; the sets $\mathscr{P}_\perp^{(5)}(A)$ and $\mathscr{P}_\perp^{(6)}(A)$ are empty.

Therefore one obtains four examples of finite orthoalgebras embeddable into a Hilbert space orthoalgebra, and one needs more rays in the E_6 case.

4.3 Non-Colourable Configurations

It is now necessary to test if the orthoalgebras corresponding to the G_2, F_4, E_7 and E_8 cases admit bivaluations. It is convenient to introduce the following definition. Let A be a finite saturated collection of rays in a d-dimensional Hilbert space $\mathscr{H}$. We say that A is *colourable* if there exists a function $v : A \to \{0, 1\}$, such that for every $B \in \mathscr{P}_\perp^{(d)}(A)$ there exists a unique $\rho \in B$ such that $v(\rho) = 1$. Otherwise A is termed *non-colourable*.

It is clear that if an orthoalgebra comes from a saturated configuration of rays A, then a bivaluation on it (if it exists), induces a colouring of A. We have the following

Proposition 4.2. *(1) The configuration corresponding to G_2 is colourable;*
(2) The configurations corresponding to F_4, E_7 and E_8 are non-colourable.

Hence we obtain three examples of finite Hilbert space suborthoalgebras not admitting bivaluations (the F_4, E_7, E_8 cases). Note that the colourability fact in the G_2 case is actually trivial: one can show that in dimension 2 a configuration A is always colourable. The proofs for the F_4, E_7 and E_8 cases can be implemented in Maple or done analytically, but we do not dwell on this subject here. We just remark, that they follow the same pattern: the whole collection A splits into a *disjoint* union of subsets $B_1, B_2, \dots, B_m \in \mathscr{P}_{\perp}^{(d)}(A)$, where d is 4, 7, or 8, respectively, and m is one half of the Coxeter number of the root system (6, 9 or 15, respectively). If the configuration A were colourable, one could select from each B_i an element r_i in such a way that $r_1, r_2, \dots, r_m$ would have been mutually non-orthogonal. A Maple computation shows that this cannot be done.

4.4 The E_6 Case

It turns out that the configuration coming from the E_6 root system can be adjoined with a finite number of rays in such a way that the result is saturated and non-colourable. The proof of this fact is quite sophisticated and we do not discuss it here, but only describe the construction. The details are discussed in [2].

It is more convenient to model the E_6 root system not on $\mathbb{R}^6$, but on a six-dimensional subspace of $\mathbb{R}^9$ formed by the vectors $(x_1, x_2, x_3; y_1, y_2, y_3; z_1, z_2, z_3) \equiv (x; y; z)$ satisfying $\sum_i x_i = \sum_i y_i = \sum_i z_i = 0$ $(i = 1, 2, 3)$. Denote this subspace by R.

Consider the following six vectors:

$$\begin{aligned} v_1 &= (1, \bar{1}, 0; 0, 0, 0; 0, 0, 0), & v_2 &= (2, \bar{1}, \bar{1}; 2, \bar{1}, \bar{1}; 2, \bar{1}, \bar{1}), \\ v_3 &= (1, \bar{1}, 0; 1, \bar{1}, 0; 0, 0, 0), & v_4 &= (2, \bar{1}, \bar{1}; \bar{2}, 1, 1; 0, 0, 0), \\ v_5 &= (\bar{4}, 2, 2; 2, \bar{1}, \bar{1}; 2, \bar{1}, \bar{1}), & v_6 &= (2, \bar{1}, \bar{1}; 0, 0, 0; 0, 0, 0), \end{aligned}$$

where the bar set over a number denotes negation (i.e. $\bar{k} = -k$, $k \in \mathbb{Z}$). Observe that if we take an element of S_3, then we can naturally act with it in four different ways on an element $(x; y; z) \in R$: one can permute $\{x_i\}_i$, or $\{y_j\}_j$, or $\{z_k\}_k$, or one can permute x, y, z in $(x; y; z)$. This defines an action of the wreath product $S_3 \wr S_3$ on R. There also exists a natural action $(\mathbb{Z}/2\mathbb{Z})^3$ on R given by negations of the x, y and z components (for example, $(1, 0, 0) \in (\mathbb{Z}/2\mathbb{Z})^3$ maps $(x; y; z)$ to $(\bar{x}; y; z)$, where $\bar{x} = (\bar{x}_1, \bar{x}_2, \bar{x}_3)$).

The E_6 root system can be realized on the subspace $R \subset \mathbb{R}^9$ as the union of the $(S_3 \wr S_3)$-orbits of the vectors v_1, $(1/3)\, v_2$ and $(-1/3)\, v_2$. Denote by O_i the orbit of v_i $(i = 1, 2, \dots, 6)$ under the action of the free product of $S_3 \wr S_3$ and $(\mathbb{Z}/2\mathbb{Z})^3$.

Theorem 4.1. *The configuration of rays A corresponding to the union of the orbits* $\cup_{i=1}^{6} O_i$ *is saturated and non-colourable.*

Note that the number of elements in this A is 558, while the projective configuration corresponding to the E_6 root system itself is much smaller, just 36 elements. We obtain another example of finite orthoalgebra without bivaluations.

4.5 Orthoalgebras Generated by E_8

In this section we focus on the orthoalgebra corresponding to the E_8 root system. It turns out that one can obtain an infinite family of orthoalgebras similar to this one.

Recall that for the E_8 case we have a saturated collection of 120 rays in the eight-dimensional space. One takes the spans of the tuples of mutually orthogonal rays and by that obtains an orthoalgebra. How can one mimic this construction if instead of rays one takes just an abstract set A?

It is necessary to consider some relation $T \subset A \times A$ thinking about it as a replacement for the orthogonality relation $\perp$. We assume that T is symmetric and anti-reflexive (just like the $\perp$ relation for the rays). For any $B \subset A$ we use the notation B^T to denote the set of all $x \in A$ such that for all $y \in B$ one has $(x,y) \in T$.

Let $\mathscr{P}_T(A)$ denote the set of all subsets $B \subset A$ such that for all $x, y \in B$, if $x \neq y$, then $(x,y) \in T$. Note that the empty subset of A, as well as all the singletons of A, are in $\mathscr{P}_T(A)$. Viewing $\mathscr{P}_T(A)$ as a partially ordered set with respect to inclusion of subsets, denote by $\mathscr{M}_T(A)$ the set of all its maximal elements. The construction proceeds as follows:

- For each $U \in \mathscr{P}_T(A)$ consider U^T; denote the set of all U^T by $\mathscr{L}_T(A)$. This set will play the role of the ground set for an orthoalgebra.
- Call the relation T *saturated* if for each $M \in \mathscr{M}_T(A)$, if one takes any $U \subset M$, then one has $U^{TT} = (M \backslash U)^T$.

Note that if we take a saturated collection of rays, then the $\perp$ relation on it is saturated. Denote by $R_T(A)$ the set of all pairs (Q, Q_1), $Q, Q_1 \in \mathscr{L}_T(A)$, such that $Q_1 \subset Q^T$. With this notation we have

Theorem 4.2. *Let A be a non-empty finite set equipped with an antireflexive, symmetric and saturated relation T. Then the map* $(Q, Q_1) \mapsto (Q \cup Q_1)^{TT}$, $Q, Q_1 \subset A$, *induces a map* $-\oplus- : R_T(A) \to \mathscr{L}_T(A)$, *which yields an orthoalgebra structure on* $\mathscr{L}_T(A)$. *The* $\mathbf{0}$ *element of this orthoalgebra is* $\emptyset$, *and* $\mathbf{1} = A$.

We will now describe a family of examples of (A, T) satisfying the conditions of this theorem. The aim is to produce an infinite family of finite orthoalgebras generalizing the E_8 case. At this moment, recall that the collection of rays represented by the roots of E_8, can be split into a disjoint union of 15 subsets consisting of eight mutually orthogonal rays each.

Let $\mathbb{F}_2$ be a finite field of two elements. Consider the vector space $V := \mathbb{F}_2^N$, where N is a positive integer; denote the standard inner product on V by $\langle -,- \rangle$. Denote by $V^\times$ the set of all non-zero vectors in V (one may identify $V^\times$ with the projective space $\mathbb{P}(V)$). Next consider $W := V \oplus V$ and view it as a symplectic vector space taking the standard symplectic structure ω given by $\omega((x,y),(x_1,y_1)) := \langle x, y_1\rangle - \langle x_1, y\rangle$. Let $b : V^\times \to \mathbb{F}_2$ and $c : V^\times \times V^\times \to \mathbb{F}_2$ be two functions. Define a set $A_N(b)$

$$A_N(b) := \{(x,y) \in W \,|\, x \in V^\times \,\&\, \langle x,y\rangle = b(x)\}. \tag{4.1}$$

Note, that $A_N(b)$ can be identified with a subset of $\mathbb{P}(W)$. Then define a relation $T_N(b,c)$ as a set of all pairs $(x,y) \neq (x_1,y_1)$ such that

$$\omega((x,y),(x_1,y_1)) = c(x,x_1) + 1. \tag{4.2}$$

Note that if $y_1 \neq y$, then $((x,y),(x,y_1)) \in T_N(b,c)$. Therefore one can view $A_N(b,c)$ as a union of mutually *disjoint* subsets $L_N(b;x)$ formed by the elements (x',y') having the first component $x' = x$. Note that if we put $b(\cdot)$ and $c(\cdot)$ to zero, then we get just a bunch of lagrangian planes.

The described construction has some nice properties. It turns out that one can adjust the parameters N, $b(\cdot)$ and $c(\cdot)$ in such a way that $T_N(b,c)$ becomes a saturated antireflexive symmetric relation. For $v = (v_1, v_2, \ldots, v_N) \in V$, denote $|v| := \#_4\{i \,|\, v_i = 1\}$, where $\#_4$ denotes the cardinality of a set modulo 4. We have

Theorem 4.3. *Let N be a positive integer divisible by 4. Let $g : \mathbb{Z}/4\mathbb{Z} \to \mathbb{F}_2$ be a function defined by $g(0) = 1$ and $g(1) = g(2) = g(3) = 0$. For $x \in V$ put $b(x) = g(|x|)$ and $c(x,x) = 0$; for $x, z \in V$, $z \neq x$, put $c(x,z) = g(|z|) + g(|x|) + g(|x+z|)$. Then:*

(1) The relation $T_N(b,c)$ is antireflexive, symmetric and saturated.

(2) The orthoalgebra corresponding to $(A_N(b,c), T_N(b,c))$ does not admit bivaluations.

(3) If $N = 4$, then this orthoalgebra is isomorphic to the orthoalgebra corresponding to the E_8 root system.

Let us make just a few remarks about one of the possible ways to establish the absence of bivaluations. Suppose we have some N, b and c, not necessarily as in the theorem above, which ensure that the corresponding relation is antireflexive, symmetric, saturated. The ground set of the corresponding orthoalgebra contains, in particular, all singletons of $A_N(b)$, and for each $x \in V^\times$ we have a decomposition of the unit element $\mathbf{1} = \oplus_y\{(x,y)\}$, where y varies over all values such that $\langle x,y\rangle = b(x)$. If our orthoalgebra admits bivaluations, then there should exist a function $\sigma : V \to V$, such that $\langle x, \sigma(x)\rangle = b(x)$, and $\langle x, \sigma(z)\rangle - \langle z, \sigma(x)\rangle = c(x,z)$, where z and x vary over $V^\times$ and $x \neq z$. This yields an *overdetermined* system on $\{\sigma(x)\}_x$, from which one can derive a condition on c:

$$c(x,z) - \sum_i x_i c(e_i, z) - \sum_j z_j c(x, e_j) + \sum_{i,j} x_i z_j c(e_i, e_j) = 0, \tag{4.3}$$

where x_i and z_j are the coordinates of x and z in the standard basis of $V = \mathbb{F}_2^N$. It is straightforward to check that if N and c are as in the theorem above, then this condition is *not* satisfied. Hence the bivaluations do not exist.

The latter equation can be reformulated in terms of "lifting property". If we have a function $\phi : V \times V \to \mathbb{F}_2$, then it is the same thing as having a bilinear map $\varphi : F \times F \to \mathbb{F}_2$, where $F := \mathbb{F}_2\langle \underline{V} \rangle$ is the free $\mathbb{F}_2$-vector space over the underlying set $\underline{V}$ of V. Now consider c and extend it to all $x, z \in V$ as follows: $c(0,z) = 0$, $c(x,0) = 0$. The map $(x,z) \mapsto c(x,z)$ gives us a bilinear function $\tilde{c} : F \times F \to \mathbb{F}_2$. Then the equation on c simply says that the function $\tilde{c}$ comes from a *bilinear* map $\beta : V \times V \to \mathbb{F}_2$, i.e. $\tilde{c} = \beta \circ \pi$, where $\pi : F \times F \to V \times V$ is induced by the canonical (summing) epimorphism $F \twoheadrightarrow V$. This is illustrated by the commutative diagram

$$\begin{array}{ccc} F \times F & \xrightarrow{\tilde{c}} & \mathbb{F}_2 \\ \Big\downarrow \pi & \nearrow \beta & \\ V \times V & & \end{array} \tag{4.4}$$

(the category is the category of finite dimensional $\mathbb{F}_2$-vector spaces).

The function $\tilde{c}$ corresponding to c from the theorem does not have the lifting property. At the same time, if one takes *any* $\tau \in \mathrm{Aut}(V) = \mathrm{GL}_N(\mathbb{F}_2)$, the function $\tilde{c}_\tau$ corresponding to $c_\tau(x,z) := c(\tau(x), \tau(z)) - c(x,z)$ *does* have it. This fact is quite important for the construction of the family of automorphisms of the orthoalgebras defined by $(A_N(b), T_N(b,c))$; these automorphisms are being used to verify that the relation $T_N(b,c)$ is saturated. In particular, for each $\tau \in \mathrm{GL}_N(\mathbb{F}_2)$, one can construct the maps $A_N(b,c) \to A_N(b,c)$ respecting $T_N(b,c)$ of the following shape $(x,y) \mapsto (\tau(x), (\tau^t)^{-1}(y) + a_\tau(x))$, where $(\cdot)^t$ denotes the transposition and $a_\tau(x)$ are some elements of V. See [3] for details.

4.6 Conclusions

The motivation for investigating the finite orthoalgebras without bivaluations comes from quantum mechanics. A family of examples of such orthoalgebras corresponds to the irreducible root systems of exceptional types. It would be interesting to investigate in analogy to these results the case of classical root systems (in particular, A_n). The E_8 system generates an infinite family of orthoalgebras. It would be interesting to generalize the corresponding construction to other root systems.

References

1. Kochen, S., Specker, E.P.: The problem of hidden variables in quantum mechanics. J. Math. Mech. **17**, 59–87 (1967)
2. Ruuge, A.E.: Exceptional and non-crystallographic root systems and the Kochen–Specker theorem. J. Phys. A. **40**(11), 2849–2859 (2007)
3. Ruuge, A.E., Van Oystaeyen, F.: New families of finite coherent orthoalgebras without bivaluations. J. Math. Phys. **47**(2), 022108-1–022108-32 (2006)

Chapter 5
A Rewriting Approach to Graph Invariants

Lars Hellström

Abstract Diagrammatic calculation is a powerful tool that gets near indispensable when one tries to manage some of the newer algebraic structures that have been popping up in the last couple of decades. Concretely, it generalises the underlying structure of expressions to being general graphs, where traditional algebraic notation only supports path- or treelike expressions. This paper demonstrates how to apply the author's Generic Diamond Lemma in diagrammatic calculations, by solving through elementary rewriting techniques the problem of classifying all multigraph invariants satisfying a linear contract–delete recursion. (As expected, this leads one to rediscover the Tutte polynomial, along with some more degenerate invariants.) In addition, a concept of "semigraph" is defined which formalises the concept of a graph-theoretical "gadget".

5.1 Background

Diagrams are abundant in both in physics and mathematics, but what is of interest here is specifically diagrams that one takes as direct objects of calculations. The best known type of such diagram is probably the Feynman diagrams – see for example [8, (6.22)] – but this is not ideal as a first example since there is so much more going on with these than just the *diagrammatic* calculations. A cleaner example is provided by the *tensor diagram notation*, which is called "birdtracks" in [5] and spin networks in [10]. A more abstract example, which however closely corresponds to the terms (formal expressions) of classical rewrite theory, is provided by the diagram shorthand for expressions in Hopf algebras [9], and more abstract still are the "string diagrams" for composite morphisms in 2-categories [2]. While the intended interpretations of these diagram types vary, there are also obvious similarities, and it

L. Hellström
Sand 216, 881 91 Sollefteå, Sweden
e-mail: Lars.Hellstrom@residenset.net

S. Silvestrov et al. (eds.), *Generalized Lie Theory in Mathematics, Physics and Beyond,*

is of this common core that diagrammatic calculations make use. Apart from some basic restrictions of a primarily syntactic character that stem from the need to have a certain interpretation, there is little that distinguishes different flavours of diagrammatic calculations from each other, and the basic principles can therefore be demonstrated with pretty much any kind of diagram.

My work in this area has been motivated primarily by problems concerning Hopf algebras, so in that sense it would have been natural to demonstrate rewriting using shorthand diagrams, but unfortunately time would not permit this; even though the syntactic structure of these diagrams is fairly simple on an intuitive level, it is still too complicated to allow a rigorous treatment in the space available here. Instead I have chosen to demonstrate rewriting on graphs, because that is what one ends up with when all syntactic restrictions are removed; other types of diagrams generally have to be formalised as graphs with extra structure satisfying various constraints, but these have little effect on the way in which the rewriting machinery works. What one loses by considering plain graphs is the direct interpretation of diagrams as expressions that can be evaluated, but evaluation is just a special case of map from diagrams to some other domain; in many cases it might be just as interesting to evaluate some quantity related to the diagram.

5.2 Graph Theory

This section is an informal review of the graph-theoretical concepts that are relevant in this paper. The methods used in Sects. 5.4–5.5 do not logically depend on this material, but it should help to explain why the given problem is relevant and how it relates to known results. For graph-theoretical concepts not defined in this text, I refer to Diestel [6]. I will mostly follow his terminology, but prefer to reserve the term *multigraph* for multigraphs without loops; if loops (and multiple edges) are allowed then the object will be called a *pseudograph*. K_n is the complete graph on n vertices. K_n^{C} is the complement of K_n, i.e., the graph with n vertices but no edges.

The *chromatic polynomial* $P_\chi(G)$ for a graph G is defined by the property that $P_\chi(G)(k)$ is the number of vertex-k-colourings of G, for any natural number k. That this function is always a polynomial is at first sight surprising, but an easy proof can be based on the delete–contract recursion for P_χ:

$$P_\chi(G)(k) = P_\chi(G-e)(k) - P_\chi(G/e)(k) \quad \text{for all } k \in \mathbb{N} \text{ and } e \in \mathrm{E}(G). \tag{5.1}$$

Here the notation $G-e$ means "the graph G with the edge e deleted", whereas G/e means "the graph G with the edge e contracted", i.e., the two endpoints of e are identified; see Fig. 5.1 for an example. The proof of this recursion is almost trivial: a k-colouring of $G-e$ either assigns different colours to the endpoints of e, and in that case it is a k-colouring of G, or assigns the same colour to the endpoints of e, and in that case it defines a k-colouring of G/e; subtract the latter, and you get the expression for the former. Since each step of the recursion decreases the size (number of edges) of the graphs involved by at least 1, one arrives after a finite

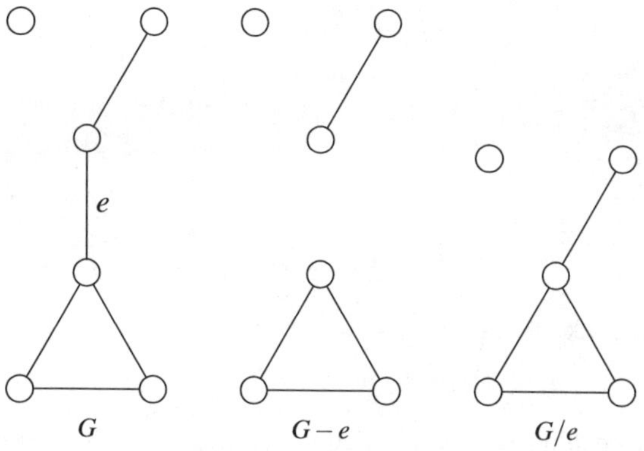

Fig. 5.1 Deletion and contraction of an edge

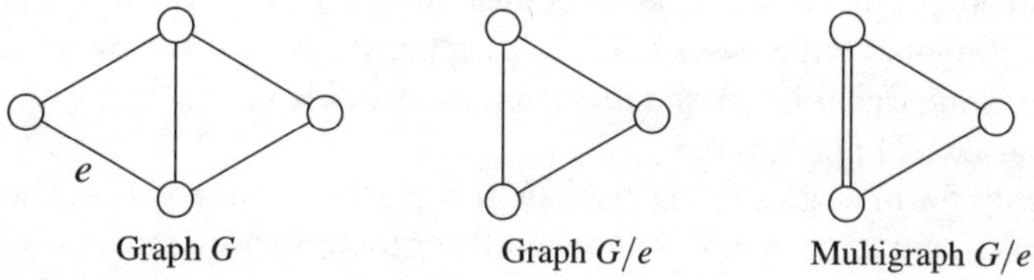

Fig. 5.2 Contracting an edge in a triangle

number of steps at a linear combination of chromatic polynomials for size 0 graphs, and these are easily found to be polynomials in k; since any assignment of colours to the vertices in K_n^{C} is a colouring, there are exactly k^n k-colourings of this graph, and hence $P_\chi(K_n^{\mathrm{C}})(x) = x^n$. The recursion (5.1) is the foremost tool for computing the chromatic polynomial, and by extension even for computing the chromatic number, of a general graph (even though relying solely on this recursion often makes the task much more laborious than it has to be).

A function of graphs which does not depend on which the vertices and edges are, but only on how they are connected, is called an *invariant*; formally a function Q is an invariant if $Q(G) = Q(H)$ whenever G and H are isomorphic. The chromatic polynomial is an invariant, and interestingly enough there are also several other graph invariants which sport similar delete–contract recursions. Hence it becomes an interesting problem to classify these invariants and perhaps find new ones. In order to do so, one must however first clarify exactly what the deletion and contraction operations should do. There is a certain amount of hindsight here, in that the choices I make are primarily dictated by the method I want to apply, but there are also more generic reasons for making these choices, and it is certainly worth while to explain them.

The difficult operation is edge contraction. The first choice (Fig. 5.2) one has to make when defining it is to decide what happens when one contracts an edge in a triangle (3-cycle): should the two remaining edges count as one edge or two? The

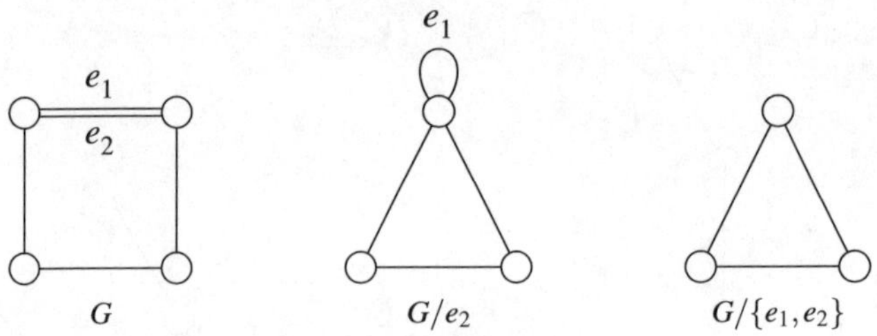

Fig. 5.3 Contracting a double edge

disadvantage of keeping two parallel edges in this case is that it means edge contraction can transform a graph into a multigraph. For the chromatic polynomial one or two edges make no difference – the endpoints are not allowed to have the same colour in either case – but for some other invariants (e.g., the flow polynomial) it is crucial to distinguish these cases, and as it happens it is then the multigraph contraction that is the right one for the recursion. Hence it is natural to let the problem concern invariants of multigraphs.

Opening up for multigraphs leads to another problem, namely what should happen when one contracts a double (or even higher multiplicity) edge, as in Fig. 5.3. Contracting one of two parallel edges will turn the other one into a loop – must one therefore extend the argument to consider also pseudographs? Although the tradition indeed is to do so, there is no compelling computational reason to take this route; loops just sit on particular vertices, so they behave as an extra set of weights that for most part can be ignored. It is possible to stay within the realm of multigraphs if one decides to always delete or contract all edges between the two endpoints in a single step, rather than removing them one at a time; this way the multigraph recursion for the chromatic polynomial proceeds exactly as with graph contraction of edges.

5.3 The Problem

Formally, the problem studied here is to classify all multigraph invariants Q, with values in a vector space W over some field $\mathcal{R}$, that for all multigraphs G satisfy an identity on the form

$$Q(G) = \alpha_{|\bar{e}|} Q(G - \bar{e}) - \beta_{|\bar{e}|} Q(G/\bar{e}) \quad \text{for all } e \in \mathrm{E}(G). \tag{5.2}$$

By $\bar{e}$ is meant the set of all edges with the same endpoints as the edge e; that this set can be regarded as the closed hull $\bar{e}$ of e is well established in matroid theory. The coefficients $\{\alpha_m\}_{m=1}^{\infty}$ and $\{\beta_m\}_{m=1}^{\infty}$ may be arbitrary elements of $\mathcal{R}$; to a great extent a class of invariants is determined by a particular parameterisation of these coefficients.

Most of the rewriting arguments in subsequent sections could just as well be carried out for coefficients from a commutative ring with unit, but the subsequent parameterisation of these coefficients for particular invariant classes becomes much easier if one can invert nonzero ring elements. As this is meant to be a pedagogical presentation, one shouldn't introduce unnecessary difficulties; those interested in the greater generality are referred to the classification of Bollobás and Riordan [4] instead.

For an algebraic treatment of graph invariants satisfying the prescribed kind of recursion, it is convenient to first fold away the condition that Q should be an invariant. A formal solution to this would be to regard Q as a function of isomorphism classes of multigraphs, as this is an alternative definition of invariant. In graph theory, such isomorphism classes are sometimes known as *unlabelled multigraphs* (multigraphs where no names have been assigned to the vertices and edges), whereas an ordinary multigraph G for which one may speak of distinct sets of vertices $\mathrm{V}(G)$ and edges $\mathrm{E}(G)$ is a *labelled multigraph.*[1] Graph theorists rarely bother about upholding this distinction, but we will have to be careful about it when we get to do algebra with graphs.

Let $\mathcal{Y}$ denote the set of all unlabelled multigraphs, which will be interpreted as a set of equivalence classes of labelled multigraphs – hence its elements will be written as $[G]$ for some labelled multigraph G. The sought class of invariants is then just the set of those functions $Q\colon \mathcal{Y} \longrightarrow W$ which satisfy

$$Q([G]) = \alpha_{|\bar{e}|} Q([G-\bar{e}]) - \beta_{|\bar{e}|} Q([G/\bar{e}]) \tag{5.3}$$

for all $e \in \mathrm{E}(G)$ and labelled multigraphs G. Next let $\mathcal{M}$ be the set of all formal $\mathcal{R}$-linear combinations of elements of $\mathcal{Y}$, i.e., the $\mathcal{R}$-vector-space with basis $\mathcal{Y}$. Any invariant $Q\colon \mathcal{Y} \longrightarrow W$ extends to a linear map $Q\colon \mathcal{M} \longrightarrow W$, and those of interest here are those that satisfy (5.3), or equivalently

$$Q\Bigl([G] - \alpha_{|\bar{e}|}[G-\bar{e}] + \beta_{|\bar{e}|}[G/\bar{e}]\Bigr) = 0 \tag{5.4}$$

for all $e \in \mathrm{E}(G)$ and labelled multigraphs G! Let $\mathcal{I} \subseteq \mathcal{M}$ be the subspace spanned by all $[G] - \alpha_{|\bar{e}|}[G-\bar{e}] + \beta_{|\bar{e}|}[G/\bar{e}]$ for $e \in \mathrm{E}(G)$ and $[G] \in \mathcal{Y}$. The invariants which satisfy the wanted recursion are exactly the linear maps that factor over $\mathcal{M}/\mathcal{I}$, so they are completely determined by their values on a basis for this quotient. Determining bases for a quotient is a standard application for the diamond lemma, and the key to start using it is to express the identity $[G] \equiv \alpha_{|\bar{e}|}[G-\bar{e}] - \beta_{|\bar{e}|}[G/\bar{e}] \pmod{\mathcal{I}}$ as a rewrite rule.

---

[1] In diagrammatic calculations, "label" is often a term used for extra information attached to an edge or vertex, but in graph theory it is just the unique identifier of that particular item. If you need both concepts, then try using "annotation" for the extra information kind of label.

5.4 Semigraphs

The recursion identity $[G] \equiv \alpha_{|\bar{e}|}[G - \bar{e}] - \beta_{|\bar{e}|}[G/\bar{e}]$ (mod $\mathfrak{I}$) has a straightforward interpretation as a rewrite rule: two vertices connected by m edges may be replaced by α_m times the same two vertices without those edges, minus β_m times the contraction of those two vertices. More symbolically, this may be written as

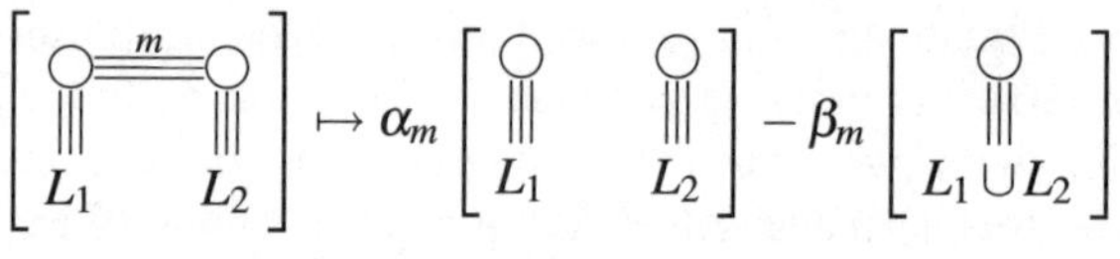

$$
\left[\begin{array}{c} \overset{m}{\circ\!=\!=\!\circ} \\ \underset{L_1}{\vert\vert\vert} \quad \underset{L_2}{\vert\vert\vert} \end{array} \right] \mapsto \alpha_m \left[\begin{array}{cc} \underset{L_1}{\overset{\circ}{\vert\vert\vert}} & \underset{L_2}{\overset{\circ}{\vert\vert\vert}} \end{array} \right] - \beta_m \left[\begin{array}{c} \overset{\circ}{\vert\vert\vert} \\ L_1 \cup L_2 \end{array} \right] \tag{5.5}
$$

where the L_1 and L_2 should be taken as the sets of edges between these two vertices and other vertices in G; it is important in the recursion that these are unaffected by the deletion and contraction operations.

In order to turn this into a rewrite rule for use with the diamond lemma of [7], the first step is to turn the diagrams in (5.5) into proper mathematical objects with which one may calculate. To that end, I now propose a generalisation of the ordinary multigraph concept, which I call *semigraph*. Intuitively, a semigraph may be thought of as an ordinary graph which someone has been placed on a hard surface and chopped in half with a cleaver. (Semi- = half.) Any edge between a vertex in one part and a vertex in the other must be split in half by such an operation, so each half-a-graph one is left with may in general have half-edges sticking out that are attached only to one vertex.

Definition 5.1. A *labelled semigraph* G is a triplet $(V(G), E(G), \phi_G)$, where $V(G)$ and $E(G)$ are finite sets and $\phi_G \colon E(G) \longrightarrow \{A \subseteq V(G) \mid 1 \leqslant |A| \leqslant 2\}$ is a map. The elements of $V(G)$ are called *vertices*, the elements of $E(G)$ are called *edges*. The elements of $\phi_G(e)$ are called *endpoints* of the edge e. An edge is said to be *internal* if it has two endpoints and *external* otherwise. External edges are also called *legs*. The set of legs of G is called the *boundary* of G and is denoted ∂G.

While the formal definition of semigraph may look very similar to a formal definition of pseudograph, the two concepts are quite different: an external edge is an edge with one endpoint, whereas a loop would be an edge with two ends, although both of them happen to be attached to the same vertex. A better characterisation would be that semigraphs are "$\{1,2\}$-uniform hypergraphs" (hypergraphs where all edges are incident with 1 or 2 vertices), but semigraphs occur often enough in practice that they deserve a simple name. In particular, the "gadgets" that graph theorists sometimes use to transform graphs or build graphs with particular properties are precisely semigraphs.

Definition 5.2. An *isomorphism* of two semigraphs G and H is a pair (α, β) of bijections $\alpha \colon V(G) \longrightarrow V(H)$ and $\beta \colon E(G) \longrightarrow E(H)$ such that

$$
\phi_H(\beta(e)) = \{\alpha(v) \mid v \in \phi_G(e)\} \qquad \text{for all } e \in E(G).
$$

Two semigraphs G and H are said to be *isomorphic*, written $G \cong H$, if there exists an isomorphism from G to H. The isomorphism (α, β) is said to be *internal* if $\beta(e) = e$ for all $e \in \partial G$. Two semigraphs G and H are said to be *internally isomorphic*, written $G \simeq H$, if there exists an internal isomorphism from G to H.

Let $\mathcal{D}$ be the set of labelled semigraphs G with $\mathrm{V}(G), \mathrm{E}(G) \subset \mathbb{Z}_{>0}$ and define $\mathcal{D}(L)$ to be the set of those $G \in \mathcal{D}$ for which $\partial G = L$. Define $\mathcal{Y}(L) = \mathcal{D}(L)/\simeq$; this is the set of *unlabelled semigraphs* with boundary L and if $G \in \mathcal{D}(L)$ then $[G]$ denotes the element of $\mathcal{Y}(L)$ containing G. Finally define $\mathcal{M}(L)$ to be the $\mathcal{R}$-vector-space with basis $\mathcal{Y}(L)$.

In this more general notation, the set of unlabelled multigraphs $\mathcal{Y}$ is the set $\mathcal{Y}(\varnothing)$ of unlabelled semigraphs with empty boundary, and similarly the vector space $\mathcal{M}$ of the previous section is $\mathcal{M}(\varnothing)$. For multigraphs the internal isomorphism concept is the same as ordinary isomorphism of multigraphs, but for more general semigraphs they are not, and consequently even unlabelled semigraphs have their external edges labelled. A natural analogy is with the names of variables in a function definition: it makes no difference which names are used, but one must use the same names in both sides of the defining equation. The rewriting will primarily operate on semigraphs with boundary. Note for example that the left hand side of (5.5) is an element of $\mathcal{Y}(L_1 \cup L_2)$ and the right hand side is an element of $\mathcal{M}(L_1 \cup L_2)$.

Lemma 5.1. *Let $L_1, L_2 \subset \mathbb{Z}_{>0}$ be given. If $\eta \colon L_1 \longrightarrow L_2$ is a bijection then it defines a bijection $\mathring{\eta} \colon \mathcal{Y}(L_1) \longrightarrow \mathcal{Y}(L_2)$ by*

$$\mathring{\eta}([G]) = \Big[\big(\mathrm{V}(G), E, \phi_G \circ \beta\big)\Big] \tag{5.6}$$

where $\beta \colon E \longrightarrow \mathrm{E}(G)$ is a bijection such that $\beta\big(\eta(e)\big) = e$ for all $e \in L_1$, and $E \subset \mathbb{Z}_{>0}$ is arbitrary such that $|E| = \big|\mathrm{E}(G)\big|$ and $L_2 \subseteq E$.

Proof. First observe that $(\mathrm{id}, \beta^{-1})$, where id denotes the identity map, is an isomorphism from G to $H = \big(\mathrm{V}(G), E, \phi_G \circ \beta\big)$. Hence if $G' \in [G]$ is some other semigraph of the same class then $H' = \big(\mathrm{V}(G), E', \phi_{G'} \circ \beta'\big)$, where $E' \subset \mathbb{Z}_{>0}$ is such that $|E'| = \big|\mathrm{E}(G)\big|$ and $L_2 \subseteq E'$, and $\beta' \colon E' \longrightarrow \mathrm{E}(G')$ is some bijection such that $\beta'\big(\eta(e)\big) = e$ for all $e \in L_1$, satisfies $H' \cong G' \simeq G \cong H$. Furthermore that isomorphism from H' to H composed in the obvious way from (id, β'), an internal isomorphism from G' to G, and $(\mathrm{id}, \beta^{-1})$ will map elements of L_2 back to themselves, and thus be an internal isomorphism. Hence $\mathring{\eta}$ is well-defined. In order to see that it is a bijection, it suffices to observe that $(\eta^{-1})^{\circ} \colon \mathcal{Y}(L_2) \longrightarrow \mathcal{Y}(L_1)$ is the inverse of $\mathring{\eta}$. □

This lemma illustrates a common awkwardness in formalising operations on labelled semigraphs (or just graphs): although the operation may be very easy to define, one usually has to make some arbitrary choice of new labels in it. This means the operations do not really have a canonical definition; a different author (or perhaps more noticeably: a different programmer) will probably make a different choice. The different choices will however usually produce isomorphic results, so in general the operation is canonical as an operation on *unlabelled* semigraphs.

Definition 5.3. Let G and H be labelled semigraphs. H is an *induced subsemigraph* of G, written $H \sqsubseteq G$, if

$$\begin{aligned} &\mathrm{V}(H) \subseteq \mathrm{V}(G), \\ &\phi_H(e) = \phi_G(e) \cap \mathrm{V}(H) \qquad \text{for all } e \in \mathrm{E}(H), \\ &\mathrm{E}(H) = \Bigl\{ e \in \mathrm{E}(G) \Bigm| 1 \leqslant \bigl|\phi_G(e) \cap \mathrm{V}(H)\bigr| \leqslant 2 \Bigr\}. \end{aligned}$$

If $X \subseteq \mathrm{V}(G)$ then $G\{X\}$ denotes the induced subsemigraph of G whose set of vertices is X.

For any $H \sqsubseteq G$, the *splice map* $G \div H \colon \mathcal{D}(\partial H) \longrightarrow \mathcal{D}(\partial G)$ is defined by

$$\begin{aligned} n &= \max\bigl(\{0\} \cup \mathrm{V}(G) \setminus \mathrm{V}(H)\bigr), \\ m &= \max\Bigl(\{0\} \cup \mathrm{E}(G) \setminus \bigl(\mathrm{E}(H) \setminus \partial H\bigr)\Bigr), \\ \mathrm{V}\bigl((G \div H)(K)\bigr) &= \bigl(\mathrm{V}(G) \setminus \mathrm{V}(H)\bigr) \cup \bigl\{ v+n \bigm| v \in \mathrm{V}(K) \bigr\}, \\ \mathrm{E}\bigl((G \div H)(K)\bigr) &= \Bigl(\mathrm{E}(G) \setminus \bigl(\mathrm{E}(H) \setminus \partial H\bigr)\Bigr) \cup \bigl\{ e+m \bigm| e \in \mathrm{E}(K) \setminus \partial K \bigr\}, \\ \phi_{(G \div H)(K)}(e) &= \begin{cases} \bigl(\phi_G(e) \setminus \mathrm{V}(H)\bigr) \cup \bigl\{ v+n \bigm| v \in \phi_K(e) \bigr\} & \text{if } e \in \partial H, \\ \bigl\{ v+n \bigm| v \in \phi_K(e-m) \bigr\} & \text{if } e > m, \\ \phi_G(e) & \text{otherwise;} \end{cases} \end{aligned}$$

the idea is to replace in G the internal parts of H by the corresponding parts of K. The integers n and m are offsets added to labels from K to avoid collisions with labels from G.

The sense in which $(G \div H)(K)$ is "G, but with the H part replaced by K" is formalised in the next lemma.

Lemma 5.2. *Let $G \in \mathcal{D}$, $H \sqsubseteq G$, and $K \in \mathcal{D}(\partial H)$ be given. If $G' = (G \div H)(K)$ and $H' = G'\bigl\{\mathrm{V}(G') \setminus \mathrm{V}(G)\bigr\}$ then $H' \simeq K$ and $G' \div H' = G \div H$. Furthermore if $F \sqsubseteq G$ is such that $\mathrm{V}(F) \cap \mathrm{V}(H) = \varnothing$ then $F \sqsubseteq G'$.*

Proof. Let $n = \max\bigl(\{0\} \cup \mathrm{V}(G) \setminus \mathrm{V}(H)\bigr)$ and $m = \max\bigl(\{0\} \cup (\mathrm{E}(G) \setminus \mathrm{E}(H)) \cup \partial H\bigr)$ be the offsets that the splice map $G \div H$ adds to vertices and internal edges of K when duplicating them in G'. Let $\alpha(v) = v+n$ for all $v \in \mathrm{V}(K)$. Let $\beta \colon \mathrm{E}(K) \longrightarrow \mathbb{Z}_{>0}$ be defined by $\beta(e) = e$ for $e \in \partial K = \partial H$ and $\beta(e) = e+m$ for $e \in \mathrm{E}(K) \setminus \partial K$. Let $X = \mathrm{V}(G') \setminus \mathrm{V}(G)$. By the definition of $G \div H$,

$$X = \bigl\{ v \in \mathrm{V}(G') \bigm| v > n \bigr\} = \bigl\{ v+n \bigm| v \in \mathrm{V}(K) \bigr\},$$

hence α is a bijection from $\mathrm{V}(K)$ to $X = \mathrm{V}(H')$. Furthermore β is a bijection from $\mathrm{E}(K)$ to $\mathrm{E}(H')$ and $\phi_{G'}\bigl(\beta(e)\bigr) \cap X = \bigl\{ \alpha(v) \bigm| v \in \phi_K(e) \bigr\}$ for all $e \in \mathrm{E}(K)$, thus $H' \simeq K$ as claimed. That $G' \div H' = G \div H$ is immediate from the definition, since $\partial H' = \partial H$, $\mathrm{E}(G') \setminus \mathrm{E}(H') = \mathrm{E}(G) \setminus \mathrm{E}(H)$, and $\mathrm{V}(G') \setminus \mathrm{V}(H') = \mathrm{V}(G) \setminus \mathrm{V}(H)$; in particular the two offsets n and m will be exactly the same in the definition of $G' \div H'$ as in the definition of $G \div H$.

For the last claim, it is clear that $\mathrm{V}(F) \subseteq \mathrm{V}(G')$. That $\mathrm{E}(F) \subseteq \mathrm{E}(G')$ follows from the observation that no edge of F may be an internal edge of H. Finally, $\phi_F(e) = \phi_G(e) \cap \mathrm{V}(F) = \phi_{G'}(e) \cap \mathrm{V}(F)$ for all $e \in \mathrm{E}(F)$ since $\phi_G(e) \setminus \phi_{G'}(e) \subseteq \mathrm{V}(H)$ and $\phi_{G'}(e) \setminus \phi_G(e) \subseteq X$, both of which are disjoint from $\mathrm{V}(F)$. □

The definition of the splice map admittedly contains a bit of arbitrariness when it comes to the assignment of vertex and edge labels – one could just as well have used $n+1$ and $m+1$ instead of n and m, since the labels are mostly irrelevant. The canonical object is instead the corresponding splice map for unlabelled semigraphs.

Lemma 5.3. *Let $G \in \mathcal{D}$ and $H \sqsubseteq G$. If $K_1, K_2 \in \mathcal{D}(\partial H)$ are such that $K_1 \simeq K_2$ then $(G \div H)(K_1) \simeq (G \div H)(K_2)$. Hence $G \div H$ is well-defined as a map $\mathcal{Y}(\partial H) \longrightarrow \mathcal{Y}(\partial G)$ and extends to a linear map $\mathcal{M}(\partial H) \longrightarrow \mathcal{M}(\partial G)$.*

Proof. Let n and m be the vertex and edge respectively offsets added by $G \div H$. Let $G_i = (G \div H)(K_i)$ and $X_i = \{v \in \mathrm{V}(G_i) \,|\, v > n\}$ for $i = 1, 2$. By Lemma 5.2, $G_1\{X_1\} \simeq K_1 \simeq K_2 \simeq G_2\{X_2\}$. Let (α, β) be such an internal isomorphism from $G_1\{X_1\}$ to $G_2\{X_2\}$. By extending α to the whole of $\mathrm{V}(G_1)$ by $\alpha(v) = v$ for $v \in \mathrm{V}(G_1) \setminus X_1$ and β to the whole of $\mathrm{E}(G_1)$ by $\beta(e) = e$ for $e \leqslant m$, one gets an internal isomorphism from G_1 to G_2. Hence $G_1 \simeq G_2$, as claimed. □

The splice maps themselves do however require some degree of semigraph labelling, as it is necessary to specify *which* induced subsemigraph is being replaced by a particular splice map. The next lemma demonstrates how different labellings of a semigraph may be used to express the same splice map.

Lemma 5.4. *Let $G_1, G_2, H_1 \in \mathcal{D}$ such that $G_1 \simeq G_2$ and $H_1 \sqsubseteq G_1$ be given. Then there exists some $H_2 \sqsubseteq G_2$ and bijection $\mathring{\eta}\colon \mathcal{Y}(\partial H_1) \longrightarrow \mathcal{Y}(\partial H_2)$ such that $H_2 \cong H_1$ and $G_1 \div H_1 = (G_2 \div H_2) \circ \mathring{\eta}$ as maps $\mathcal{Y}(\partial H_1) \longrightarrow \mathcal{Y}(\partial G_1)$.*

Proof. Let (α, β) be the internal isomorphism from G_1 to G_2. The restrictions of α and β to $\mathrm{V}(H_1)$ and $\mathrm{E}(H_1)$ provide an isomorphism from H_1 to $H_2 := G_2\{\alpha(\mathrm{V}(H_1))\}$. Let η be the restriction of β to ∂H_1 and let $\mathring{\eta}$ be the corresponding bijection as defined in Lemma 5.1. The task is now to show that for any $K_1 \in \mathcal{D}(\partial H_1)$ there is some $K_2 \in \mathring{\eta}([K_1])$ such that $(G_1 \div H_1)(K_1) \simeq (G_2 \div H_2)(K_2)$.

Let n_1 and m_1 be the offsets $G_1 \div H_1$ adds to vertex and edge respectively labels. Let n_2 and m_2 be the offsets $G_2 \div H_2$ adds to vertex and edge respectively labels. Let $E = \partial H_2 \cup \{e + m_2 \,|\, e \in \mathrm{E}(K_1) \setminus \partial K_1\}$ and define $\gamma\colon E \longrightarrow \mathrm{E}(K_1)$ by $\gamma(\eta(e)) = e$ for all $e \in \partial K_1$ and $\gamma(e + m_2) = e$ for all $e \in \mathrm{E}(K_1) \setminus \partial K_1$. Then γ is a bijection and $(\mathrm{id}, \gamma^{-1})$ is an isomorphism from K_1 to $K_2 := (\mathrm{V}(K_1), E, \phi_{K_1} \circ \gamma)$, which implies $K_2 \in \mathring{\eta}([K_1])$. The needed isomorphism (α', β') from $(G_1 \div H_1)(K_1)$ to $(G_2 \div H_2)(K_2)$ is defined by

$$\alpha'(v) = \begin{cases} \alpha(v) & \text{if } v \leqslant n_1, \\ v - n_1 + n_2 & \text{if } v > n_1, \end{cases}$$

$$\beta'(e) = \begin{cases} \beta(e) & \text{if } e \leqslant m_1, \\ \gamma^{-1}(e - m_1) + m_2 = e - m_1 + 2m_2 & \text{if } e > m_1 \end{cases}$$

for all $v \in \mathrm{V}\big((G_1 \div H_1)(K_1)\big)$ and $e \in \mathrm{E}\big((G_1 \div H_1)(K_1)\big)$. □

Another useful property of the splice maps is that the order of replacing disjoint parts of a semigraph is not important.

Lemma 5.5. *Let $G \in \mathcal{D}$ and $H_1, H_2 \sqsubseteq G$ be such that $\mathrm{V}(H_1) \cap \mathrm{V}(H_2) = \varnothing$. Then for any $K_1 \in \mathcal{D}(\partial H_1)$ and $K_2 \in \mathcal{D}(\partial H_2)$,*

$$\big((G \div H_1)(K_1) \div H_2\big)(K_2) \simeq \big((G \div H_2)(K_2) \div H_1\big)(K_1). \tag{5.7}$$

It follows that there is a map $G \div H_1 \div H_2 : \mathcal{Y}(\partial H_1) \times \mathcal{Y}(\partial H_2) \longrightarrow \mathcal{Y}(\partial G)$ defined by

$$(G \div H_1 \div H_2)\big([K_1],[K_2]\big) = \Big[\big((G \div H_1)(K_1) \div H_2\big)(K_2)\Big] \tag{5.8}$$

that extends to a bilinear map $\mathcal{M}(\partial H_1) \times \mathcal{M}(\partial H_2) \longrightarrow \mathcal{M}(\partial G)$.

Proof. By Lemma 5.2, $H_2 \sqsubseteq (G \div H_1)(K_1)$ and $H_1 \sqsubseteq (G \div H_2)(K_2)$, hence both sides of (5.7) are well-defined. Let n_i and m_i be the offsets added to vertex and edge respectively labels of K_i in the left hand side of (5.7), and let n_i' and m_i' be the offsets added to vertex and edge respectively labels of K_i in the right hand side of (5.7), for $i = 1,2$. The wanted isomorphism (α, β) from left hand side to right hand side is then given by

$$\alpha(v) = \begin{cases} v - n_2 + n_2' & \text{if } v > n_2, \\ v - n_1 + n_1' & \text{if } n_2 \geqslant v > n_1, \\ v & \text{if } n_1 \geqslant v, \end{cases} \quad \beta(e) = \begin{cases} e - m_2 + m_2' & \text{if } e > m_2, \\ e - m_1 + m_1' & \text{if } m_2 \geqslant e > m_1, \\ e & \text{if } m_1 \geqslant e. \end{cases}$$

It follows from Lemma 5.3 that the right hand side of (5.8) depends only on the internal isomorphism class $[K_2]$ of K_2, and not on its exact labelling. (5.8) is furthermore by (5.7) equivalent to

$$(G \div H_1 \div H_2)\big([K_1],[K_2]\big) = \Big[\big((G \div H_2)(K_2) \div H_1\big)(K_1)\Big]$$

and hence the values of $G \div H_1 \div H_2$ is indeed determined only by the internal isomorphism classes of its arguments. □

The next lemma treats the opposite case of composition of splice maps: what happens when one part is contained in another. The proof is left as an exercise to the reader.

Lemma 5.6. *Let $F \in \mathcal{D}$ and $H \sqsubseteq G \sqsubseteq F$ be given. For any $K \in \mathcal{D}(\partial H)$,*

$$(F \div H)(K) \simeq (F \div G)\big((G \div H)(K)\big). \tag{5.9}$$

More generally, if $G_1 \in \mathcal{D}(\partial G)$ and $H_1 \sqsubseteq G_1$ then there exist $F_2 \in \mathcal{D}(\partial F)$ and $H_2 \in \mathcal{D}(\partial H_1)$ such that $H_2 \sqsubseteq F_2$ and for all $K \in \mathcal{D}(\partial H_1)$ it holds that

$$(F_2 \div H_2)(K) \simeq (F \div G)\big((G_1 \div H_1)(K)\big). \tag{5.10}$$

The previous lemmas have all concerned fairly generic properties of isomorphism and splicing in diagrams, in the sense that some variant on these properties should hold for diagrams no matter what how they are formalised, and they are part of a general toolbox for defining reductions on $\mathcal{M}(L)$. The utility of the next lemma is more specific to the concrete rewriting system studied.

Lemma 5.7. *For any $G \in \mathcal{D}$, $H \sqsubseteq G$, and $K \in \mathcal{D}(\partial H)$,*

$$\Big|\mathrm{E}\big((G \div H)(K)\big)\Big| = \big|\mathrm{E}(G)\big| - \big|\mathrm{E}(H)\big| + \big|\mathrm{E}(K)\big|. \tag{5.11}$$

Proof. By the construction of the edge offset factor m in Definition 5.3, the union in the definition of $\mathrm{E}\big((G \div H)(K)\big)$ is disjoint. Hence

$$\Big|\mathrm{E}\big((G \div H)(K)\big)\Big| = \big|\mathrm{E}(G)\big| - \Big(\big|\mathrm{E}(H)\big| - |\partial H|\Big) + \Big(\big|\mathrm{E}(K)\big| - |\partial K|\Big) \tag{5.12}$$

and the claim follows from $\partial K = \partial H$. □

The splice maps for unlabelled graphs may be compared to maps multiplying by a constant monomial, in the sense that they play the same role in the next section as multiplication by a monomial does in Gröbner basis theory or Bergman's diamond lemma [3]. This may seem an overly modest foundation to build an algebraic theory on, but it does suffice, and it handles many issues – in particular the construction of simple reductions – quite elegantly. It is however not the only possibility for a "multiplication structure" on semigraphs.

A natural product of unlabelled semigraphs is to join up common external edges; formally one may define $G \cdot H$ for $G, H \in \mathcal{D}$ by

$$\begin{aligned}
n &= \max\big(\{0\} \cup \mathrm{V}(G)\big),\\
m &= \max\big(\{0\} \cup \mathrm{E}(G) \cup \partial H\big),\\
\mathrm{V}(G \cdot H) &= \mathrm{V}(G) \cup \big\{v+n \,\big|\, v \in \mathrm{V}(H)\big\},\\
\mathrm{E}(G \cdot H) &= \mathrm{E}(G) \cup \partial H \cup \big\{e+m \,\big|\, e \in \mathrm{E}(H) \setminus \partial H\big\},\\
\phi_{G \cdot H}(e) &= \begin{cases} \big\{v+n \,\big|\, n \in \phi_H(e-m)\big\} & \text{if } e > m,\\ \big\{v+n \,\big|\, n \in \phi_H(e)\big\} & \text{if } e \in \partial H \setminus \partial G,\\ \phi_G(e) \cup \big\{v+n \,\big|\, n \in \phi_H(e)\big\} & \text{if } e \in \partial H \cap \partial G,\\ \phi_G(e) & \text{otherwise.} \end{cases}
\end{aligned}$$

If G and H are multigraphs, then this $G \cdot H$ is simply the disjoint union of G and H. It should be no surprise that $G_1 \cdot H_1 \simeq G_2 \cdot H_2$ whenever $G_1 \simeq G_2$ and $H_1 \simeq H_2$, so this multiplication is well-defined also for unlabelled semigraphs. Slightly more surprising is perhaps that

$$G\big\{\mathrm{V}(G) \setminus \mathrm{V}(H)\big\} \cdot K \simeq (G \div H)(K), \tag{5.13}$$

meaning the "quotients" $G \div H$ can actually themselves be identified with semigraphs. Defining $\mathcal{A} = \bigoplus_{\text{finite } L \subset \mathbb{Z}_{>0}} \mathcal{M}(L)$, one has even produced an "$\mathcal{R}$-algebra of unlabelled semigraphs", which turns out to be graded by the group of finite subsets of $\mathbb{Z}_{>0}$ under symmetric difference, since $\partial(G \cdot H) = (\partial G \setminus \partial H) \cup (\partial H \setminus \partial G)$! This view has many nice features, but there is also a downside.

One problem with the semigraph algebra $\mathcal{A}$ is that its multiplicative structure is rather far from what is common in rewriting theories: there is no unique factorisation, not even cancellation (the definition of splice map $G \div H$ requires G and H to be labelled), and consequently $\mathcal{A}$ is rich with zero divisors. A more important problem is however that the semigraph multiplication doesn't generalise well to diagrams with more structure. In a directed semigraph, it wouldn't be possible to join an external edge e of G with the external edge e of H if both ends are heads or both ends are tails. In a directed acyclic semigraph, which is something found underneath a PROP (or symmetric monoidal category), it need not be allowed to join an output of G to an input of H at the same time as one joins an output of H to an input of G, as this could create a cycle. If one is working with plane diagrams, then joining external edges with equal labels is likely to produce a non-planar diagram. And so on.

The key concept is "replacing part of a diagram with another diagram of the same sort", and that is what the splice maps do. Multiplication of diagrams assumes that "a diagram with a part removed" (which is the essential meaning of some $G \div H$) can be identified with another diagram, but that is often not possible. A plane diagram has to be a plane semigraph with all external edges in the outer facet, but a plane diagram with a piece missing may have external edges also in some inner facet, where the extra piece would be plugged in. Similarly a tree with a piece missing is generally not a tree, but rather a forest. Inserting any tree (connected acyclic semigraph) with the right boundary will turn the forest back into a tree however, so splicing on trees is a well-behaved operation in that it avoids going outside the set of trees. The same holds for the other types of diagrams mentioned above.

A more traditional formalism for graph rewriting, in that it avoids semigraphs and half-edges, can be found in [1]. The conceptual difference is mostly that the roles of vertices and edges are swapped, so that vertices act as connectors between edges rather than vice versa, but of course either way will work.

5.5 Applying the Diamond Lemma

This section is all about applying the machinery from [7] to the present semigraph rewriting problem. Not only results, but also definitions and notation from that paper will be used extensively and without restraint. The conclusion can be understood without first having read [7], but the proof of it probably cannot, although it should be clear how the diagrammatic calculations lead to the condition (5.20).

The basic framework for the diamond lemma is of the kind that can be set up using [7, Construction 7.2], but not all bells and whistles of that are actually needed.

Furthermore several parts of the framework have already been set up, although a review may be in order:

- The index set I – the set of sorts – will be the set of all finite subsets of $\mathbb{Z}_{>0}$, i.e., the set of all boundaries of semigraphs in $\mathcal{D}$.
- The abelian group of all expressions of sort $L \in I$ will be the vector space $\mathcal{M}(L)$ of Definition 5.2.
- Similarly, the set of all "monomials" of sort $L \in I$ will be the set $\mathcal{Y}(L)$ of unlabelled semigraphs with boundary L from that same definition.

In addition, there are some advanced aspects of the framework which we won't have to worry about:

- The set $R(L)$ of endomorphisms of $\mathcal{M}(L)$ is simply going to be the set of actions of $\mathcal{R}$ on $\mathcal{M}(L)$, so an $R(L)$-module is just an $\mathcal{R}$-vector-space.
- The topology used will be the discrete topology, so $B_n(L) = \{0\}$ for all $n \geqslant 1$ and $L \in I$. This also implies that $\overline{\mathcal{M}}(L) = \mathcal{M}(L)$ for all $L \in I$, that all maps $\mathcal{M}(L) \longrightarrow \mathcal{M}(L)$ are equicontinuous, and that the topological descending chain condition is the same thing as the ordinary descending chain condition (any subset contains a minimal element).

Simple reductions are defined by putting rewrite rules into all possible contexts.
For all $m \in \mathbb{Z}_{>0}$ and $L_1, L_2 \in I$ such that $L_1 \cap L_2 = \varnothing$ define

$$
s(L_1, m, L_2) := \left(\left[\begin{array}{c} \overset{m}{\rule{0pt}{0pt}} \\ L_1 \quad L_2 \end{array} \right], \alpha_m \left[\begin{array}{cc} \\ L_1 & L_2 \end{array} \right] - \beta_m \left[\begin{array}{c} \\ L_1 \cup L_2 \end{array} \right] \right), \quad (5.14)
$$

which is an element of $\mathcal{Y}(L_1 \cup L_2) \times \mathcal{M}(L_1 \cup L_2)$. For all $L \in I$, let $S(L)$ be the set of all $s(L_1, m, L \setminus L_1)$ for $m \in \mathbb{Z}_{>0}$ and $L_1 \subseteq L$, and let $S = \bigcup_{L \in I} S(L)$. It is convenient to introduce a less spacious notation for unlabelled semigraphs without internal edges, as the results once rewriting is complete will involve a lot of these. Therefore let $[L_1, \ldots, L_n]$ denote the element $[G] \in \mathcal{Y}(L_1 \cup \cdots \cup L_n)$ where $V(G) = \{1, \ldots, n\}$, $E(G) = \bigcup_{k=1}^{n} L_k$, and $\phi_G(e) = \{k\}$ for all $e \in L_k$, i.e., L_1 through L_n are the sets of external edges incident with vertices 1 through n respectively. (The only aspect of this notation that is not uniquely determined by the underlying element of $\mathcal{Y}(L)$ is the order of the L_k sets.) For $(\mu, a) = s(L_1, m, L_2)$ this means a can be expressed as $\alpha_m[L_1, L_2] - \beta_m[L_1 \cup L_2]$.

For any finite $L, L' \in I$ define

$$
V(L, L') = \{ G \div H \colon \mathcal{M}(L') \longrightarrow \mathcal{M}(L) \,|\, G \in \mathcal{D}(L), H \sqsubseteq G, \partial H = L' \}. \quad (5.15)
$$

Note that by Lemma 5.6, if $v \in V(L, L')$ and $w \in V(L', L'')$, then $v \circ w \in V(L, L'')$. For any $v \in V(L, L')$ and $s = (\mu_s, a_s) \in S(L')$ define $t_{v,s} \colon \mathcal{M}(L) \longrightarrow \mathcal{M}(L)$ to be the linear map which satisfies

$$
t_{v,s}(\lambda) = \begin{cases} v(a_s) & \text{if } \lambda = v(\mu_s), \\ \lambda & \text{otherwise,} \end{cases} \quad \text{for all } \lambda \in \mathcal{Y}(L). \quad (5.16)
$$

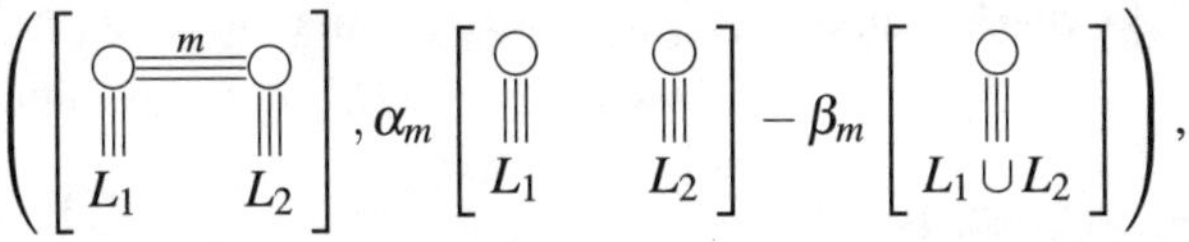

Finally let $T_1(S)(L) = \{t_{v,s} \,|\, v \in V(L,L'), s \in S(L'), L' \in I\}$ be the set of simple reductions on $\mathcal{M}(L)$, for every $L \in I$. With this in place, the derived sets $\mathcal{I}(S)(L)$, $\mathrm{Irr}(S)(L)$, etc. are defined for all $L \in I$. Also note that the maps in $V(L,L')$ are all advanceable with respect to $T(S)(L')$ and $T(S)(L)$ by [7, Lemma 7.3].

In order to verify that this has anything do with the problem that was posed in Sect. 5.3, it must be established that $\mathcal{I} = \mathcal{I}(S)(\varnothing)$, or in more elementary language that the rules replace two adjacent vertices by the wanted linear combination of their delete–contract counterparts. By [7, Lemma 3.7], $\mathcal{I}(S)(\varnothing)$ is the set spanned by all $\lambda - t(\lambda)$ such that $\lambda \in \mathcal{Y}(\varnothing)$ and $t \in T_1(S)(\varnothing)$, i.e., the set spanned by all $\lambda - t_{v,s}(\lambda)$ such that $\lambda \in \mathcal{Y}(\varnothing)$, $v \in V(\varnothing, L)$, and $s \in S(L)$ for some $L \in I$. By the definition of $t_{v,s}$, $\lambda - t_{v,s}(\lambda) = 0$ unless $\lambda = v(\mu_s)$, and in that case $\lambda - t_{v,s}(\lambda) = v(\mu_s) - v(a_s)$.

Every $v \in V(\varnothing, L)$ is of the form $G \div H$ for some $G \in \mathcal{D}(\varnothing)$ and $H \sqsubseteq G$ such that $\partial H = L$. Similarly every $s = (\mu_s, a_s) \in S(L)$ is such that $v(\mu_s) = \bigl[(G \div H)(K)\bigr]$ for some $K \in \mathcal{D}(L)$ such that $\mathrm{V}(K) = \{u_1, u_2\}$, and if one furthermore defines $L_i = \{e \in L \,|\, \phi_K(e) = \{u_i\}\}$ for $i = 1,2$ and $m = \bigl|\mathrm{E}(K) \setminus L\bigr|$ then $a_s = \alpha_m[L_1, L_2] - \beta_m[L]$. Letting $G' = (G \div H)(K)$ and $e \in \mathrm{E}(G') \setminus \mathrm{E}(G)$, it follows that $|\bar{e}| = m$, $[G' - \bar{e}] = (G \div H)\bigl([L_1, L_2]\bigr)$, and $[G'/\bar{e}] = (G \div H)\bigl([L]\bigr)$, whence $v(\mu_s) - v(a_s) = [G'] - \alpha_m[G' - \bar{e}] + \beta_m[G'/\bar{e}] \in \mathcal{I}$. Hence $\mathcal{I}(S)(\varnothing) \subseteq \mathcal{I}$.

Conversely, for any labelled multigraph $G \in \mathcal{D}(\varnothing)$ and every edge $e \in \mathrm{E}(G)$, there are two endpoints $\{u_1, u_2\} = \phi_G(e)$. Let H be the induced subsemigraph $G\{\phi_G(e)\}$, let $L_i = \{e \in \partial H \,|\, \phi_H(e) = \{u_i\}\}$ for $i = 1,2$, and let $m = |\bar{e}|$. Then $s(L_1, m, L_2) \in S(\partial H)$ satisfies $\mu_{s(L_1,m,L_2)} = [H]$ and $G \div H \in V(\varnothing, \partial H)$. Furthermore $[G - \bar{e}] = (G \div H)\bigl([L_1, L_2]\bigr)$ and $[G/\bar{e}] = (G \div H)\bigl([L_1 \cup L_2]\bigr)$, hence

$$
\begin{aligned}
&[G] - \alpha_m[G - \bar{e}] + \beta_m[G/\bar{e}] = \\
&\quad = (G \div H)\bigl([H]\bigr) - \alpha_m (G \div H)\bigl([L_1, L_2]\bigr) + \beta_m (G \div H)\bigl([L_1 \cup L_2]\bigr) = \\
&\qquad = (G \div H)(\mu_{s(L_1,m,L_2)}) - (G \div H)(a_{s(L_1,m,L_2)}) \in \mathcal{I}(S)(\varnothing)
\end{aligned}
$$

and thus $\mathcal{I} \subseteq \mathcal{I}(S)(\varnothing)$.

The next necessary step is to define a suitable partial order on the unlabelled multigraphs, but for resolving ambiguities it is convenient to have corresponding partial orders defined for unlabelled semigraphs of all sorts. Constructing these things can be quite complicated for some diagrammatical calculation problems, but in the present case it is sufficient to compare elements by size (number of edges). Thus for every sort $L \in I$, let the partial order $P(L)$ on $\mathcal{Y}(L)$ be defined by $[G] < [H]$ in $P(L)$ iff $\bigl|\mathrm{E}(G)\bigr| < \bigl|\mathrm{E}(H)\bigr|$. Since there are only finitely many possible values for $\bigl|\mathrm{E}(G)\bigr|$ if $[G] < [H]$ in $P(L)$ for some given $[H]$, it follows that all $P(L)$ satisfy the descending chain condition.

For the issue of whether the simple reductions are compatible with these partial orders, one should first observe that for all $v \in V(L, L')$ and $\mu, \lambda \in \mathcal{Y}(L')$, it follows from Lemma 5.7 that $v(\mu) < v(\lambda)$ in $P(L)$ if and only if $\mu < \lambda$ in $P(L')$, and hence

$$
v\Bigl(\mathrm{DSM}\bigl(\lambda, P(L')\bigr)\Bigr) \subseteq \mathrm{DSM}\bigl(v(\lambda), P(L)\bigr). \tag{5.17}
$$

Thus the issue of whether $t_{v,s}$ is compatible with $P(L)$ reduces to the issue of whether $a_s \in \mathrm{DSM}\bigl(\mu_s, P(L')\bigr)$, and that is easily verified by considering the explicit form $s(L_1,m,L_2)$ for the rule s: any $H \in \mu_{s(L_1,m,L_2)}$ has $\bigl|\mathrm{E}(H)\bigr| = |L_1|+m+|L_2|$ whereas $H \in [L_1,L_2]$ or $H \in [L_1 \cup L_2]$ has $\bigl|\mathrm{E}(H)\bigr| = |L_1|+|L_2|$. The general conditions for the diamond lemma are thus fulfilled, and one may conclude that:

- $\mathrm{Irr}(S)(\varnothing)$ is the subspace of $\mathcal{M}(\varnothing)$ spanned by the unlabelled graphs without edges [7, Theorem 5.6].
- $\mathcal{M}(\varnothing) = \mathrm{Irr}(S)(\varnothing) \oplus \mathcal{I}(S)(\varnothing)$ if and only if all ambiguities of $T_1(S)(\varnothing)$ are resolvable relative to $P(\varnothing)$ [7, Theorem 5.11].

It is when verifying that the ambiguities are resolvable that one ends up making diagrammatic calculations, but most of them are about semigraphs of other sorts than $\varnothing$, as the critical part of an ambiguity of $T_1(S)(\varnothing)$ is typically a much smaller ambiguity of some other $T_1(S)(L)$. For the purpose of analysing ambiguities, the following lemma is very convenient.

Lemma 5.8. *Let $L \in I$, $t \in T_1(S)(L)$, and $\lambda \in \mathcal{Y}(L)$ such that t acts nontrivially on λ be given. For every $G \in \lambda$ there exists some $H \sqsubseteq G$ and $s \in S(\partial H)$ such that $H \in \mu_s$ and $t = t_{G \div H,s}$.*

Proof. Since t acts nontrivially on λ it must be the case that $t = t_{v,s'}$, where $v(\mu_{s'}) = \lambda$, for some $v \in V(L,L')$, $s' \in S(L')$, and $L' \in I$. By Lemma 5.2, this means $v = G' \div K$ for some $G' \in \lambda$ and $K \in \mu_{s'}$ such that $K \sqsubseteq G'$. By Lemma 5.4 and since $G' \simeq G$, there exists some $H \sqsubseteq G$ and $\mathring{\eta} \colon \mathcal{Y}(\partial K) \longrightarrow \mathcal{Y}(\partial H)$ such that $H \cong K$ and $v = G' \div K = (G \div H) \circ \mathring{\eta}$. Letting $\mathring{\eta}$ act on both parts of s', which must be on the form $s(L,m,J)$ for some $L, J \in I$ and $m \in \mathbb{Z}_{>0}$, one finds that

$$\mathring{\eta}(\mu_{s(L,m,J)}) = \left[\vphantom{\int}\right.$$
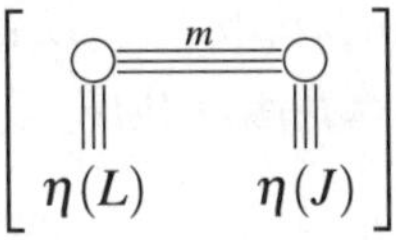

$$\left.\vphantom{\int}\right] = \mu_{s(\eta(L),m,\eta(J))},$$

$$\begin{aligned}\mathring{\eta}(a_{s(L,m,J)}) = \mathring{\eta}\bigl(\alpha_m[L,J] - \beta_m[L \cup J]\bigr) &= \alpha_m \mathring{\eta}\bigl([L,J]\bigr) - \beta_m \mathring{\eta}\bigl([L \cup J]\bigr) = \\ &= \alpha_m\bigl[\eta(L),\eta(J)\bigr] - \beta_m\bigl[\eta(L) \cup \eta(J)\bigr] = \\ &= a_{s(\eta(L),m,\eta(J))}.\end{aligned}$$

Hence for $s = s(\eta(L),m,\eta(J))$, the wanted result is obtained. □

Let an ambiguity (t_1,μ,t_2) of $T_1(S)(\varnothing)$ be given, and fix some $G \in \mu$. The lemma then implies that (t_1,μ,t_2) is of the form $(t_{G \div H_1,s_1}, [G], t_{G \div H_2,s_2})$, where $H_i \sqsubseteq G$ satisfies $H_i \in \mu_{s_i}$ for $i = 1,2$. This description of ambiguities is concrete enough that a resolution can be computed. There are essentially three ambiguity cases in this system, and these may be distinguished by the number of vertices that H_1 and H_2 have in common. Since $\bigl|\mathrm{V}(H_1)\bigr| = \bigl|\mathrm{V}(H_2)\bigr| = 2$, the first case is that $\mathrm{V}(H_1) = \mathrm{V}(H_2)$, so that both reductions act on exactly the same part of G. Due to the symmetry of the rules ($s(L,m,J) = s(J,m,L)$ for all rules $s(L,m,J)$), all such cases have $t_1 = t_2$ and are thus trivially resolvable.

A not at all trivial case occurs if $|V(H_1) \cap V(H_2)| = 1$. Consider first what happens in $H = G\{V(H_1) \cup V(H_2)\}$. This semigraph has the form

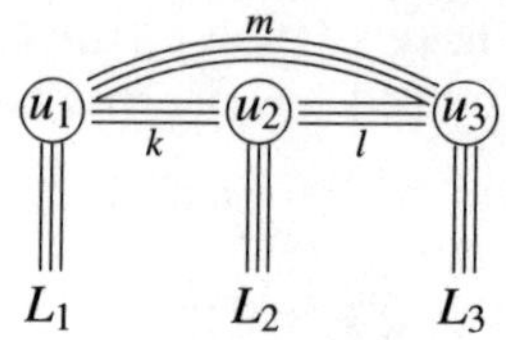

where $u_1 \in V(H_1) \setminus V(H_2)$, $u_2 \in V(H_1) \cap V(H_2)$, $u_3 \in V(H_2) \setminus V(H_1)$, and $L_i = \{e \in \partial H \mid \phi_H(e) = \{u_i\}\}$ for $i = 1, 2, 3$. $k \geqslant 1$ is the number of edges between u_1 and u_2, $l \geqslant 1$ is the number of edges between u_2 and u_3, whereas $m \geqslant 0$ is the number of edges between u_3 and u_1. This is the site of the ambiguity $\left(t_{H \div H_1, s_1}, [H], t_{H \div H_2, s_2} \right)$ of $T_1(S)(\partial H)$, which is resolved as follows. On the $t_{H \div H_1, s_1}$ side,

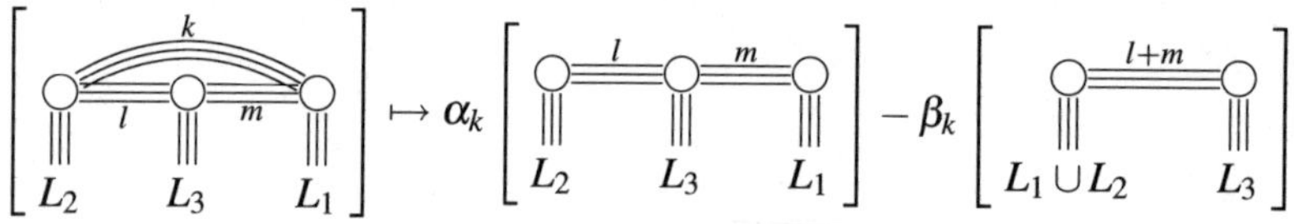

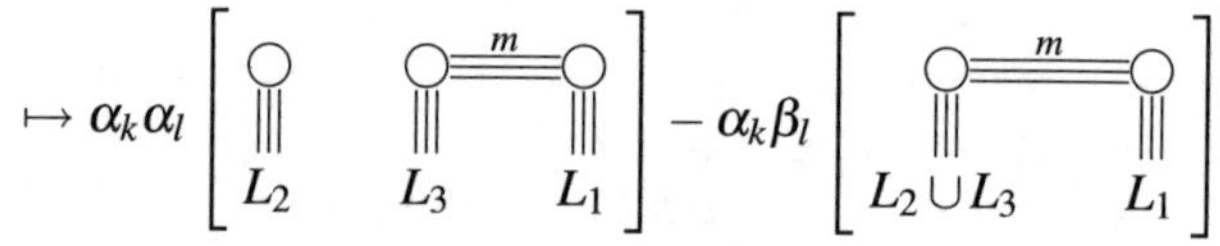

$$\begin{bmatrix} \raisebox{0pt}{k} \\ L_2 \quad L_3 \quad L_1 \end{bmatrix} \mapsto \alpha_k \begin{bmatrix} l \quad m \\ L_2 \quad L_3 \quad L_1 \end{bmatrix} - \beta_k \begin{bmatrix} l+m \\ L_1 \cup L_2 \quad L_3 \end{bmatrix}$$

$$\mapsto \alpha_k \alpha_l \begin{bmatrix} m \\ L_2 \quad L_3 \quad L_1 \end{bmatrix} - \alpha_k \beta_l \begin{bmatrix} m \\ L_2 \cup L_3 \quad L_1 \end{bmatrix}$$

$$- \beta_k \alpha_{l+m}[L_1 \cup L_2, L_3] + \beta_k \beta_{l+m}[L_1 \cup L_2 \cup L_3]$$

$$\mapsto \alpha_k \alpha_l \alpha_m [L_1, L_2, L_3] - \alpha_k \alpha_l \beta_m [L_1 \cup L_3, L_2]$$

$$- \alpha_k \beta_l \alpha_m [L_1, L_2 \cup L_3] - \beta_k \alpha_{l+m}[L_1 \cup L_2, L_3]$$

$$+ (\alpha_k \beta_l \beta_m + \beta_k \beta_{l+m})[L_1 \cup L_2 \cup L_3], \tag{5.18}$$

and on the $t_{H \div H_2, s_2}$ side,

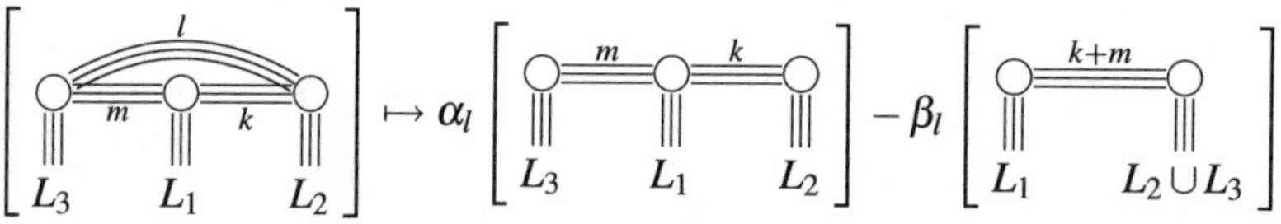

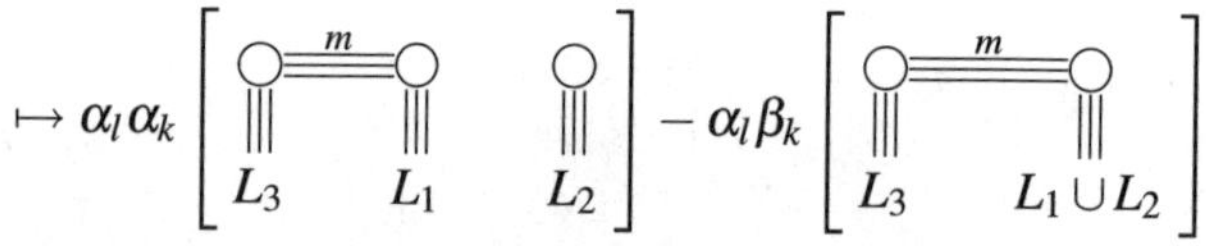

$$\begin{bmatrix} l \\ L_3 \quad L_1 \quad L_2 \end{bmatrix} \mapsto \alpha_l \begin{bmatrix} m \quad k \\ L_3 \quad L_1 \quad L_2 \end{bmatrix} - \beta_l \begin{bmatrix} k+m \\ L_1 \quad L_2 \cup L_3 \end{bmatrix}$$

$$\mapsto \alpha_l \alpha_k \begin{bmatrix} m \\ L_3 \quad L_1 \quad L_2 \end{bmatrix} - \alpha_l \beta_k \begin{bmatrix} m \\ L_3 \quad L_1 \cup L_2 \end{bmatrix}$$

$$- \beta_l \alpha_{k+m}[L_1, L_2 \cup L_3] + \beta_l \beta_{k+m}[L_1 \cup L_2 \cup L_3]$$

$$\mapsto \alpha_l \alpha_k \alpha_m [L_3, L_1, L_2] - \alpha_l \alpha_k \beta_m [L_1 \cup L_3, L_2]$$

$$- \alpha_l \beta_k \alpha_m [L_1 \cup L_2, L_3] - \beta_l \alpha_{k+m}[L_1, L_2 \cup L_3]$$

$$+ (\alpha_l \beta_k \beta_m + \beta_l \beta_{k+m})[L_1 \cup L_2 \cup L_3]. \tag{5.19}$$

The $[L_1, L_2, L_3]$ and $[L_1 \cup L_3, L_2]$ terms of these reductions are always equal, but the $[L_1 \cup L_2, L_3]$ terms are only equal if $\beta_k \alpha_{l+m} = \beta_k \alpha_l \alpha_m$, the $[L_1, L_2 \cup L_3]$ terms are only equal if $\alpha_k \beta_l \alpha_m = \beta_l \alpha_{k+m}$, and the $[L_1 \cup L_2 \cup L_3]$ terms are only equal if $\alpha_k \beta_l \beta_m + \beta_k \beta_{l+m} = \alpha_l \beta_k \beta_m + \beta_l \beta_{k+m}$. These are thus the conditions under which this ambiguity is resolvable.

There is a slight formal complication in that the reduction steps removing edges between u_1 and u_3 are not really allowed if $m = 0$, as that means there aren't any edges there to remove, but if one sets $\alpha_0 = 1$ and $\beta_0 = 0$ then the reduction step carried out above is equivalent to applying the identity reduction id. Hence the $m = 0$ cases can be resolved using a calculation of the same form as the $m > 0$ cases, although a more direct approach would be to make a separate calculation for the $m = 0$. The given equations furthermore turn out to be trivially fulfilled for $m = 0$ if $\alpha_0 = 1$ and $\beta_0 = 0$, so it can be argued that $m = 0$ really represents a different type of ambiguity than the $m > 0$ cases.

All of this has been about an ambiguity of $T_1(S)(\partial H)$, however – what about the ambiguity (t_1, μ, t_2) that we began with? By [7, Lemma 6.3] this ambiguity is resolvable relative to $P(\varnothing)$, since it is a shadow of the ambiguity at $[H]$ that was resolved above. The key to making this claim is Lemma 5.6, which implies that $(G \div H)\bigl(t_{H \div H_i, s_i}(\lambda)\bigr) = t_i\bigl((G \div H)(\lambda)\bigr)$ for all $\lambda \in \mathcal{Y}(\partial H_i)$ and $i = 1, 2$, and as explained above more generally implies that $G \div H \in V(\partial G, \partial H)$ is an advanceable map.

What remains is therefore the case $\mathrm{V}(H_1) \cap \mathrm{V}(H_2) = \varnothing$, which again is practically trivial (although the theory is somewhat involved). What happens in this case is that (t_1, μ, t_2) is a montage of the two pieces $\bigl([H_1], t_{\mathrm{id}, s_1}\bigr)$ and $\bigl([H_2], t_{\mathrm{id}, s_2}\bigr)$ under the composition map $w = G \div H_1 \div H_2$ as detailed in Lemma 5.5; the biadvanceability of w is immediate from $w\bigl([K_1], \cdot\bigr) = (G \div H_1)(K_1) \div H_2 \in V(\partial G, \partial H_2)$ and $w\bigl(\cdot, [K_2]\bigr) = (G \div H_2)(K_2) \div H_1 \in V(\partial G, \partial H_1)$. Since these expressions via (5.17) also demonstrate that the conditions in [7, Lemma 6.7] are fulfilled, it follows that all ambiguities of this last class are resolvable relative to $P(\varnothing)$.

Since $\mathcal{M}(\varnothing) = \mathrm{Irr}(S)(\varnothing) \oplus \mathcal{I}(S)(\varnothing)$ implies $\mathcal{M}(\varnothing) / \mathcal{I}(S)(\varnothing) \cong \mathrm{Irr}(S)(\varnothing)$, it has thus been shown that:

Lemma 5.9. *If the coefficients* $\{\alpha_m\}_{m=1}^{\infty}, \{\beta_m\}_{m=1}^{\infty} \subseteq \mathcal{R}$ *satisfy*

$$(\alpha_k \alpha_m - \alpha_{k+m}) \beta_l = 0, \tag{5.20a}$$

$$\alpha_k \beta_l \beta_m + \beta_k \beta_{l+m} = \alpha_l \beta_k \beta_m + \beta_l \beta_{k+m} \tag{5.20b}$$

for all $k, l, m \in \mathbb{Z}_{>0}$ *then a* W*-valued multigraph invariant* Q *which satisfies* (5.2) *is uniquely determined by its values for graphs with no edges, and every assignment of values to the graphs with no edges extends to an invariant* Q *for all multigraphs.*

The part about the invariant being uniquely determined by its values on graphs with no edges is pretty easy to arrive at by elementary methods, but the part that every possible assignment of values to these graphs is allowed for such invariants is not. Likewise, it is fairly easy to show that (5.20) are necessary conditions for such an invariant, but harder to show that they are also sufficient.

5.6 Classification of Invariants

The two main cases in the classification are (1) invariants for which the conditions of Lemma 5.9 are fulfilled and (2) invariants for which these conditions are not fulfilled. In the latter case, it is possible to derive additional relations which lead to simpler recursions or classifications, which means these on the whole tend to have fewer degrees of freedom. The former case is more interesting, so let us begin with that.

Denote by q^n the value of $Q(K_n^{\mathrm{C}})$. (Even if W is not formally required to be a space of polynomials, it turns out to be a very natural identification, and there is no loss of information as long as one preserves the coefficients of the polynomial.) Also let $|G|$ denote the order (number of vertices) of the multigraph G, let $\|G\|$ denote the size (number of edges) of the multigraph G, and let $c(G)$ denote the number of components (connected nonempty induced subsemigraphs without external edges) in G. Some invariants turn out to be explainable in terms of these elementary invariants alone.

Invariant class 1. $\beta_m = 0$ for all $m \in \mathbb{Z}_{>0}$. In this case (5.20) is fulfilled for all $\{\alpha_m\}_{m=1}^{\infty} \subseteq \mathcal{R}$, and so these may be chosen arbitrarily.

Since $Q(G) = \alpha_m Q(G - \bar{e})$ if $|\bar{e}| = m$, the coefficient α_m is essentially a weight attributed to edges of multiplicity m, and apart from that Q only keeps track of the number of vertices.

If $\beta_l \neq 0$ for some l then (5.20a) implies $\alpha_{k+m} = \alpha_k \alpha_m$ for all k and m. This has the unique solution $\alpha_m = \alpha_1^m$, and so the values of α's in the remaining cases are determined by the values of α_1.

Invariant class 2. $\beta_m \neq 0$ for some m and $\alpha_m = 0$ for all m. (5.20b) simplifies to $\beta_k \beta_{l+m} = \beta_l \beta_{k+m}$, which for $l = 1$ and $k = m+1$ reads $\beta_{m+1}^2 = \beta_1 \beta_{2m+1}$. Since $\beta_m \neq 0$ for some m, it follows that $\beta_1 \neq 0$. Setting $k = 2$ and $l = 1$ one gets $\beta_2 \beta_{m+1} = \beta_1 \beta_{m+2}$ for all $m \in \mathbb{Z}_{>0}$, from which follows that

$$\beta_k = \beta_1^{2-k} \beta_2^{k-1} = \beta_1 (\beta_1^{-1} \beta_2)^{k-1} \qquad \text{for all } k \in \mathbb{Z}_{>0}. \tag{5.21}$$

Hence the parameters of this invariant class are β_1, β_2/β_1, and $\{q^n\}_{n=0}^{\infty}$.

As with the previous invariant class, $Q(G)$ is always a single q^n times some weight factors, but in this case n will be the number of components $c(G)$. The exponent on β_1 will be the number of contractions made, i.e., $|G| - c(G)$ and the exponent on β_2/β_1 will be the number of edges minus the number of contractions. Hence the value of this invariant is completely determined by $c(G)$, $|G|$, and $\|G\|$.

Now assume $\beta_l \neq 0$ for some l and $\alpha_k = \alpha_1^k \neq 0$ for all k. Setting $l = 1$ and $k = m+1$ in (5.20b) leads to $\alpha_1^{m+1} \beta_1 \beta_m + \beta_{m+1}^2 = \alpha_1 \beta_{m+1} \beta_m + \beta_1 \beta_{2m+1}$, or from collecting terms $\beta_1 (\alpha_1^{m+1} \beta_m - \beta_{2m+1}) = \beta_{m+1} (\alpha_1 \beta_m - \beta_{m+1})$. If $\beta_1 = \beta_m = 0$ then this implies $\beta_{m+1} = 0$ too, and in particular $\beta_1 = 0$ implies $\beta_2 = 0$. As it was assumed some $\beta_m \neq 0$, it follows that $\beta_1 \neq 0$.

Having established that, it is possible to fix

$$\beta_m = \alpha_1^m \cdot (\beta_1/\alpha_1) \cdot \gamma_m \tag{5.22}$$

for some new family of parameters $\{\gamma_m\}_{m=1}^{\infty} \subseteq \mathcal{R}$ where $\gamma_1 = 1$. Inserting this into (5.20b) yields the homogeneous equation system

$$\gamma_l \gamma_m + \gamma_k \gamma_{l+m} = \gamma_k \gamma_m + \gamma_l \gamma_{k+m} \qquad \text{for all } k,l,m \in \mathbb{Z}_{>0}. \tag{5.23}$$

Let $z = \gamma_2 - 1$, $l = 1$, and $k = 2$. This equation then becomes $\gamma_m + (1+z)\gamma_{m+1} = (1+z)\gamma_m + \gamma_{m+2}$, or equivalently $z(\gamma_{m+1} - \gamma_m) = \gamma_{m+2} - \gamma_{m+1}$. Hence $\gamma_{m+1} - \gamma_m = z^m$ and $\gamma_m = \sum_{k=0}^{m-1} z^k$ – the so-called z-natural numbers.

Invariant class 3 (The Tutte polynomial). If some β_k and α_k are both nonzero then the parameters of the invariant are α_1, β_1/α_1, z, and $\{q^n\}_{n=0}^{\infty}$, where

$$\alpha_m = \alpha_1^m, \tag{5.24a}$$

$$\beta_m = \alpha_1^m \cdot (\beta_1/\alpha_1) \cdot \sum_{k=0}^{m-1} z^k. \tag{5.24b}$$

It should be observed that α_1 is essentially a weight on edges and β_1/α_1 is a weight on vertices, where the latter however has some interaction with the q^n.

If one denotes by Q' an invariant of this class with values in $\mathcal{R}[q]$ and $\alpha_1 = \beta_1 = 1$ but the same value of z as a particular $\mathcal{R}[q]$-valued Q, then these two are related by

$$Q(G)(q) = \alpha_1^{\|G\|} \cdot (\beta_1/\alpha_1)^{|G|} \cdot Q'(G)(q\alpha_1/\beta_1),$$

so any nontrivial information from Q has been encoded into this simpler Q'. What is this invariant Q'?

For $z = 0$ one recovers the chromatic polynomial, but $Q'(G)$ seen as a polynomial in the two variables q and z also happens to be a known graph invariant: it is the *Tutte polynomial* – in a variant of q-state Potts model variables; see [11] for a nice overview of different forms of the Tutte polynomial – and a closed form expression for it is

$$Q'(G) = (-1)^{|G|} \sum_{F \subseteq \mathrm{E}(G)} (-q)^{c(G-F)} (z-1)^{\|G-F\| + c(G-F) - |G|}. \tag{5.25}$$

In order to verify that this satisfies the recursion (5.2), one may collect the terms of the sum depending on whether $F \supseteq \bar{e}$ or $F \not\supseteq \bar{e}$, where the former group turns out to sum to $Q'(G - \bar{e})$ and the latter group sums to $-\gamma_{|\bar{e}|} Q'(G/\bar{e})$.

For the invariants whose coefficients do not satisfy (5.20), one may return to the identity

$$(\beta_k\alpha_{l+m} - \beta_k\alpha_l\alpha_m)[L_1 \cup L_2, L_3] + (\alpha_k\beta_l\alpha_m - \beta_l\alpha_{k+m})[L_1, L_2 \cup L_3] + \\ + (\alpha_k\beta_l\beta_m + \beta_k\beta_{l+m} - \alpha_l\beta_k\beta_m - \beta_l\beta_{k+m})[L_1 \cup L_2 \cup L_3] \in \\ \in \mathcal{I}(S)(L_1 \cup L_2 \cup L_3) \quad (5.26)$$

that in (5.18) and (5.19) was derived for all $k,l,m \in \mathbb{Z}_{>0}$ and disjoint $L_1, L_2, L_3 \in I$. It is convenient to consider the special case $L_2 = \varnothing$ (which arises when the common vertex u_2 have no neighbours other than u_1 and u_3), as the identity then simplifies to the two terms

$$(\beta_k\alpha_{l+m} - \beta_k\alpha_l\alpha_m + \alpha_k\beta_l\alpha_m - \beta_l\alpha_{k+m})[L_1, L_3] + \\ + (\alpha_k\beta_l\beta_m + \beta_k\beta_{l+m} - \alpha_l\beta_k\beta_m - \beta_l\beta_{k+m})[L_1 \cup L_3] \in \mathcal{I}(S)(L_1 \cup L_3).$$

Invariant class 4. If $\beta_k\alpha_{l+m} - \beta_k\alpha_l\alpha_m + \alpha_k\beta_l\alpha_m - \beta_l\alpha_{k+m} \neq 0$ for some $k,l,m \in \mathbb{Z}_{>0}$ then the invariant Q besides (5.2) also satisfies a recursion on the form

$$Q(G) = \gamma_0 Q(G/xy) \qquad \text{for all nonadjacent } x,y \in \mathrm{V}(G), \qquad (5.27)$$

and from applying this to (5.2) they can both be combined into

$$Q(G) = \gamma_{|\overline{xy}|} Q(G/xy) \qquad \text{for all } x,y \in \mathrm{V}(G), \qquad (5.28)$$

where $\gamma_m = \gamma_0\alpha_m - \beta_m$ and $\overline{xy} = \{e \in \mathrm{E}(G) \,|\, \phi_G(e) = \{x,y\}\}$ denotes the set of edges between x and y.

Invariants of class 4 are subject to the same uncertainties as those satisfying (5.2) in general as to whether these really are the simplest possible recursions, or whether there are some still simpler identities that can be derived. Since the formal process of checking this exactly mirrors what was done in Sect. 5.5, except that the calculations are a bit simpler, it seems appropriate to leave this as an exercise for the reader. It may be noted, however, that $q^n = \gamma_0^{n-1} q$, which indeed limits the degrees of freedom quite considerably.

Invariant class 5. If $\beta_k\alpha_{l+m} - \beta_k\alpha_l\alpha_m + \alpha_k\beta_l\alpha_m - \beta_l\alpha_{k+m} = 0$ for all $k,l,m \in \mathbb{Z}_{>0}$ but $\alpha_k\beta_l\beta_m + \beta_k\beta_{l+m} - \alpha_l\beta_k\beta_m - \beta_l\beta_{k+m} \neq 0$ for some $k,l,m \in \mathbb{Z}_{>0}$ then the invariant Q becomes really trivial, since there is then an identity stating that

$$(\alpha_k\beta_l\beta_m + \beta_k\beta_{l+m} - \alpha_l\beta_k\beta_m - \beta_l\beta_{k+m})Q(G) = 0 \qquad (5.29)$$

if G has at least one vertex. In other words, Q may at most distinguish between the empty graph (no vertices or edges) and all other graphs.

For a generic choice of parameters for an invariant of class 4, it is very likely that one ends up with a parameter behaving as the ones in class 5. The final class of invariants is a bit more interesting.

Invariant class 6. The only other possibility is that $\beta_k\alpha_{l+m} - \beta_k\alpha_l\alpha_m + \alpha_k\beta_l\alpha_m - \beta_l\alpha_{k+m} = 0$ and $\alpha_k\beta_l\beta_m + \beta_k\beta_{l+m} - \alpha_l\beta_k\beta_m - \beta_l\beta_{k+m} = 0$ for all $k,l,m \in \mathbb{Z}_{>0}$, but $\beta_k\alpha_{l+m} \neq \beta_k\alpha_l\alpha_m$ for some $k,l,m \in \mathbb{Z}_{>0}$. Equation (5.26) then takes on the form

$$(\beta_k \alpha_{l+m} - \beta_k \alpha_l \alpha_m)\big([L_1 \cup L_2, L_3] - [L_1, L_2 \cup L_3]\big) \in \mathcal{I}(S)(L_1 \cup L_2 \cup L_3), \quad (5.30)$$

which in plain English means Q cannot tell the difference between G and G' if they differ only in that G has some set L_2 of edges attached to a vertex u_1 whereas G' has them attached to the vertex u_3, where u_1 and u_3 are non-adjacent. It is furthermore easy to get rid of the non-adjacency condition by going via some additional vertex – if necessary, such a vertex can be manufactured by running the recursion (5.2) backwards: $G = G''/e'$, where one endpoint of e' is a new leaf, so that $\beta_1 Q(G) = \alpha_1 Q(G'' - e) - Q(G'')$.

In other words, this kind of invariant does not care which vertices an edge is incident with, so it can at most keep track of the size and order of a graph.

References

1. Michel Bauderon and Bruno Courcelle: Graph expressions and graph rewritings. *Math. Syst. Theory* **20**, 83–127 (1987).
2. John C. Baez and Aaron D. Lauda: Higher-dimensional algebra V: 2-Groups. *Theory Appl. Categ.* **12**, 423–491 (2004); arXiv: math.QA/0307200v3.
3. George M. Bergman: The diamond lemma for ring theory. *Adv. Math.* **29**, 178–218 (1978).
4. Bela Bollobás and Oliver Riordan: A Tutte polynomial for coloured graphs. *Combin. Probab. Comput.* **8**, 45–93 (1999).
5. Predrag Cvitanović: *Group Theory – Tracks, LieÕs, and Exceptional Groups*. Web book (modification of March 17, 2004), `http://chaosbook.org/GroupTheory/`
6. Reinhard Diestel: *Graph Theory* (Graduate Texts in Mathematics **173**), Springer, Berlin, 1997; ISBN 0-387-98211-6.
7. Lars Hellström: *A Generic Framework for Diamond Lemmas*, 2007; arXiv:0712.1142v1 [math.RA].
8. Gerardus 't Hooft and Martinus J. G. Veltman: *Diagrammar*, CERN, Geneva, 1973.
9. Shahn Majid: Cross products by braided groups and bosonization. *J. Algebra* **163**, 165–190 (1994).
10. Seth A. Major: *Spin Network Primer*, (1999); arXiv:gr-qc/9905020v2.
11. Alan D. Sokal: *The multivariate Tutte polynomial (alias Potts model) for graphs and matroids*, arXiv:math.CO/0503607v1.

Part II
Non-Commutative Deformations, Quantization, Homological Methods, and Representations

Chapter 6
Graded q-Differential Algebra Approach to q-Connection

Viktor Abramov

Abstract We propose a concept of a q-connection, where q is a Nth primitive root of unity, which is constructed by means of a graded q-differential algebra $\mathfrak{B}$ with N-differential satisfying $d^N = 0$. Having proved that the Nth power of a q-connection is the endomorphism of the left $\mathfrak{B}$-module $\mathfrak{F} = \mathfrak{B} \otimes_{\mathfrak{A}} \mathfrak{E}$, where $\mathfrak{A}$ is the subalgebra of elements of grading zero of $\mathfrak{B}$, and $\mathfrak{E}$ is a left $\mathfrak{A}$-module, we give the definition of the curvature of a q-connection. We prove that the curvature satisfies the Bianchi identity. Assuming that $\mathfrak{E}$ is a free finitely generated module we associate to a q-connection the matrix of this connection and the curvature matrix. We calculate the expression for the curvature matrix in terms of the entries of the matrix of q-connection. We also find the form of the Bianchi identity in terms of the curvature matrix and the matrix of a q-connection.

6.1 Introduction

Within the framework of a non-commutative geometry there has arisen an interest towards a possible generalization of an exterior calculus such that $d^N = 0, N \geq 2$, where d is an exterior differential. It seems that there are two possible directions to construct a generalization of an exterior calculus with $d^N = 0$. The first one could be called a geometric approach when we construct an analog of differential forms on some geometric space with exterior differential satisfying $d^N = 0$. This line has been elaborated in the series of papers [1, 2, 9, 10]. The second direction could be called an algebraic approach, and it is based on the fact that from an algebraic point of view the algebra of differential forms on a smooth manifold is a graded differential algebra. This second direction has led to a concept of a graded q-differential algebra [6–8] which is a natural generalization of a graded differential algebra. In

V. Abramov
Institute of Mathematics, University of Tartu, Liivi 2, Tartu, Estonia
e-mail: viktor.abramov@ut.ee

S. Silvestrov et al. (eds.), *Generalized Lie Theory in Mathematics, Physics and Beyond,*

this paper we make use of a notion of a graded q-differential algebra to construct a q-connection which may be viewed as a generalization of a classical connection. We consider the tensor product $\mathfrak{F} = \mathfrak{B} \otimes_{\mathfrak{A}} \mathfrak{E}$, where $\mathfrak{B}$ is a graded q-differential algebra, $\mathfrak{A} \subset \mathfrak{B}$ is the subalgebra of $\mathfrak{B}$ of elements of grading zero and $\mathfrak{E}$ is a left $\mathfrak{A}$-module. We define a q-connection (Definition 6.1) as a linear operator of degree one on the $\mathfrak{F}$. We study the structure of a q-connection, give the definition of its curvature (Definition 6.2) and prove the Bianchi identity. In the last section we study the analog of a local structure of a q-connection. We introduce the matrix of a q-connection and the curvature matrix. We also find the expression for the curvature matrix and the form of the Bianchi identity in terms of the entries of the matrix of a q-connection. It should be pointed that the expression for the curvature of a q-connection in the case of $N = 3$ is very similar to the Chern–Simons term on three dimensional manifold, and this suggests a possible relation of q-connection to Chern–Simons theory. The approach we use to define and study a q-connection is an algebraic one, and an interesting question is to construct a q-connection on a geometric space by means of analogs of differential forms. It should be noted that a first attempt to construct a q-connection on a classical smooth manifold was made in [9], where the author used the differential forms with exterior differential satisfying $d^N = 0$. Unfortunately later it was shown that the algebraic structure of the differential forms is not self-consistent and algebra degenerates. We think that the Theorem 6.1 given at the end of the second section can be used to construct a graded q-differential algebra of analogs of differential forms on a generalized Clifford algebra which may be viewed as a non-commutative space, and this will lead to geometric realization for a q-connection.

6.2 Graded q-Differential Algebra

In this section we remind a notion of a graded q-differential algebra, where q is a primitive N-th root of unity. Given an associative unital graded algebra $\mathfrak{B}$ over the complex numbers $\mathbb{C}$ containing an element w of grading one satisfying $w^N = ae$, where $a \in \mathbb{C}$ and e is the identity element of $\mathfrak{B}$, we construct the graded q-differential algebra with the help of a graded q-commutator.

We remind that an associative unital graded algebra is said to be a graded differential algebra if it is equipped with a linear mapping d of degree one satisfying the graded Leibniz rule and the nilpotency condition $d^2 = 0$. A linear mapping d of degree one is called a differential of a corresponding graded differential algebra. This notion of a graded differential algebra has a natural generalization called a graded q-differential algebra which has been proposed within the framework of a non-commutative geometry [6]. Let $N \geq 2$ be an integer, q be a primitive N-th root of unity and $\mathfrak{B} = \oplus_i \mathfrak{B}^i$ be an associative unital graded algebra over the complex numbers. Let us denote the grading of a homogeneous element u of $\mathfrak{B}$ by $|u|$. An algebra $\mathfrak{B}$ is said to be a graded q-differential algebra if it is endowed with a linear mapping d of degree one, i.e. $d : \mathfrak{B}^i \to \mathfrak{B}^{i+1}$, satisfying the graded q-Leibniz rule

$d(uv) = d(u)\,v + q^{|u|}\,u\,d(v)$, where $u, v \in \mathfrak{B}$, and the N-nilpotency condition $d^N = 0$. According to the terminology elaborated in [8] we shall call d the N-differential of a graded q-differential algebra $\mathfrak{B}$. It is easy to see that a graded differential algebra is the particular case of a graded q-differential algebra when $N = 2, q = -1$. Hence a concept of a graded q-differential algebra can be viewed as a generalization of a graded differential algebra.

Let $\mathfrak{B}$ be a graded q-differential algebra. Obviously the subspace $\mathfrak{B}^0 \subset \mathfrak{B}$ of elements of grading zero is the subalgebra of an algebra $\mathfrak{B}$. A graded q-differential algebra $\mathfrak{B}$ is said to be an N-differential calculus over an associative unital algebra $\mathfrak{A}$ provided that $\mathfrak{A} = \mathfrak{B}^0$. If $u, v \in \mathfrak{B}^0$ and $w \in \mathfrak{B}^i$ then the mappings $\mathfrak{B}^0 \times \mathfrak{B}^i \to \mathfrak{B}^i$ and $\mathfrak{B}^i \times \mathfrak{B}^0 \to \mathfrak{B}^i$ defined by $(u, w) \mapsto uw, (w, v) \mapsto wv$ determine the structure of a bimodule over the algebra $\mathfrak{B}^0$ on the subspace $\mathfrak{B}^i \subset \mathfrak{B}$ of elements of grading i. Consequently every subspace $\mathfrak{B}^i$ can be considered as the bimodule over the algebra $\mathfrak{B}^0$. Thus we have the following sequence

$$\ldots \xrightarrow{d} \mathfrak{B}^{i-1} \xrightarrow{d} \mathfrak{B}^i \xrightarrow{d} \mathfrak{B}^{i+1} \xrightarrow{d} \ldots \tag{6.1}$$

of bimodules over the algebra $\mathfrak{B}^0$. If we consider the part of this sequence which is $d : \mathfrak{B}^0 \to \mathfrak{B}^1$ and take into account that in this case d satisfies the Leibniz rule we conclude that the pair $(\mathfrak{B}^1, d)$ is the differential calculus on the algebra $\mathfrak{B}^0$. If we consider each $\mathfrak{B}^i$ in (6.1) as a vector space over the complex numbers and take into account the N-nilpotency of d then the sequence (6.1) is an N-complex of vector spaces [8]. The generalized cohomologies $H^i_{(k)}(\mathfrak{B})$ of this N-complex are defined by the formula

$$H^i_{(k)}(\mathfrak{B}) = \{\mathrm{Ker}(d^k) : \mathfrak{B}^i \to \mathfrak{B}^{i+k}\}/\{\mathrm{Im}(d^{N-k}) : \mathfrak{B}^{i-N+k} \to \mathfrak{B}^i\}.$$

Given a graded q-differential algebra $\mathfrak{B}$ one can associate to it the generalized cohomologies $H^i_{(k)}(\mathfrak{B})$ of the corresponding N-complex of vector spaces (6.1).

Let $\mathfrak{B} = \oplus_{i \in \mathbb{Z}_N} \mathfrak{B}^i$ be an associative unital $\mathbb{Z}_N$-graded algebra over the complex numbers, and e is the identity element of this algebra. If u, v are homogeneous elements of $\mathfrak{B}$ then the graded q-commutator of u, v is defined by the formula $[u, v]_q = uv - q^{|u||v|}vu$. The following theorem [3] shows the way one can construct the structure of a graded q-differential algebra on an algebra $\mathfrak{B}$.

Theorem 6.1. *If there exists an element $w \in \mathfrak{B}$ of grading one such that $w^N = a\,e$, where $a \in \mathbb{C}$, then an algebra $\mathfrak{B}$ equipped with the linear mapping $d : \mathfrak{B} \to \mathfrak{B}$ defined by the formula $d(v) = [w, v]_q, v \in \mathfrak{B}$ is the $\mathbb{Z}_N$-graded q-differential algebra, and d is its N-differential.*

6.3 *q*-Connection and Its Curvature

The aim of this section is to introduce a notion of a q-connection and its curvature. We use an algebraic approach based on a graded q-differential algebra which plays the part of an analog of algebra of differential forms. In the classical theory of connections the algebra of differential forms on a manifold has $\mathbb{Z}_2$-graded structure and assuming that a vector bundle over this manifold has also $\mathbb{Z}_2$-graded structure (superbundle) one can construct a $\mathbb{Z}_2$-connection or superconnection [11]. Following the same scheme and making use of a $\mathbb{Z}_N$-graded structure of a graded q-differential algebra we give a definition of a q-connection and study its structure.

Let $\mathfrak{B}$ be an N-differential calculus over an algebra $\mathfrak{A}$, i.e. $\mathfrak{B}$ is a graded q-differential algebra with N-differential d and $\mathfrak{A} = \mathfrak{B}^0$. Let $\mathfrak{E}$ be a left $\mathfrak{A}$-module. Since an algebra $\mathfrak{B}$ can be considered as the $(\mathfrak{B},\mathfrak{A})$-bimodule the tensor product $\mathfrak{F} = \mathfrak{B} \otimes_{\mathfrak{A}} \mathfrak{E}$ is the left $\mathfrak{B}$-module [5]. It should be mentioned here that an algebra $\mathfrak{B}$ may be also viewed as the $(\mathfrak{A},\mathfrak{A})$-bimodule and in this case the tensor product $\mathfrak{F}$ has the structure of a left $\mathfrak{A}$-module. Taking into account that an algebra $\mathfrak{B}$ can be viewed as the direct sum of $(\mathfrak{A},\mathfrak{A})$-bimodules $\mathfrak{B}^i$ we can split the left $\mathfrak{A}$-module $\mathfrak{F}$ into the direct sum of the left $\mathfrak{A}$-modules $\mathfrak{F}^i = \mathfrak{B}^i \otimes_{\mathfrak{A}} \mathfrak{E}$ [5], i.e. $\mathfrak{F} = \oplus_i \mathfrak{F}^i$, which means that the left $\mathfrak{A}$-module $\mathfrak{F}$ is the graded left $\mathfrak{A}$-module. The tensor product $\mathfrak{F}$ is also the vector space over $\mathbb{C}$ where this vector space is the tensor product of the vector spaces $\mathfrak{B}$ and $\mathfrak{E}$. It is evident that $\mathfrak{F}$ is a graded vector space, i.e. $\mathfrak{F} = \oplus_i \mathfrak{F}^i$, where $\mathfrak{F}^i$ is the tensor product of the vector spaces $\mathfrak{B}^i$ and $\mathfrak{E}$. Keeping in mind this structure of a vector space of $\mathfrak{F}$ we can use such notions as a linear operator on $\mathfrak{F}$, bilinear forms and so on. Let us denote the vector space of linear operators on $\mathfrak{F}$ by $\mathrm{Lin}(\mathfrak{F})$. The structure of the graded vector space of $\mathfrak{F}$ induces the structure of a graded vector space on $\mathrm{Lin}(\mathfrak{F})$, and we shall denote the subspace of homogeneous linear operators of degree i by $\mathrm{Lin}^i(\mathfrak{F})$.

Definition 6.1. A q-connection on the left $\mathfrak{B}$-module $\mathfrak{F}$ is a linear operator $D : \mathfrak{F} \to \mathfrak{F}$ of degree one satisfying the condition

$$D(u\xi) = d(u)\xi + q^{|u|}\, uD(\xi), \tag{6.2}$$

where $u \in \mathfrak{B}, \xi \in \mathfrak{F}$, and d is the N-differential of a graded q-differential algebra $\mathfrak{B}$.

Making use of the previously introduced notations we can write $D \in \mathrm{Lin}^1(\mathfrak{F})$. Let us note that if $N = 2$ then $q = -1$, and in this particular case the Definition 6.1 gives us the notion of a classical connection. Hence a concept of a q-connection can be viewed as a generalization of a classical connection.

Proposition 6.1. *The N-th power of any q-connection D is the endomorphism of degree N of the left $\mathfrak{B}$-module $\mathfrak{F}$.*

The proof of this proposition is based on the following formula

$$D^k(u\xi) = \sum_{m=0}^{k} q^{m|u|} \begin{bmatrix} k \\ m \end{bmatrix}_q d^{k-m}(u)\, D^m(\xi), \tag{6.3}$$

where u is a homogeneous element of $\mathfrak{B}$, $\xi \in \mathfrak{F}$, and $\begin{bmatrix} k \\ m \end{bmatrix}_q$ are the q-binomial coefficients . As d is the N-differential of a graded q-differential algebra $\mathfrak{B}$ we have

$d^N(u) = 0$. Taking into account that $\begin{bmatrix} N \\ m \end{bmatrix}_q = 0$ for $1 \leq m \leq N-1$ we see that in the case of $k = N$ (6.3) takes on the form $D^N(u\xi) = q^{N|u|} u D^N(\xi) = u D^N(\xi)$ and this proves that D^N is the endomorphism of the left $\mathfrak{B}$-module $\mathfrak{F}$.

Definition 6.2. The endomorphism $F_D = D^N$ of degree N of the left $\mathfrak{B}$-module $\mathfrak{F}$ is said to be the curvature of a q-connection D.

The graded vector space $\mathrm{Lin}(\mathfrak{F})$ can be endowed with the structure of a graded algebra if one takes the product $A \circ B$ of two linear operators A, B of the vector space $\mathfrak{F}$ as an algebra multiplication. Let us extend a q-connection D to the linear operator on the vector space $\mathrm{Lin}(\mathfrak{F})$ by means of the graded q-commutator as follows

$$D(A) = [D, A]_q = D \circ A - q^{|A|} A \circ D, \tag{6.4}$$

where A is a homogeneous linear operator. Obviously D is the linear operator of degree one on the vector space $\mathrm{Lin}(\mathfrak{F})$, i.e. $D : \mathrm{Lin}^i(\mathfrak{F}) \to \mathrm{Lin}^{i+1}(\mathfrak{F})$, and D satisfies the graded q-Leibniz rule with respect to the algebra structure of $\mathrm{Lin}(\mathfrak{F})$. It follows from the definition of the curvature of a q-connection that F_D can be viewed as the linear operator of degree N on the vector space $\mathfrak{F}$, i.e. $F_D \in \mathrm{Lin}^N(\mathfrak{F})$. Consequently one can act on F_D by D, and it holds that

Proposition 6.2. *The curvature F_D of any q-connection D on $\mathfrak{F}$ satisfies the Bianchi identity $D(F_D) = 0$.*

We have $D(F_D) = [D, F_D]_q = D \circ F_D - q^N F_D \circ D = D^{N+1} - D^{N+1} = 0$.

As it was noted earlier the concept of a q-connection can be viewed as a generalization of a classical connection. It is well known that slightly modifying the construction of a connection in the classical case by assuming that a vector bundle is a superbundle one can construct a superconnection or $\mathbb{Z}_2$-graded connection [11]. The key role plays the fact that the algebra of differential forms has the $\mathbb{Z}_2$-structure determined by the forms of odd and even degree. It turns out that the notion of a q-connection suits well to construct in analogy with the classical case a $\mathbb{Z}_N$-graded q-connection because of a natural $\mathbb{Z}_N$-graded structure of a graded q-differential algebra. The $\mathbb{Z}_N$-graded q-connection was defined and studied in [4].

6.4 Matrix of a q-Connection

It is well known that a connection on the vector bundle of finite rank over a finite dimensional smooth manifold can be studied locally by choosing a local trivialization of the vector bundle and this leads to the basis for the module of sections of this vector bundle. Our next aim is to study an analog of the local structure of a q-connection D and this can be achieved if we assume that $\mathfrak{E}$ is a finitely generated free $\mathfrak{A}$-module. Let $\mathfrak{e} = \{\mathfrak{e}_\mu\}$, where $1 \leq \mu \leq r$, be a finite $\mathfrak{A}$-basis for a module $\mathfrak{E}$. In what follows we shall use the Einstein summation convention with respect to the indices induced by a basis $\mathfrak{e}$. For any element $\xi \in \mathfrak{E}$ we have $\xi = \xi^\mu \mathfrak{e}_\mu$, where $\xi^\mu \in \mathfrak{A}$. It is well known [5] that the left $\mathfrak{A}$-modules $\mathfrak{F}^0 = \mathfrak{A} \otimes_{\mathfrak{A}} \mathfrak{E}$ and $\mathfrak{E}$ are isomorphic, where the isomorphism $\phi : \mathfrak{E} \longrightarrow \mathfrak{F}^0$ can be defined by $\phi(\xi) = e \otimes \xi$, where e is the identity element of the algebra $\mathfrak{A}$. Hence $\mathfrak{f} = \{\mathfrak{f}_\mu\}$, where $\mathfrak{f}_\mu = e \otimes \mathfrak{e}_\mu$, is the $\mathfrak{A}$-basis for the left $\mathfrak{A}$-module $\mathfrak{F}^0$, and for any $\xi \in \mathfrak{F}^0$ we have $\xi = \xi^\mu \mathfrak{f}_\mu$. Analogously we can express any element of the $\mathfrak{F}^i$ as a linear combination of $\mathfrak{f}_\mu$ with coefficients from $\mathfrak{B}^i$. Indeed let $\omega \otimes \xi$ be an element of $\mathfrak{F}^i = \mathfrak{B}^i \otimes_{\mathfrak{A}} \mathfrak{E}$. Then

$$\begin{aligned} \omega \otimes \xi &= (\omega e) \otimes (\xi^\mu \mathfrak{e}_\mu) = (\omega e \xi^\mu) \otimes \mathfrak{e}_\mu \\ &= (\omega \xi^\mu e) \otimes \mathfrak{e}_\mu = \omega \xi^\mu (e \otimes \mathfrak{e}_\mu) = \omega^\mu \mathfrak{f}_\mu, \end{aligned}$$

where $\omega^\mu = \omega \xi^\mu \in \mathfrak{B}^i$.

Let $\xi = \xi^\mu \mathfrak{f}_\mu \in \mathfrak{F}^0$. Obviously $D\xi \in \mathfrak{F}^1$, and making use of (6.2) we can express the element $D\xi$ as follows

$$D\xi = D(\xi^\mu \mathfrak{f}_\mu) = d\xi^\mu \mathfrak{f}_\mu + \xi^\mu D\mathfrak{f}_\mu. \tag{6.5}$$

Since any element of $\mathfrak{F}^1$ can be expressed in terms of the basis $\{\mathfrak{f}_\mu\}$ with coefficients from $\mathfrak{B}^1$ we have

$$D\mathfrak{f}_\mu = \theta_\mu^\nu \mathfrak{f}_\nu, \tag{6.6}$$

where $\theta_\mu^\nu \in \mathfrak{B}^1$. In analogy with the classical theory of connections in differential geometry of fibre bundles we shall call the $r \times r$-matrix $\hat{\theta} = (\theta_\mu^\nu)$, whose entries θ_μ^ν are the elements of $\mathfrak{B}^1$, the matrix of a q-connection D with respect to the $\mathfrak{A}$-basis $\mathfrak{f}$ of the left $\mathfrak{A}$-module $\mathfrak{F}^0$. Let $\mathrm{Mat}_r(\mathfrak{B})$ be the vector space of the $r \times r$-matrices with entries from a graded q-differential algebra $\mathfrak{B}$. If each entry of a matrix $\hat{\omega} = (\omega_\mu^\nu)$ is an element of a homogeneous subspace $\mathfrak{B}^i$, i.e. $\omega_\mu^\nu \in \mathfrak{B}^i$ then we shall call $\hat{\omega}$ a homogeneous matrix of grading i and denote the vector space of such matrices by $\mathrm{Mat}_r(\mathfrak{B}^i)$. Obviously $\mathrm{Mat}_r(\mathfrak{B}) = \oplus_i \mathrm{Mat}_r(\mathfrak{B}^i)$ and $\hat{\theta} \in \mathrm{Mat}_r(\mathfrak{B}^1)$. We define the product of two matrices $\hat{\omega} = (\omega_\mu^\nu), \hat{\eta} = (\eta_\mu^\nu) \in \mathrm{Mat}_r(\mathfrak{B})$ as follows

$$(\hat{\omega}\,\hat{\eta})_\mu^\nu = \omega_\mu^\sigma\, \eta_\sigma^\nu. \tag{6.7}$$

It is obvious that $\mathrm{Mat}_r(\mathfrak{B})$ is a graded associative algebra with respect to the multiplication (6.7). If $\hat{\omega}, \hat{\eta} \in \mathrm{Mat}_r(\mathfrak{B})$ are homogeneous matrices then we define the graded q-commutator by $[\hat{\omega}, \hat{\eta}] = \hat{\omega}\,\hat{\eta} - q^{|\hat{\omega}||\hat{\eta}|}\hat{\eta}\,\hat{\omega}$. We extend the N-differential d of a graded q-differential algebra $\mathfrak{B}$ to the algebra $\mathrm{Mat}_r(\mathfrak{B})$ as follows $d\hat{\omega} = d(\omega_\mu^\nu) = (d\omega_\mu^\nu)$.

From (6.5) and (6.6) it follows that

$$D\xi = (d\xi^\mu + \xi^\nu \theta_\nu^\mu)\, \mathfrak{f}_\mu. \tag{6.8}$$

In order to express the curvature F_D of a q-connection D in the terms of the entries of the matrix $\hat{\theta}$ of a q-connection D we should express the kth power of a q-connection D, where $1 \leq k \leq N$, in the terms of the entries of the matrix $\hat{\theta}$. It can be calculated that the kth power of D has the following form

$$
\begin{aligned}
D^k\xi &= \sum_{l=0}^{k} \begin{bmatrix} k \\ l \end{bmatrix}_q d^{k-l}\xi^{\mu}\, \Theta_{\mu}^{(l,k)\nu}\, \mathfrak{f}_\nu \\
&= (d^k\xi^{\mu}\,\Theta_{\mu}^{(0,k)\nu} + [k]_q d^{k-1}\xi^{\mu}\,\Theta_{\mu}^{(1,k)\nu} + \ldots + \xi^{\mu}\,\Theta_{\mu}^{(k,k)\nu})\,\mathfrak{f}_\nu, \qquad (6.9)
\end{aligned}
$$

where $\Theta_{\mu}^{(l,k)\nu} \in \mathfrak{B}^l$ are polynomials on the entries θ_{ν}^{μ} of the matrix $\hat{\theta}$ of a q-connection D and their differentials. We can calculate the polynomials $\Theta_{\mu}^{(l,k)\nu}$ by means of the following recursion formula

$$
\Theta_{\mu}^{(l,k)\nu} = d\Theta_{\mu}^{(l-1,k)\nu} + q^{l-1}\,\Theta_{\mu}^{(l-1,k)\sigma}\,\theta_{\sigma}^{\nu}, \qquad (6.10)
$$

or in the matrix form

$$
\hat{\Theta}^{(l,k)} = d\hat{\Theta}^{(l-1,k)} + q^{l-1}\,\hat{\Theta}^{(l-1,k)}\,\hat{\theta}, \qquad (6.11)
$$

where we begin with the polynomial $\Theta_{\mu}^{(0,k)\nu} = \delta_{\mu}^{\nu}\, e \in \mathfrak{A}$, and e is the identity element of $\mathfrak{A} \subset \mathfrak{B}$. For example the first four polynomials in the expansion (6.9) obtained with the help of the recursion formula (6.10) have the form

$$
\begin{aligned}
\Theta_{\mu}^{(1,k)\nu} &= \theta_{\mu}^{\nu}, \qquad (6.12)\\
\Theta_{\mu}^{(2,k)\nu} &= d\theta_{\mu}^{\nu} + q\,\theta_{\mu}^{\sigma}\theta_{\sigma}^{\nu}, \qquad (6.13)\\
\Theta_{\mu}^{(3,k)\nu} &= d^2\theta_{\mu}^{\nu} + (q+q^2)\,d\theta_{\mu}^{\sigma}\theta_{\sigma}^{\nu} + q^2\,\theta_{\mu}^{\sigma}\,d\theta_{\sigma}^{\nu} + q^3\,\theta_{\mu}^{\tau}\theta_{\tau}^{\sigma}\theta_{\sigma}^{\nu}. \qquad (6.14)\\
\Theta_{\mu}^{(4,k)\nu} &= d^3\theta_{\mu}^{\nu} + (q+q^2+q^3)\,d^2\theta_{\mu}^{\sigma}\theta_{\sigma}^{\nu} + q^3\,\theta_{\mu}^{\sigma}\,d^2\theta_{\sigma}^{\nu}\\
&\quad + (q^3+q^4+q^5)\,d\theta_{\mu}^{\tau}\theta_{\tau}^{\sigma}\theta_{\sigma}^{\nu} + (q^4+q^5)\,\theta_{\mu}^{\tau}d\theta_{\tau}^{\sigma}\theta_{\sigma}^{\nu}\\
&\quad + q^5\,\theta_{\mu}^{\tau}\theta_{\tau}^{\sigma}d\theta_{\sigma}^{\nu} + (q^2+q^3+q^4)\,d\theta_{\mu}^{\sigma}d\theta_{\sigma}^{\nu} + q^6\,\theta_{\mu}^{\tau}\theta_{\tau}^{\sigma}\theta_{\sigma}^{\rho}\theta_{\rho}^{\nu}. \qquad (6.15)
\end{aligned}
$$

From (6.9) it follows that if $k = N$ then the first term $d^N\xi^{\mu}\,\Theta_{\mu}^{(0,N)\nu}$ in this expansion vanishes because of the N-nilpotency of the N-differential d, and the next terms corresponding to the l values from 1 to $N-1$ also vanish because of the well known property of q-binomial coefficients $\begin{bmatrix} N \\ l \end{bmatrix}_q = 0$ provided q is a primitive Nth root of unity. Hence if $k = N$ then (6.9) takes on the form

$$
D^N\xi = \xi^{\mu}\,\Theta_{\mu}^{(N,N)\nu}\,\mathfrak{f}_\nu. \qquad (6.16)
$$

In order to simplify the notations and assuming that N is fixed we shall denote $\Theta_{\mu}^{(N)\nu} = \Theta_{\mu}^{(N,N)\nu}$. We shall call the $(r\times r)$-matrix $\hat{\Theta}^{(N)} = (\Theta_{\mu}^{(N)\nu})$, whose entries are the elements of grading N of a graded q-differential algebra $\mathfrak{B}$, the curvature matrix of a q-connection D. Obviously $\hat{\Theta}^{(N)} \in \mathrm{Mat}_r(\mathfrak{B}^N)$. In new notations (6.16) can be written as follows $D^N\xi = \xi^{\mu}\,\Theta_{\mu}^{(N)\nu}\,\mathfrak{f}_\nu$, and it clearly demonstrates that D^N is the endomorphism of degree N of the left $\mathfrak{B}$-module $\mathfrak{F}$.

Let us consider the form of the curvature matrix of a q-connection in two special cases. If $N=2$ then $q=-1$, and $\mathfrak{B}$ is a graded differential algebra with differential d satisfying $d^2=0$. This is a classical case, and if we assume that $\mathfrak{B}$ is the algebra of differential forms on a smooth manifold M with exterior differential d and exterior multiplication $\wedge$, $\mathfrak{E}$ is the module of smooth sections of a vector bundle $\pi: E \to M$ over M, D is a connection on E, $\mathfrak{e}$ is a local frame of a vector bundle E then $\hat{\theta}$ is the matrix of 1-forms of a connection D and (6.13) gives the expression for the curvature 2-form

$$\Theta_\mu^{(2)\nu} = d\theta_\mu^\nu + q\,\theta_\mu^\sigma\theta_\sigma^\nu = d\theta_\mu^\nu - \theta_\mu^\sigma \wedge \theta_\sigma^\nu = d\theta_\mu^\nu + \theta_\sigma^\nu \wedge \theta_\mu^\sigma,$$

in which we immediately recognize the classical expression for the curvature.

If $N=3$ then $q=\exp(\frac{2\pi i}{3})$ is the cubic root of unity satisfying the relations $q^3=1,\ 1+q+q^2=0$. This is a first non-classical case of a q-connection, and (6.14) gives the following expression for the curvature of a q-connection

$$\begin{aligned}\Theta_\mu^{(3)\nu} &= d^2\theta_\mu^\nu + (q+q^2)\,d\theta_\mu^\sigma\theta_\sigma^\nu + q^2\,\theta_\mu^\sigma\,d\theta_\sigma^\nu + q^3\,\theta_\mu^\tau\theta_\tau^\sigma\theta_\sigma^\nu \\ &= d^2\theta_\mu^\nu - d\theta_\mu^\sigma\theta_\sigma^\nu + q^2\,\theta_\mu^\sigma\,d\theta_\sigma^\nu + \theta_\mu^\tau\theta_\tau^\sigma\theta_\sigma^\nu \\ &= d^2\theta_\mu^\nu - (d\theta_\mu^\sigma\theta_\sigma^\nu - q^2\,\theta_\mu^\sigma\,d\theta_\sigma^\nu) + \theta_\mu^\tau\theta_\tau^\sigma\theta_\sigma^\nu \end{aligned} \tag{6.17}$$

We can put this expression into the form

$$\Theta_\mu^{(3)\nu} = d^2\theta_\mu^\nu + \theta_\mu^\sigma\Theta_\sigma^{(2)\nu} - \Theta_\mu^{(2)\sigma}\theta_\sigma^\nu + q\,\theta_\mu^\tau\theta_\tau^\sigma\theta_\sigma^\nu. \tag{6.18}$$

Taking into account that $\Theta_\mu^{(2)\sigma} = d\theta_\mu^\sigma + q\,\theta_\mu^\tau\theta_\tau^\sigma$ can be viewed as a q-deformed classical curvature 2-form we see that expression (6.18) resembles the Chern–Simons term in dimension three, and this suggests a possible relation of the curvature of q-connection in the case of $N=3$ to the Chern–Simons theory. It is useful to write the expression (6.17) for the curvature in a matrix form

$$\hat{\Theta}^{(3)} = d^2\hat{\theta} - [d\hat{\theta},\hat{\theta}]_q + \hat{\theta}^3. \tag{6.19}$$

If $N=4$ then $q=i$ is the fourth root of unity satisfying relations $1+q^2=0, q^2=-1$. The expression (6.15) for curvature in this case takes on the form

$$\begin{aligned}\Theta_\mu^{(4)\nu} &= d^3\theta_\mu^\nu - (d^2\theta_\mu^\sigma\theta_\sigma^\nu - q^3\,\theta_\mu^\sigma\,d^2\theta_\sigma^\nu) + d\theta_\mu^\tau\theta_\tau^\sigma\theta_\sigma^\nu \\ &\quad -(q^2+q^3)\,\theta_\mu^\tau d\theta_\tau^\sigma\theta_\sigma^\nu + q\,\theta_\mu^\tau\theta_\tau^\sigma d\theta_\sigma^\nu - q\,d\theta_\mu^\sigma d\theta_\sigma^\nu - \theta_\mu^\tau\theta_\tau^\sigma\theta_\sigma^\rho\theta_\rho^\nu. \end{aligned} \tag{6.20}$$

This expression can be put into a matrix form as follows

$$\hat{\Theta}^{(4)} = d^3\hat{\theta} - [d^2\hat{\theta},\hat{\theta}]_q + [[d\hat{\theta},\hat{\theta}]_q,\hat{\theta}]_q - q\,(d\hat{\theta})^2 - \hat{\theta}^4. \tag{6.21}$$

It should be mentioned that in the case of $N=4$ it holds that

$$[d\hat{\theta},\hat{\theta}]_q = d\hat{\theta}\;\hat{\theta} - q^2\,\hat{\theta}\,d\hat{\theta} = d\hat{\theta}\;\hat{\theta} + \hat{\theta}\,d\hat{\theta} = [\hat{\theta},d\hat{\theta}]_q, \tag{6.22}$$

and the graded q-commutator degenerates in the case of $d\hat{\theta}$, i.e. $[d\hat{\theta}, d\hat{\theta}] = d\hat{\theta}\, d\hat{\theta} - q^4 d\hat{\theta}\, d\hat{\theta} = 0$.

Let us remind that it was proved in the previous section that the curvature of a q-connection satisfies the Bianchi identity. Straightforward computation shows that in both cases of (6.19) and (6.21) this identity in terms of the matrices of the curvature and a connection takes on the form

$$d\hat{\Theta} = [\hat{\Theta}, \hat{\theta}]_q. \tag{6.23}$$

Acknowledgements The author gratefully acknowledges the financial support of his research by the Estonian Science Foundation under the Grant Nr. 6206.

References

1. V. Abramov, R. Kerner and B. Le Roy, Hypersymmetry: A $\mathbb{Z}_3$-generalization of supersymmetry, *J. Math. Phys.*, **38**(3), 1650–1669 (1997).
2. V. Abramov and R. Kerner, Exterior differentials of higher order and their covariant generalization, *J. Math. Phys.*, **41**(8), 5598–5614 (2000).
3. V. Abramov, On a graded q-differential algebra, *J. Nonlinear Math. Phys.*, **13**(Suppl.), 1–8 (2006).
4. V. Abramov, Generalization of superconnection in non-commutative geometry, *Proc. Estonian Acad. Sci. Phys. Math.*, **55**(1), 3–15 (2006).
5. C. Curtis and I. Reiner, Representation Theory of Finite Groups and Associative Algebras, Interscience, New York, 1962.
6. M. Dubois-Violette and R. Kerner, Universal q-differential calculus and q-analog of homological algebra, *Acta Math. Univ. Comenian.*, **65**, 175–188 (1996).
7. Dubois-Violette M. Lectures on graded differential algebras and noncommutative geometry, In *Noncommutative differential geometry and its applications to physics: Proceedings of the workshop (Maeda, Yoshiaki., eds.)*. Kluwer, Dordrecht. *Math. Phys. Stud.*, **23**, 245–306 (2001).
8. M. Dubois-Violette, Lectures on differentials, generalized differentials and on some examples related to theoretical physics, *math.QA*/0005256.
9. M. Kapranov, On the q-analog of homological algebra, *math.QA*/9611005.
10. R. Kerner and V. Abramov, On certain realizations of q-deformed exterior differential calculus, *Rep. Math. Phys.*, **43**(1–2), 179–194 (1999).
11. V. Mathai and D. Quillen, Superconnections, Thom classes and equivariant differential forms, *Topology*, **25**, 85–110 (1986).

Chapter 7
On Generalized N-Complexes Coming from Twisted Derivations

Daniel Larsson and Sergei D. Silvestrov

Abstract Inspired by a result of V. Abramov [1] on q-differential graded algebras, we prove a theorem, analogous to Abramov's result but in a slightly different set-up, using a σ- (twisted) derivation as the differential-like map. As an application, we construct a generalized N-complex based on the ring of Laurent polynomials.

7.1 Introduction

In [1] V. Abramov effectively constructs an N-complex from a $\mathbb{Z}$-graded, unital $\mathbb{C}$-algebra via a generalized commutator. More precisely, he considers a $\mathbb{Z}$-graded associative $\mathbb{C}$-algebra $D = \bigoplus_{n\in\mathbb{Z}} D_n$ with unity together with the (graded) q-commutator with $q \in \mathbb{C}$, defined by

$$\langle a,b\rangle_q := ab - q^{\deg(a)\deg(b)}ba, \quad \text{for} \quad a \in D_{\deg(a)}, \quad \text{and} \quad b \in D_{\deg(b)},$$

and where $\deg(\cdot)$ is the graded degree-function. Notice that this is undefined for non-homogenous elements and that this definition uses more than the fact that $\mathbb{Z}$ is a group: it uses the fact that $\mathbb{Z}$ is a ring!

From definition it follows easily that

$$\langle a,bc\rangle_q = \langle a,b\rangle_q c + q^{\deg(a)\deg(b)} b\langle a,c\rangle_q,$$

D. Larsson
Department of Mathematics, Uppsala University, Box 480, 751 06 Uppsala, Sweden
e-mail: daniel.larsson@math.uu.se

S.D. Silvestrov
Centre for Mathematical Sciences, Box 118, 221 00 Lund, Sweden
e-mail: sergei.silvestrov@math.lth.se

S. Silvestrov et al. (eds.), *Generalized Lie Theory in Mathematics, Physics and Beyond,*

that is, the mapping $\mathrm{d}_a(b) := \langle x, \cdot\rangle_q(b) = \langle a, b\rangle_q$ is a q-differential on D. Notice, however, that d_a is linear only on homogeneous components. This is due to the involvement of the factor $q^{\deg(a)\deg(b)}$ and the fact that $\deg(\cdot)$ is not linear.

Abramov's main result can now be formulated as

Theorem 7.1 (Abramov [1]). *Suppose $D = \bigoplus_{n\in\mathbb{Z}} D_n$ and that $N \geq 2$ is given such that q is a primitive N^{th}-root of unity. Assume further that $a \in D_1$ and that $a^N = u\mathbf{1}_D \in D_0$, for $u \in \mathbb{C}$. Then $\mathrm{d}_a^N(b) = 0$ for all $b \in A$.*

7.1.1 Our Main Result

The aim of this short note is to prove an analogous result where Abramov's q-differential is replaced by a so-called σ- (twisted) derivation (see Sect. 7.2 for the definition).

Let k be a commutative, associative ring with unity and A an associative k-algebra with unity. Furthermore, let G be a subset of A and form $k[\mathrm{G}]$, the k-algebra generated by G. Let $\sigma : k[\mathrm{G}] \to k[\mathrm{G}]$ be a k-algebra homomorphism defined by $\sigma(b) = \phi(a,b)b$ for a fixed $a \in k[\mathrm{G}]$ and all $b \in k[\mathrm{G}]$, where ϕ is a map $\phi : \{a\} \times k[\mathrm{G}] \to \mathrm{Z}(k[\mathrm{G}])$. Here $\mathrm{Z}(k[\mathrm{G}])$ denotes the center of $k[\mathrm{G}]$.

Put $\Delta(b) := [a, \cdot\rangle(b) = [a, b\rangle = ab - \sigma(b)a$, for $b \in k[\mathrm{G}]$. This is a σ-derivation on $k[\mathrm{G}]$ with values in A. If $k[\mathrm{G}]$ is a two-sided ideal in A (for example the whole A), then $\Delta(k[\mathrm{G}]) \subseteq k[\mathrm{G}]$ and hence Δ becomes a σ-derivation of $k[\mathrm{G}]$.

Compare this with Abramov's q-differential $\mathrm{d}_a = \langle a, \cdot\rangle_q$.

Theorem 7.2. *Let $a \in k[\mathrm{G}]$ be such that $a^N \in \mathrm{Z}(k[\mathrm{G}])$ for some $N \geq 2$. Suppose that $\phi(a,a)$ is a primitive N^{th}-root of unity and that $\phi(a,b)^N = \mathbf{1}$ for all $b \in k[\mathrm{G}]$. Then $\Delta^N(b) = 0$ for all $b \in k[\mathrm{G}]$.*

7.2 General Framework of (σ, τ)-Derivations

Let A be a k-algebra and N an A-bimodule. A *module derivation on A* is a k-linear map $\mathscr{D} : A \to N$ satisfying $\mathscr{D}(ab) = \mathscr{D}(a)b + a\mathscr{D}(b)$ for $a, b \in A$. Furthermore, let Γ and M be left A-modules (in particular k-modules). Then Γ is said to act on M if there is a k-linear map $\mu : \Gamma \otimes_k M \to M$. We write $\gamma.x$ for $\mu(\gamma \otimes_k x)$. A *general derivation* on (A, Γ, M) is a quadruple $(\sigma, \tau, \Delta, \mathscr{D})$ (see [4]) where

- $\sigma, \Delta : \Gamma \to \Gamma$
- $\tau, \mathscr{D} : M \to M$

are all k-linear maps such that

$$\mathscr{D}(\gamma.x) = \Delta(\gamma).\tau(x) + \sigma(\gamma).\mathscr{D}(x). \tag{7.1}$$

Definition 7.1. If $\Gamma = M = A$ and $\mathscr{D} = \Delta$, then a general derivation $(\sigma, \tau, \Delta, \mathscr{D})$ is said to be a *(σ, τ)-derivation* on A and when $\tau = \mathrm{id}_M$ it is usually called a *σ-derivation*. Here we simply write this as Δ.

Assume that A is a k-algebra equipped with a k-algebra endomorphism σ. Define the operator $[a, \cdot\rangle : A \to A$, for each $a \in A$, by:

$$\Delta(b) := [a, \cdot\rangle(b) := ab - \sigma(b)a, \tag{7.2}$$

i.e., $\Delta := [a, \cdot\rangle$. Clearly Δ is k-linear since σ is. It is easy to see that

$$[a, bc\rangle = [a, b\rangle c + \sigma(b)[a, c\rangle.$$

In other words, $[a, \cdot\rangle$ is a σ-twisted derivation for each $a \in A$ and algebra endomorphism σ. In fact, $[a, \cdot\rangle$ is called *σ-inner* in analogy with the classical case $\sigma = \mathrm{id}_A$.

From now on we fix $a \in k[\mathrm{G}]$ and assume that σ given by $\sigma(b) = \sigma_a(b) := \phi(a,b)b$ is a k-algebra morphism on $k[\mathrm{G}]$ with $\phi : \{a\} \times k[\mathrm{G}] \to \mathrm{Z}(k[\mathrm{G}])$. For $b, c \in k[\mathrm{G}]$ we have

$$0 = \sigma_a(bc) - \sigma_a(b)\sigma_a(c) = (\phi(a,bc) - \phi(a,b)\phi(a,c))bc$$

and so if bc is not a (right) zero divisor $\phi(a,bc) = \phi(a,b)\phi(a,c)$.

We introduce the notation $\phi^{(\ell)}(a,b) := \phi(a, \phi(a, \ldots, \phi(a,b)))$ (ℓ appearances of ϕ for integer $\ell \geq 1$). For instance, $\phi^{(3)}(a,b) = \phi(a, \phi(a, \phi(a,b)))$. Also, it is convenient to interpret $\phi^{(0)}(a,b)$ as b.

Lemma 7.1. *The following identities hold for $b \in k[\mathrm{G}]$:*

(i) $\sigma_a(\phi^{(\ell)}(a,b)) = \phi^{(\ell+1)}(a,b)\phi^{(\ell)}(a,b)$,

(ii) $\sigma_a^{\ell}(b) = \prod_{j=0}^{\ell} \phi^{(\ell-j)}(a,b)^{\binom{\ell}{j}}$.

Proof. Identity (i) follows immediately from definition. The second one is proved by induction where the case $\ell = 1$ is $\sigma_a^1(b) = \sigma_a(b) = \phi(a,b)b$ which is (ii) for $\ell = 1$. Assume now that (ii) holds for ℓ. Then

$$\begin{aligned}\sigma_a^{\ell+1}(b) &= \sigma_a(\sigma_a^{\ell}(b)) = \sigma_a\Big(\prod_{j=0}^{\ell} \phi^{(\ell-j)}(a,b)^{\binom{\ell}{j}}\Big) = \prod_{j=0}^{\ell} \sigma_a(\phi^{(\ell-j)}(a,b))^{\binom{\ell}{j}} = \\ &= \prod_{j=0}^{\ell} \phi^{(\ell+1-j)}(a,b)^{\binom{\ell}{j}} \phi^{(\ell-j)}(a,b)^{\binom{\ell}{j}} = \prod_{j=0}^{\ell+1} \phi^{(\ell+1-j)}(a,b)^{\binom{\ell+1}{j}},\end{aligned}$$

where we have used identity (i) and after re-arranging the product, the Pascal identity $\binom{\ell}{j} + \binom{\ell}{j+1} = \binom{\ell+1}{j+1}$. (Notice that we used that $\phi^{(i)}(a,b) \in \mathrm{Z}(A)$ and that σ_a is multiplicative.) □

Lemma 7.2. *For $a \in k[\mathrm{G}]$ we have $\phi(a,a)\Delta \circ \sigma = \sigma \circ \Delta$.*

Proof. This follows from the following simple computation:

$$\begin{aligned}\sigma \circ \Delta(b) &= \sigma(ab - \sigma(b)a) = \sigma(a)\sigma(b) - \sigma(\sigma(b))\sigma(a) = \\ &= \phi(a,a)a\sigma(b) - \sigma(\sigma(b))\phi(a,a)a = \phi(a,a)(a\sigma(b) - \sigma(\sigma(b))a) = \\ &= \phi(a,a)\Delta \circ \sigma(b).\end{aligned}$$

This completes the proof. □

Compare this with [2] wherein we have the reversed order, i.e., $\Delta \circ \sigma = \delta\sigma \circ \Delta$, for $\delta \in A$ (in [2] A was supposed to be commutative as well). In fact, adopting the order from the above Lemma in [2] leads to analogous result and so we have a connection to the theory developed in [2].

7.2.1 Main Result

Assume that k is an integral domain and let Σ denote the maximal subalgebra of $\mathrm{Z}(k[\mathrm{G}])$ such that $\sigma_a|_\Sigma = \mathrm{id}_A$ and such that Σ is an integral domain as well. From now on (unless stated otherwise) we suppose $\phi : \{a\} \times k[\mathrm{G}] \to \Sigma$. This implies that if $s \in \Sigma$ then $\phi(a,s) = \mathbf{1}$ since, on the one hand, $\sigma_a(s) = s$, and on the other, $\sigma_a(s) = \phi(a,s)s$. Also, by construction σ_a satisfies $\sigma_a(sb) = s\sigma_a(b)$ for $s \in \Sigma$. This is all sufficient to have $\Delta(\sigma_a(b)) = \Delta(\phi(a,b)b) = \phi(a,b)\Delta(b)$, for instance. In general $\Delta(sb) = s\Delta(b)$ for $s \in \Sigma$.

Let $a,b \in k[\mathrm{G}]$ and put $\varepsilon_a := \phi(a,a)$ and $\varepsilon_b := \phi(a,b)$. Formally, for $q \in \Sigma^* := \Sigma \setminus \{0\}$, we denote by $\{n\}_q \in \Sigma$ the polynomial $\mathbf{1} + q + q^2 + \cdots + q^{n-1}$ for $n \in \mathbb{N}^+ := \mathbb{N} \cup \{0\}$, defining $\{0\}_q := 0$. Note that we do not exclude the possibility of $\{\ell\}_q = \mathbf{1} + q + q^2 + \cdots + q^{\ell-1}$ being zero for some $\ell \in \mathbb{N}^+$. Define the q-binomial coefficient as the (unique) solution to the q-Pascal recurrence relation:

$$\binom{n+1}{j+1}_q = q^{n-j}\binom{n}{j}_q + \binom{n}{j+1}_q \tag{7.3}$$

or 0 either if $j+1 < 0$ or $j+1 > n+1$ and 1 if $j+1 = 0$ or $j+1 = n+1$. It can be proven [3] that $\binom{n}{j}_q$ is a polynomial in q for all n and j. Also, in analogy with the classical case, it can be shown that if neither of the involved products in the denominator is zero, we have $\binom{n}{j}_q := \frac{\{n\}_q!}{\{j\}_q!\{n-j\}_q!}$.

An element $q \in \Sigma^*$ is an n-th root of unity if $q^n = \mathbf{1}$ and a primitive n-th root of unity if $q^n = \mathbf{1}$, and $\{\ell\}_q \neq 0$ for $\ell < n$. The property of q being an n-th root of unity, i.e., $q^n - \mathbf{1} = 0$, is equivalent to

$$(\mathbf{1} + q + q^2 + \cdots + q^{n-1})(q - \mathbf{1}) = \{n\}_q(q - \mathbf{1}) = 0.$$

If $q \neq \mathbf{1}$, this is equivalent to $\{n\}_q = 0$ since Σ is a domain.

Proposition 7.1. *For $a,b \in k[\mathrm{G}]$ with $\varepsilon_a \neq 0$ we have*

$$\Delta^\ell(b) = \sum_{j=0}^{\ell} (-1)^j \varepsilon_a^{\frac{j(j-1)}{2}} \varepsilon_b^j \binom{\ell}{j}_{\varepsilon_a} a^{\ell-j} b a^j. \tag{7.4}$$

Proof. The Proposition is verified for $\ell = 1,2,3$ without difficulty. Assume that (7.4) is true for ℓ. Then

$$\Delta^{\ell+1}(b) = \Delta(\Delta^{\ell}(b)) = \sum_{j=0}^{\ell}(-1)^{j}\varepsilon_a^{\frac{j(j-1)}{2}}\varepsilon_b^{j}\binom{\ell}{j}_{\varepsilon_a}\Delta(a^{\ell-j}ba^{j}). \tag{7.5}$$

We have

$$\begin{aligned}\Delta(a^{\ell-j}ba^{j}) &= [a, a^{\ell-j}ba^{j}\rangle = a^{\ell-j+1}ba^{j} - \sigma_a(a)^{\ell-j}\sigma_a(b)\sigma_a(a)^{j}a = \\ &= a^{\ell-j+1}ba^{j} - \varepsilon_a^{\ell}\varepsilon_b a^{\ell-j}ba^{j+1}.\end{aligned}$$

This means that

$$\begin{aligned}\Delta^{\ell+1}(b) = &\sum_{j=0}^{\ell}(-1)^{j}\varepsilon_a^{\frac{j(j-1)}{2}}\varepsilon_b^{j}\binom{\ell}{j}_{\varepsilon_a}a^{\ell-j+1}ba^{j} + \\ &+\sum_{j=0}^{\ell}(-1)^{j+1}\varepsilon_a^{\frac{j(j-1)}{2}+\ell}\varepsilon_b^{j+1}\binom{\ell}{j}_{\varepsilon_a}a^{\ell-j}ba^{j+1}.\end{aligned}$$

Write the first sum as

$$\sum_{j=1}^{\ell}(-1)^{j}\varepsilon_a^{\frac{j(j-1)}{2}}\varepsilon_b^{j}\binom{\ell}{j}_{\varepsilon_a}a^{\ell-j+1}ba^{j} + a^{\ell+1}b = \mathsf{S}_1 + a^{\ell+1}b$$

and the second as

$$\begin{aligned}&\sum_{j=0}^{\ell-1}(-1)^{j+1}\varepsilon_a^{\frac{j(j-1)}{2}+\ell}\varepsilon_b^{j+1}\binom{\ell}{j}_{\varepsilon_a}a^{\ell-j}ba^{j+1} + (-1)^{\ell+1}\varepsilon_a^{\frac{\ell(\ell+1)}{2}}\varepsilon_b^{\ell+1}ba^{\ell+1} = \\ &\qquad = \mathsf{S}_2 + (-1)^{\ell+1}\varepsilon_a^{\frac{\ell(\ell+1)}{2}}\varepsilon_b^{\ell+1}ba^{\ell+1}.\end{aligned}$$

The S_1-term can be written as

$$\sum_{j=0}^{\ell-1}(-1)^{j+1}\varepsilon_a^{\frac{j(j+1)}{2}}\varepsilon_b^{j+1}\binom{\ell}{j+1}_{\varepsilon_a}a^{\ell-j}ba^{j+1}.$$

Adding S_1 and S_2 we get:

$$\mathsf{S}_1 + \mathsf{S}_2 = \sum_{j=0}^{\ell-1}(-1)^{j+1}\varepsilon_b^{j+1}\Big(\varepsilon_a^{\frac{j(j+1)}{2}}\binom{\ell}{j+1}_{\varepsilon_a} + \varepsilon_a^{\frac{j(j-1)}{2}+\ell}\binom{\ell}{j}_{\varepsilon_a}\Big)a^{\ell-j}ba^{j+1}.$$

Note that $\frac{j(j-1)}{2} = \frac{j(j+1)}{2} - j$ so the parentheses becomes

$$\varepsilon_a^{\frac{j(j+1)}{2}}\Big(\binom{\ell}{j+1}_{\varepsilon_a} + \varepsilon_a^{\ell-j}\binom{\ell}{j}_{\varepsilon_a}\Big).$$

Using (7.3) this is the same as $\varepsilon_a^{\frac{j(j+1)}{2}}\binom{\ell+1}{j+1}_{\varepsilon_a}$. Then $\mathsf{S}_1 + \mathsf{S}_2$ add up to

$$\sum_{j=0}^{\ell-1}(-1)^{j+1}\varepsilon_a^{\frac{j(j+1)}{2}}\varepsilon_b^{j+1}\binom{\ell+1}{j+1}_{\varepsilon_a}a^{\ell-j}ba^{j+1} =$$

$$=\sum_{j=1}^{\ell}(-1)^{j}\varepsilon_a^{\frac{j(j-1)}{2}}\varepsilon_b^{j}\binom{\ell+1}{j}_{\varepsilon_a}a^{\ell+1-j}ba^{j}.$$

Putting everything together yields

$$a^{\ell+1}b+\mathsf{S}_1+\mathsf{S}_2+(-1)^{\ell+1}\varepsilon_a^{\frac{\ell(\ell+1)}{2}}\varepsilon_b^{\ell+1}ba^{\ell+1} =$$

$$=\sum_{j=0}^{\ell+1}(-1)^{j}\varepsilon_a^{\frac{j(j-1)}{2}}\varepsilon_b^{j}\binom{\ell+1}{j}_{\varepsilon_a}a^{\ell+1-j}ba^{j}$$

and the proof is complete.

Suppose $\varepsilon_a \neq \mathbf{1}$ and $\varepsilon_a \in \Sigma \subseteq \mathsf{Z}(k[\mathsf{G}])$ is a primitive n-th root of unity. Then, $\{n\}_{\varepsilon_a} = \mathbf{1}+\varepsilon_a+\varepsilon_a^2+\cdots+\varepsilon_a^{n-1}=0$, and $\binom{n}{j}_{\varepsilon_a}=0$ for $j\neq 0,n$. Hence

$$\Delta^n(b)=a^nb+(-1)^n\varepsilon_a^{\frac{n(n-1)}{2}}\varepsilon_b^n ba^n.$$

Assuming that a^n and b commute (if $a^n \in \mathsf{Z}(k[\mathsf{G}])$, for instance), we get

$$\Delta^n(b)=(\mathbf{1}+(-1)^n\varepsilon_a^{\frac{n(n-1)}{2}}\varepsilon_b^n)a^nb.$$

From this follows that $\Delta^n(b)=(\mathbf{1}-\varepsilon_b^n)a^nb$, if n is odd, and

$$\Delta^n(b)=(\mathbf{1}+(\varepsilon_a^{\frac{n}{2}})^{n-1}\varepsilon_b^n)a^nb,$$

if n is even. However, since ε_a is a primitive n-th root of unity $\varepsilon_a^{\frac{n}{2}}=-\mathbf{1}$ and so both these cases are the same.

Corollary 7.1. *If, in addition to the above assumptions, $\varepsilon_b^n=\mathbf{1}$ then $\Delta^n(b)=0$.*

7.3 Generalized N-Complexes and an Example

A *generalized N-complex*, $N\geq 0$, is a sequence of objects $\{C_i\}_{i\in\mathbb{Z}}$, in an abelian category A together with a sequence of morphisms $\mathrm{d}_i \in \mathrm{Hom}(C_i, C_{i+p})$ for some (fixed) $p\in\mathbb{Z}$ and such that

$$\mathrm{d}^N := \mathrm{d}_{i+(N-1)p}\circ\mathrm{d}_{i+(N-2)p}\circ\cdots\circ\mathrm{d}_{i+p}\circ\mathrm{d}_i = 0 : C_i \to C_{i+Np}.$$

The case $N=0$ is interpreted as there being no vanishing condition at all on the differential and $N=1$ means $\mathrm{d}=0$. We write a generalized N-complex as $(C_n,\mathrm{d}_n)_{n\in\mathbb{Z}}^{N,p}$.

If $p = 1$ we get the class of N-complexes and if in addition $N = 2$ we get the ordinary complexes from ordinary homological algebra. Of course we could have defined $\mathrm{d}_i \in \mathrm{Hom}(C_i, C_{i+p_i})$ for some family of p_i's but such a definition would drown in indices so we refrain from explicitly stating it.

In this paper we are considering only the case when $\mathsf{A} = \mathsf{Mod}(k)$, the abelian category of k-(bi-)modules.

7.3.1 An Elaborated Example

Here we assume that A is the k-algebra of Laurent polynomials over k, i.e., $A = k[t,t^{-1}]$. This is a $\mathbb{Z}$-graded k-algebra generated over k by $\{\mathbf{1},t,t^{-1}\}$. Put $\mathsf{G} = \{\mathbf{1},t,t^{-1}\}$.

The most general algebra endomorphism σ on A is one on the form $\sigma(t) = q_1 t^{s_1}$ and $\sigma(t^{-1}) = q_2 t^{s_2}$. But this choice should respect $tt^{-1} = t^{-1}t = \mathbf{1}$. So if σ is multiplicative we have to condition $q_2 = q_1^{-1} =: q$ and $s_2 = -s_1 =: s$. We then have $\sigma(t) = qt^s = \phi(a,t)t$ and so $\phi(a,t) = qt^{s-1}$. From this follows $\phi(a,t)\phi(a,t^{-1}) = 1$, i.e., $\phi(a,t)^{-1} = \phi(a,t^{-1})$ by the uniqueness of inverses. Extend σ to A by the obvious $\sigma(u_1 t^n + u_2 t^m) := u_1\sigma(t^n) + u_2\sigma(t^m)$ for $u_1, u_2 \in k$, $n,m \in \mathbb{Z}$.

Take $a \in A$ and form $\Delta := a(\mathrm{id}_A - \sigma)$. We know that Δ is a σ-derivation since A is commutative. Applying Δ to a homogeneous component A_n we find

$$\Delta(ut^n) = a(\mathrm{id}_A - \sigma)(ut^n) = au(t^n - \phi(a,t)^n t^n) = au(\mathbf{1} - \phi(a,t)^n)t^n.$$

The degree of Δ is therefore in general undefined since $\mathbf{1}$ and $\phi(a,t)^n$ will belong to different graded components; indeed, $\phi(a,t)^n \notin A_0 \approx k$ in general. However, if $\phi(a,t) \in A_0$ then $\phi(a,t)^n \in A_0$ for all $n \in \mathbb{Z}$ since A_0 is a subalgebra. Accordingly, we assume from now on that $\phi(a,t) \in k$. Then $\Delta(ut^n) = au(\mathbf{1} - \phi(a,t)^n)t^n \in A_{n+\deg(a)}$ with $u \in k$.

Thus, we have a generalized complex $(A_n, \Delta)_{n\in\mathbb{Z}}^{0,\deg(a)}$, where $\Delta : A_n \to A_{n+\deg(a)}$, for each $a \in A_{\deg(a)}$.

From Proposition 7.1 we have

$$\Delta^\ell(b) = \sum_{j=0}^{\ell} (-1)^j \phi(a,a)^{\frac{j(j-1)}{2}} \phi(a,b)^j \binom{\ell}{j}_{\phi(a,a)} a^\ell b.$$

Suppose $\phi(a,a)^\ell = \mathbf{1}$ and $\phi(a,a)^m \neq \mathbf{1}$ for $m < \ell$, i.e., $\phi(a,a)$ is a primitive ℓ^{th}-root of unity and suppose $\phi(a,b)^\ell = \mathbf{1}$ for all $b \in A$. Then we are in the situation of Corollary 7.1:

$$\Delta^\ell(b) = 0, \quad \text{for all} \quad b \in A = k[t,t^{-1}],$$

and so we have constructed an ℓ-complex.

Acknowledgements We are grateful to the Crafoord foundation, the Swedish Foundation for International Cooperation in Research and Higher Education (STINT), the Royal Swedish Academy of Sciences, the Royal Physiographic Society in Lund, and the Swedish Research Council for support of this work.

References

1. Abramov V., *On a graded q-differential algebra*, J. Nonlinear Math. Phys., **13**(Suppl.), 1–8 (2006). (math.QA/0509481).
2. Hartwig, J.T., Larsson, D., Silvestrov, S.D., *Deformations of Lie algebras using σ-derivations*, J. Algebra **295**, 314–361 (2006).
3. Hellström, L., Silvestrov, S.D., *Commuting Elements in q-Deformed Heisenberg Algebras*, World Scientific, Singapore, 2000, 256 pp.
4. Laksov, D., Thorup, A., *These are the differentials of order n*, Trans. Am. Math. Soc. **351**(4), 1293–1353 (1999).

Chapter 8
Remarks on Quantizations, Words and R-Matrices

Hilja L. Huru

Abstract We consider the monoidal category of modules graded by the monoid of words made from a finite alphabet. The associativity constraints, braidings and quantizations related to the grading are described explicitly. By quantizations of R-matrices in the same manner as braidings we obtain new R-matrices that by construction still satisfy the Yang–Baxter equation.

8.1 Introduction

Throughout this paper let M be a monoid, written multiplicatively, let k be a commutative ring and $U(k)$ be the set of units in k.

We extend the results of [2], [3], [4] and [5] concerning quantizations of monoid graded modules, define the multiplicative cohomology of monoids that not necessarily are commutative, and show that in the monoidal category of modules over a commutative ring k graded by a monoid M the non-trivial associativity constraints and quantizations are in one-to-one correspondence with, respectively, the third and second cohomology groups of M with values in the group of units of k. Braidings of this category are also investigated. Specifically, we investigate the quantizations in the category of modules graded by words generated by some finite alphabet A, that is, by a free monoid and the graded module is the tensor algebra of a finite number of vector spaces.

Furthermore, we find quantizations by considering the commutative grading given by the map from the monoid of words to $\mathbb{Z}^n$ that counts the number of times each of the n letters of A occurs in a word. The non-trivial quantizations are representatives of the second cohomology groups of $\mathbb{Z}$ with values in the group of units of k.

H.L. Huru
Department of Mathematics and Statistics, University of Tromsø, 9037 Tromsø, Norway
e-mail: hilja.huru@matnat.uit.no

S. Silvestrov et al. (eds.), *Generalized Lie Theory in Mathematics, Physics and Beyond,*

Braidings and quantizations of braidings satisfy the Yang–Baxter diagram, see [6, 8]. This is exploited to develop a method where we from an R-matrix easily generate new R-matrices which all satisfy the Y–B equation. In other words, after finding the quantizations for some graded vector spaces we consider any R-matrix as a braiding and quantize to produce new R-matrices.

As an example we look at the case when the vector spaces are isomorphic to each other. For an R-matrix given by B. A. Kupershmidt in [7] we assume that the vector space is finitely graded, and quantize.

8.2 Multiplicative Cohomologies of Monoids

We define the cohomology of monoids as defined for groups in [1], where we assume that the monoid M acts trivially on $U(k)$. $U(k)$ can be replaced by any commutative group.

A *normalized n-cochain of M with coefficients in* $U(k)$ is a function θ of n variables, $m_1, \dots, m_n$,

$$\theta : M \times \cdots \times M \to U(k), \tag{8.1}$$

that satisfies the normalization condition, that is, if one or more of the entries $m_i, i = 1, \dots, n$ are equal to $0 \in M$ then $\theta(m_1, \dots, m_n) = 1$. Let us denote by $C^n(M, U(k))$ the set of normalized n-cochains.

The product of two n-cochains θ and θ' is defined by

$$(\theta \cdot \theta')(m_1, \dots, m_n) = \theta(m_1, \dots, m_n) \cdot \theta'(m_1, \dots, m_n) \tag{8.2}$$

where the product on the right hand side is the product in $U(k)$, and is again an n-cochain. With this multiplication the n-cochains form an abelian group.

Define the *coboundary operator*

$$\delta : C^n(M, U(k)) \to C^{n+1}(M, U(k)), \tag{8.3}$$

and the coboundary $\delta(\theta)$ of $\theta \in C^n(M, U(k))$ as follows,

$$\delta(\theta)(m_0, m_1, \dots, m_n) \tag{8.4}$$
$$= \theta(m_1, m_2, \dots, m_n)\, \theta^{-1}(m_0 m_1, m_2, \dots, m_n)\, \theta(m_0, m_1 m_2, m_3, \dots, m_n) \cdots \tag{8.5}$$
$$\cdots \theta^{(-1)^n}(m_0, m_1, \dots, m_{n-2}, m_{n-1} m_n)\, \theta^{(-1)^{n+1}}(m_0, m_1, \dots, m_{n-1}). \tag{8.6}$$

The operator δ has the following properties; $\delta(\theta)$ is an $(n+1)$-cochain,

$$\delta(\theta \cdot \theta') = \delta(\theta) \cdot \delta(\theta'), \delta(\delta(\theta)) = 1, \tag{8.7}$$

for $\theta, \theta' \in C^n(M, U(k))$, hence δ is a group-homomorphism from the group of n-cochains to the group of $(n+1)$-cochains.

Definition 8.1. The subgroup of $C^n(M,U(k))$ of *n-cocycles is*

$$Z^n(M,U(k)) = \{\theta \in C^n(M,U(k)) \mid \delta(\theta) = 1\}. \tag{8.8}$$

Definition 8.2. The subgroup of $C^n(M,U(k))$ of *n-coboundaries* is

$$B^n(M,U(k)) = \left\{\theta \in C^n(M,U(k)) \mid \theta = \delta\theta', \theta' \in C^{n-1}(M,U(k))\right\}. \tag{8.9}$$

By the property $\delta^2 = 1$, $B^n(M,U(k))$ is also a subgroup of $Z^n(M,U(k))$.

Definition 8.3. The factor group

$$H^n(M,U(k)) = Z^n(M,U(k))/B^n(M,U(k)) \tag{8.10}$$

is called the *n-th cohomology group of the monoid M with coefficients in* $U(k)$.

It can easily be shown that $H^n(M,U(k))$ is an abelian group.
The cup product is defined as in [1];

Definition 8.4. Let $\theta \in C^s(M,U(k))$ and $\theta' \in C^t(M,U(k))$. The cup product is

$$\theta \cup \theta'(m_1,\ldots,m_{s+t}) = \theta(m_1,\ldots,m_s)\,\theta'(m_{s+1},\ldots,m_{s+t}) \in C^{s+t}(M,U(k)) \tag{8.11}$$

where the product on the right hand side is the multiplication in k.

The cup product has the following properties; it is additive in each variable, associative and

$$\delta(\theta \cup \theta') = (\delta\theta \cup \theta') + (-1)^s(\theta \cup \delta\theta'). \tag{8.12}$$

Definition 8.5. Let θ and θ' be representatives of cohomology classes in $H^s(M,U(k))$ and $H^t(M,U(k))$ respectively. We have the cup product $[\theta] \cup [\theta'] = [\theta \cup \theta']$ where $[\theta \cup \theta'] \in H^{s+t}(M,U(k))$.

By the property (8.12) this is well defined. With the cup product and the addition of functions $H^*(M,U(k)) = \sum_n H^n(M,U(k))$ is an $\mathbb{N}$-graded k-algebra.

8.2.1 Cohomology of Free Groups

Let F be a free group generated by a set X. To find quantizations we need to find the multiplicative cohomology $H^2(F,U(k))$, see Sect. 8.3.1. Let D be an $F-F$-bimodule. The additive cohomology group is found in [11], $H^n(F,D) = 0, n \neq 0,1$. Assume that $k = \mathbb{C}$. We have the short exact sequence of coefficients $0 \to 2\pi i\mathbb{Z} \to \mathbb{C} \to \mathbb{C}^* \to 0$, and hence the long exact sequence

$$\cdots \to H^n(F,\mathbb{Z}) \to H^n(F,\mathbb{C}) \to H^n(F,\mathbb{C}^*) \to H^{n+1}(F,\mathbb{Z}) \to \cdots. \tag{8.13}$$

We are interested in the multiplicative cohomology, $H^n(F,\mathbb{C}^*)$, for $n = 2$. However, $H^2(F,\mathbb{C})$ and $H^3(F,\mathbb{Z})$ are trivial, hence, so is $H^2(F,\mathbb{C}^*)$.

Remark 8.1. Despite this result the second cohomology of a free abelian monoid or of a free abelian group may be non-trivial. The quantization considered in Sect. 8.4.2 is an example of such.

8.3 Graded Modules

Denote by $mod_k(M)$ the strict monoidal category, see [10], of M-graded k-modules, $X = \oplus_{m\in M} X_m$. Denote the grading of a homogeneous element $x \in X$ either by $|x| \in M$ or write x_m, $m \in M$. We define the tensor product $X \otimes_k X'$ of two objects in $mod_k(M)$ by,

$$(X \otimes_k X')_m = \oplus_{ij=m}(X_i \otimes_k X'_j). \tag{8.14}$$

Remark 8.2. The tensor product may not be defined like this for all M (the sum needs to be finite). However, for a free monoid with a finite number of generators this definition is ok.

8.3.1 Associativity Constraints, Quantizations and Braidings

For the monoidal category of M-graded k-modules, we get an explicit description of the associativity constraints, braidings and quantizations in terms of the grading. The procedure to show this is the same as for the finite commutative case, see [5] and [2], we state the results only.

In a monoidal category C an associativity constraint α is a natural isomorphism

$$\alpha_{X,Y,Z} : X \otimes (Y \otimes Z) \to (X \otimes Y) \otimes Z, \tag{8.15}$$

for objects X, Y and Z in C, that satisfies the coherence condition, see [10].

Proposition 8.1. *Any associativity constraint* α *in the monoidal category of* M*-graded* k*-modules is a normalized* 3*-cocycle of* M *with values in* $\mathbf{U}(k)$*, the group of unit elements of* k*, and is of the form*

$$\alpha_{X,Y,Z} : x \otimes (y \otimes z) \longmapsto \alpha(|x|,|y|,|z|)(x \otimes y) \otimes z, \tag{8.16}$$

for homogeneous elements $x \in X, y \in Y, z \in Z$*, where* X,Y,Z *are objects in* $mod_k(M)$*.*

The cocycle condition follows from the coherence condition for associativity constraints.

Parts of this result can be generalized.

Proposition 8.2. *Any unit-preserving natural isomorphism* γ *from one* n*-functor* F *to another* n*-functor* F' *in* $mod_k(M)$ *can be represented as a normalized* n*-cochain of* M *with coefficients in* $U(k)$, $\gamma : M \times \cdots \times M \to U(k)$*, and is of the form*

$$\gamma_{X_1,\ldots,X_n} : F(x_1,\ldots,x_n) \longmapsto \gamma(|x_1|,\ldots,|x_n|)\, F'(x_1,\ldots,x_n) \tag{8.17}$$

for homogeneous $x_i \in X_i$ *where* $X_i \in Obj(mod_k(M))$ *and* $i = 1,\ldots,n$.

The following theorem gives a complete description of associativity constraints in the category of M-graded k-modules.

Theorem 8.1. *The orbits of all associativity constraints in the monoidal category* $mod_k(M)$ *under the action of natural isomorphisms of the tensor bifunctor are in one-to-one correspondence with the* 3rd *cohomology group* $H^3(M, \mathbf{U}(k))$.

In a monoidal category C a quantization is a natural isomorphism

$$q_{X,Y} : X \otimes Y \to X \otimes Y, \tag{8.18}$$

for objects X, Y in C that satisfies the coherence condition for quantizations, see [9], which means that any associativity constraint is preserved under quantizations.

The following theorem gives a complete description of quantizations in the category of M-graded k-modules.

Theorem 8.2. *The orbits of all quantizations in the monoidal category* $mod_k(M)$ *under the action of unit-preserving natural isomorphisms of the identity functor are in one-to-one correspondence with the* 2nd *cohomology group,* $H^2(M, \mathbf{U}(k))$. *For homogeneous elements* $x \in X$, $y \in Y$ *in the* M*-graded modules* X *and* Y *a quantization has the form*

$$q_{X,Y} : x \otimes y \longmapsto q(|x|, |y|)\, x \otimes y, \tag{8.19}$$

where q *is a* 2*-cocycle representing a class in* $H^2(M, \mathbf{U}(k))$.

Recall that the 2-cocycle condition is,

$$q(i,j)\, q^{-1}(i,jl)\, q(ij,l)\, q^{-1}(j,l) = 1, i,j,l \in M. \tag{8.20}$$

A braiding σ in a monoidal category C is a natural isomorphism

$$\sigma_{X,Y} : X \otimes Y \to Y \otimes X, \tag{8.21}$$

for objects X and Y that satisfies the coherence conditions for braidings, see [10]. A braiding is a symmetry if $\sigma_{Y,X} \circ \sigma_{X,Y} = Id$, and a monoidal category equipped with a symmetry is called symmetric.

Quantizations act on the set of braidings as follows $(\sigma_q)_{X,Y} = q^{-1}_{Y,X} \circ \sigma_{X,Y} \circ q_{X,Y}$, and σ_q is a braiding too.

The coherence condition on braidings gives certain properties for the braidings of graded modules, summarized in the following theorem.

Theorem 8.3. *A braiding* σ *in the monoidal category* $mod_k(M)$ *has the form*

$$\sigma_{X,Y} : x \otimes y \longmapsto \sigma(|x|, |y|)\, y \otimes x \tag{8.22}$$

for homogeneous $x \in X, y \in Y$, *where* $\sigma : M \times M \to \mathbf{U}(k)$ *is a normalized* 2*-cochain satisfying*

$$\sigma(ij,l) = \frac{\alpha(i,l,j)}{\alpha(i,j,l)\,\alpha(l,i,j)}\sigma(i,l)\,\sigma(j,l), \tag{8.23}$$

$$\sigma(i,jl) = \frac{\alpha(i,j,l)\,\alpha(j,l,i)}{\alpha(j,i,l)}\frac{\sigma(i,j)}{\sigma(l,i)}, \tag{8.24}$$

and putting $j = 0$ in (8.24) yields $\sigma(i,l)\,\sigma(l,i) = 1$, i.e. any braiding in $mod_k(M)$ is a symmetry. Furthermore, if the associativity constraint α is trivial, we get the bihomomorphism conditions for σ,

$$\sigma(ij,l) = \sigma(i,l)\,\sigma(j,l), \sigma(i,jl) = \sigma(i,j)\,\sigma(i,l), i,j,l \in M. \tag{8.25}$$

8.4 Letters and Words

Let $A = \{a_1,\ldots,a_n\}$ be an alphabet with n letters. Let $W(A)$ be the set of all possible words generated by letters in A. Clearly, $W(A)$ is a non-commutative monoid.

8.4.1 Generators of Monoidal Categories

Let C be the monoidal category of finite dimensional k-vector spaces. Assume that there is a set of vector spaces $\mathfrak{A} = \langle V_1,\ldots,V_n\rangle$ such that any object V in C can be written as a direct sum of tensors of objects in $\mathfrak{A}$, $V \simeq \oplus V_{l_1} \otimes \cdots \otimes V_{l_s}, l_i \in A$. One can say that V is graded by the monoid of words, where the component $V_w = V_{l_1} \otimes \cdots \otimes V_{l_s}$ of V has the corresponding grading $w = l_1 \cdots l_s$. Hence, there is an isomorphism between this monoidal category C and the category of modules graded by the monoid of words. We will consider all tensors of the members of the family $\mathfrak{A}$, $T(\mathfrak{A})$, as our example of one such module. The associativity constraints, braidings and quantizations of C and $T(\mathfrak{A})$ depend only on the grading by $W(A)$, as described in Sect. 8.3.1. We will show the quantizations. They are of the following form

$$q_{V\otimes V'} : V_w \otimes V'_{w'} \mapsto q\left(w,w'\right) V_w \otimes V'_{w'}, \tag{8.26}$$

where $V = \oplus_w V_w$, $V' = \oplus_{w'} V'_{w'}$, $w = l_1 \cdots l_{|w|}, w' = l'_1 \cdots l'_{|w'|} \in W(A)$, $|w|$ and $|w'|$ are the lengths of the words, and $q : W(A) \times W(A) \to U(k)$, is the corresponding 2-cocycle of $W(A)$ with coefficients in $U(k)$. Assume that a quantization q of the family $\mathfrak{A}$ is a bihomomorphism. Then

$$\begin{aligned} q\left(w,w'\right) &= q\left(l_1,l'_1\right) q\left(l_1,l'_2\right) \cdots q\left(l_{|w|},l'_{|w'|-1}\right) q\left(l_{|w|},l'_{|w'|}\right) \\ &= q\left(a_1,a_1\right)^{|a_1|_w |a_1|_{w'}}\ q\left(a_1,a_2\right)^{|a_1|_w |a_2|_{w'}} \cdots q\left(a_n,a_n\right)^{|a_n|_w |a_n|_{w'}}, \end{aligned}$$

where $|a_i|_w$ is the number of times the letter a_i occurs in the word w. It is easy to see that (8.20) is satisfied.

If q is exponential, $q(a_i,a_j) = \exp(q_{ij})$, for some q_{ij} depending on $a_i,a_j \in A$, then $q(w,w') = \exp\left(\left\langle \left\|q_{ij}\right\|^T |A|_w, |A|_{w'}\right\rangle\right)$, where $\left\|q_{ij}\right\|$ is the matrix of all q_{ij} and $|A|_w = \begin{pmatrix} |a_1|_w \\ \vdots \\ |a_n|_w \end{pmatrix}$. This is easily shown,

$$\begin{aligned}
q(w,w') &= q(l_1,l'_1)\,q(l_1,l'_2)\cdots q\left(l_{|w|},l'_{|w'|-1}\right)q\left(l_{|w|},l'_{|w'|}\right) \\
&= \exp\begin{pmatrix} |a_1|_w|a_1|_{w'}\,q_{11} + |a_1|_w|a_2|_{w'}\,q_{12} + \cdots \\ +|a_1|_w|a_n|_{w'}q_{1n} + |a_2|_w|a_1|_{w'}q_{21} + \cdots + |a_n|_w|a_n|_{w'}q_{nn} \end{pmatrix} \\
&= \exp\left(\left\langle \begin{bmatrix} q_{11} & q_{21} & \cdots & \cdots & q_{n1} \\ q_{12} & \ddots & & & \vdots \\ \vdots & & \ddots & & \vdots \\ \vdots & & & \ddots & q_{n(n-1)} \\ q_{1n} & \cdots & \cdots & q_{(n-1)n} & q_{nn} \end{bmatrix} |A|_w, |A|_{w'}\right\rangle\right) \\
&= \exp\left(\left\langle \left\|q_{ij}^{W(A)}\right\|^T |A|_w, |A|_{w'}\right\rangle\right).
\end{aligned}$$

8.4.2 *Grading by a Lattice*

We consider complex vector spaces. The map $\mathbb{A} : w \mapsto \mathbb{A}(w)$ that counts the number of times each letter occurs in a word is a monoid homomorphism $\mathbb{A} : W(A) \to \mathbb{Z}^n$ and we use it to get quantizations of $T(\mathfrak{A})$. More precisely, there is a $\mathbb{N}^n$-grading on $T(\mathfrak{A})$ where each $V_w = V_{l_1} \otimes \cdots \otimes V_{l_{|w|}} \in T(\mathfrak{A})$ is given the grading element $|A|_w \in \mathbb{N}^n$. We can extend this grading to $\mathbb{Z}^n$ where the components graded by vectors not in $\mathbb{N}^n$ are $T(\mathfrak{A})_{z \notin \mathbb{N}^n} = 0$.

A quantization $q_{\mathbb{Z}^n}$ of $T(\mathfrak{A})$ as a $\mathbb{Z}^n$-graded module depends only on this grading and is of the form

$$(q_{\mathbb{Z}^n})_{V \otimes V'} : V_w \otimes V_{w'} \longmapsto q_{\mathbb{Z}^n}\left(|A|_w, |A|_{w'}\right) V_w \otimes V_{w'}, \tag{8.27}$$

$V_w, V_{w'} = T(\mathfrak{A})$, where $q_{\mathbb{Z}^n}$ is a representative of the second cohomology group of $\mathbb{Z}^n$ with coefficients in $\mathbb{C}^*$, $q(w,w') = q_{\mathbb{Z}^n}(|A|_w, |A|_{w'}) = \exp(\langle Q|A|_w, |A|_{w'}\rangle)$, and Q is a skew-symmetric matrix with entries in $\mathbb{C}^*$.

8.5 Quantizations of *R*-Matrices

Let V be a k-vector space graded by a monoid M. We consider R-matrices

$$R : X \otimes Y \to Y \otimes X, R_{ij} : X_i \otimes Y_j \to Y_j \otimes X_i, \tag{8.28}$$

that satisfy the (quantum) Yang–Baxter equation,

$$R_{jl} \circ R_{il} \circ R_{ij} = R_{ij} \circ R_{il} \circ R_{jl} : X_i \otimes Y_j \otimes Z_l \to Z_l \otimes Y_j \otimes X_i, \tag{8.29}$$

for $i, j, l \in M$, $R_{ij} = R \otimes 1$, $R_{jl} = 1 \otimes R$ and graded objects X, Y and Z.

Notice that any braiding σ in a monoidal category satisfies the Yang–Baxter (Y–B) equation, see [6, 8]. This is given by the coherence conditions for braidings. For the monoidal category of graded modules the Y–B takes the form

$$\sigma(j,l)\,\sigma(i,l)\,\sigma(i,j) = \sigma(i,j)\,\sigma(i,l)\,\sigma(j,l), \tag{8.30}$$

and since each $\sigma(i, j) \in U(k)$ this is clearly satisfied.

We find quantizations of the R-matrix as for symmetries,

$$R_q = q^{-1}_{Y \otimes X} \circ R \circ q_{X \otimes Y} : X \otimes Y \to Y \otimes X, \tag{8.31}$$

and for homogeneous elements $x \in X, y \in Y$, R_q takes the form

$$R_q(x \otimes y) = q^{-1}(|y|, |x|)\, q(|x|, |y|)\, R(x \otimes y), \tag{8.32}$$

where $|x|, |y| \in M$ and $q : M \times M \to U(k)$ is a 2-cocycle. It is easy to check that (8.29) is satisfied for R_q. Note that $q^{-1}_{Y \otimes X} \circ \tau \circ q_{X \otimes Y}$, i.e. a quantization of the twist, is an unitary solution of the Y–B equation.

If we consider the family of vector spaces $\mathfrak{A}$ and consider R-matrices $R_{ij} : V_{a_i} \otimes V_{a_j} \to V_{a_j} \otimes V_{a_i}$, we can quantize using grading by words as above.

8.5.1 Non-Unitary R-Matrices

Let $V_{a_i} \simeq V$, for all $a_i \in A$ and some vector space V. Let r be a classical r-matrix as in [7], $r : V \otimes V \to V \otimes V$, such that r satisfies the unitary condition if

$$\tau \circ r \circ \tau = -r, \tag{8.33}$$

where τ is the twist $\tau(x \otimes y) = y \otimes x$, $x, y \in V$. A quantization of r is a family of operators

$$R = R(h) = \tau + hr + O\left(h^2\right) : V \otimes V \to V \otimes V, \tag{8.34}$$

called R-matrices that satisfy (8.29). The unitary condition of R corresponding to (8.33) is $R^2 = 1$, and with the vocabulary we have used R is a symmetry.

In [7] the following counterexample is found showing that an R-matrix is not necessarily unitary even though the classical r is; $R = R(h; \lambda) : V \otimes V \to V \otimes V$ with $\dim(V) = 2$, basis $\{e_0, e_1\}$ and

$$R(e_0 \otimes e_0) = e_0 \otimes e_0, R(e_0 \otimes e_1) = (e_1 + he_0) \otimes e_0,$$
$$R(e_1 \otimes e_0) = e_0 \otimes (e_1 - he_0), R(e_1 \otimes e_1) = e_1 \otimes e_1 + \lambda h^2 e_0 \otimes e_0,$$

which is not unitary unless the arbitrary constant $\lambda = 0$.

Assume that V is a $\mathbb{Z}_2$-graded vector space over $k = \mathbb{R}$, with e_0 and e_1 of grading 0 and 1 respectively. We consider the following quantization of $\mathbb{Z}_2$-graded modules from [5],

$$q_{\mathbb{Z}_2}(i,j) = (-1)^{ij}, \tag{8.35}$$

$i,j \in \mathbb{Z}_2$. The quantization of R, $R_{q_{\mathbb{Z}_2}}$, is the same as R except when applied to $e_1 \otimes e_1$,

$$R_{q_{\mathbb{Z}_2}}(e_1 \otimes e_1) = e_1 \otimes e_1 - \lambda h^2 e_0 \otimes e_0. \tag{8.36}$$

By construction is (8.29) satisfied. We see that $R_{q_{\mathbb{Z}_2}}$ is still a non-unitary solution,

$$\begin{aligned}
&R^2_{q_{\mathbb{Z}_2}}(e_1 \otimes e_1) \\
&= R_{q_{\mathbb{Z}_2}}\left(q_{\mathbb{Z}_2}(1,1)\left(\frac{1}{q_{\mathbb{Z}_2}(1,1)} e_1 \otimes e_1 + \frac{1}{q_{\mathbb{Z}_2}(0,0)} \lambda h^2 e_0 \otimes e_0\right)\right) \\
&= R_{q_{\mathbb{Z}_2}}\left(-\left(-e_1 \otimes e_1 + \lambda h^2 e_0 \otimes e_0\right)\right) \\
&= q_{\mathbb{Z}_2}(1,1)\left(\frac{1}{q_{\mathbb{Z}_2}(1,1)} e_1 \otimes e_1 + \frac{1}{q_{\mathbb{Z}_2}(0,0)} \lambda h^2 e_0 \otimes e_0\right) - \lambda h^2 \frac{q_{\mathbb{Z}_2}(0,0)}{q_{\mathbb{Z}_2}(0,0)}(e_0 \otimes e_0) \\
&= e_1 \otimes e_1 - 2\lambda h^2 e_0 \otimes e_0.
\end{aligned}$$

Nor is the Hecke condition satisfied. This is easily checked.

We give one more example of how R-matrices may be quantized when there is a grading present. Assume V is of dimension 4 and graded by $M = (\mathbb{Z}_2)^2$ over $\mathbb{R}$ with the basis $\{e_0, e'_0, e_1, e'_1\}$ with the gradings $(0,0), (1,1), (1,0)$ and $(0,1)$ respectively. Note that $e'_0 = e_1 e'_1$. Consider more or less the same R-matrix, where $R \circ \otimes$ acts as follows

$R \circ \otimes$	e_0	e'_0	e_1	e'_1
e_0	$e_0 \otimes e_0$	$e'_0 \otimes e_0$	$(e_1 + he_0) \otimes e_0$	$(e'_1 + he'_0) \otimes e_0$
e'_0	$e_0 \otimes e'_0$	$e'_0 \otimes e'_0$	$(e_1 + he_0) \otimes e'_0$	$(e'_1 + he'_0) \otimes e'_0$
e_1	$e_0 \otimes (e_1 - he_0)$	$e'_0 \otimes (e_1 - he_0)$	$e_1 \otimes e_1 + \lambda h^2 e_0 \otimes e_0$	$e'_1 \otimes e_1 + \lambda h^2 e'_0 \otimes e_0$
e'_1	$e_0 \otimes (e'_1 - he'_0)$	$e'_0 \otimes (e'_1 - he'_0)$	$e_1 \otimes e'_1 + \lambda h^2 e_0 \otimes e'_0$	$e'_1 \otimes e'_1 + \lambda h^2 e'_0 \otimes e'_0$

(i.e. we only care about subscripts and clearly (8.29) is satisfied). We have for example the following quantization of $(\mathbb{Z}_2)^2$-graded modules as in [5],

$$q_{(\mathbb{Z}_2)^2}(i,j) = (-1)^{f_1(i)f_2(j) + f_1(i)f_1(j) + f_2(i)f_2(j)}, \tag{8.37}$$

where $f_1(i) = \begin{cases} 1, & i = |e_1|, |e'_0| \\ 0, & i \neq |e_1|, |e'_0| \end{cases}$, $f_2(i) = \begin{cases} 1, & i = |e'_1|, |e'_0| \\ 0, & i \neq |e'_1|, |e'_0| \end{cases}$, are the coordinate functions $(\mathbb{Z}_2^+)^2$ to $\mathbb{Z}_2^+$. The quantization $R_{q_{(\mathbb{Z}_2)^2}} \circ \otimes$ acts as follows,

	e_0	e'_0	e_1	e'_1
e_0	$e_0 \otimes e_0$	$e'_0 \otimes e_0$	$(e_1 + he_0) \otimes e_0$	$(e'_1 + he'_0) \otimes e_0$
e'_0	$e_0 \otimes e'_0$	$e'_0 \otimes e'_0$	$-(e_1 + he_0) \otimes e'_0$	$-(e'_1 + he'_0) \otimes e'_0$
e_1	$e_0 \otimes (e_1 - he_0)$	$-e'_0 \otimes (e_1 + he_0)$	$e_1 \otimes e_1 - \lambda h^2 e_0 \otimes e_0$	$-e'_1 \otimes e_1 - \lambda h^2 e'_0 \otimes e_0$
e'_1	$e_0 \otimes (e'_1 - he'_0)$	$-e'_0 \otimes (e'_1 + he'_0)$	$-e_1 \otimes e'_1 + \lambda h^2 e_0 \otimes e'_0$	$e'_1 \otimes e'_1 + \lambda h^2 e'_0 \otimes e'_0$

$R_{q_{(\mathbb{Z}_2)^2}}$ is non-unitary and satisfies the Y–B equation, see (8.29).

Remark 8.3. In the two examples above the R-matrices do not preserve the grading, hence the quantizations are of the larger category of vector spaces.

Acknowledgements I would like to express my gratitude to Volodya Rubtsov for getting me started on this paper and to Valentin Lychagin for suggestions improving the result.

References

1. S. Eilenberg, S. Mac Lane. *Cohomology theory in abstract groups 1.* Ann. Math., Vol. 48, No.1, 51–78, 1947.
2. H. L. Huru. *Associativity constraints, braidings and quantizations of modules with grading and action.* Lobachevskii J. Math., Vol. 23, 5–27, 2006, http://ljm.ksu.ru/vol23/110.html.
3. H. L. Huru. *Quantization of braided algebras. 1. Monoidal categories.* Lobachevskii J. Math., Vol.24, 13–42, 2006, http://ljm.ksu.ru/vol24/121.html.
4. H. L. Huru. *Quantization of braided algebras. 2. Graded Modules.* Lobachevskii J. Math., Vol. 25, 131–160, 2007, http://ljm.ksu.ru/vol25/145.html.
5. H. L. Huru, V. V. Lychagin. *Quantization and classical non-commutative and non-associative algebras.* J. Gen. Lie Theory Appl., Vol 2, No 1, 35–44, 2008.
6. A. Joyal, R. Street. *The geometry of tensor calculus.* Adv. Math., Vol. 88, 55–112, 1991.
7. B. A. Kupershmidt. *Remarks on quantization of classical r-matrices.* J. Nonlinear Math. Phys., Vol 6, No 3, 269–272, 1999.
8. V. V. Lychagin. *Differential operators and quantizations*, Preprint series in Pure Mathematics, Matematisk institutt, Universitetet i Oslo, No. 44, 1993.
9. V. V. Lychagin. *Quantizations of differential equations.* Pergamon Nonlinear Analysis, Vol. 47, 2621–2632, 2001.
10. Saunders Mac Lane. Categories for the working mathematician, *Graduate Texts in Mathematics*, Vol 5. Springer, Berlin, 1998.
11. C. A. Weibel. An introduction to homological algebra, *Cambridge studies in advanced mathematics*, Vol 38. Cambridge University Press, Cambridge, 1994.

Chapter 9
Connections on Modules over Singularities of Finite and Tame CM Representation Type

Eivind Eriksen and Trond Stølen Gustavsen

Abstract Let R be the local ring of a singular point of a complex analytic space, and let M be an R-module. Under what conditions on R and M is it possible to find a connection on M? To approach this question, we consider maximal Cohen–Macaulay (MCM) modules over CM algebras that are isolated singularities, and review an obstruction theory implemented in the computer algebra system Singular. We report on results, with emphasis on singularities of finite and tame CM representation type.

9.1 Introduction

Let R be the local ring of a complex analytic space, and let M be an R-module. In this paper we discuss the existence of a connection on M, i.e. an R-linear map $\nabla : \mathrm{Der}_C(R) \to \mathrm{End}_C(M)$ that satisfies the Leibniz rule. These connections are related to the topology of the singularity via the Riemann–Hilbert correspondence. In the case of normal surface singularities, this relationship is particularly strong, and it is described explicitly in Gustavsen and Ile [16]. However, it is a delicate problem to describe these connections, or even to determine when such a connection exists.

To our knowledge, Kahn was the first to consider this question, see Chap. 5 in Kahn [18]. Using known properties of vector bundles on elliptic curves and an analysis of the representations of the local fundamental group of a simple elliptic surface singularity, he was able to show that any maximal Cohen–Macaulay module

E. Eriksen
Oslo University College, P.O. Box 4, St Olavs Plass, 0130 Oslo, Norway
e-mail: eeriksen@hio.no

T. S. Gustavsen
BI Norwegian School of Management, 0442 Oslo, Norway
e-mail: trond.s.gustavsen@bi.no

S. Silvestrov et al. (eds.), *Generalized Lie Theory in Mathematics, Physics and Beyond,*

on such a singularity admits an integrable connection. In the present paper, the emphasis is on more algebraic methods which enable us to treat a more diverse class of singularities.

After some preliminary material in Sect. 2, we develop an algebraic approach to the question of existence of connections in Sect. 3. This approach is given in terms of an obstruction theory, which we have implemented as the library CONN.LIB [9] in the computer Algebra system SINGULAR 3.0 [14]. We focus on the case of maximal Cohen–Macaulay (MCM) modules over CM algebras that are isolated singularities, and discuss some results that we have obtained. In the case of normal surface singularities, we show how the MCM modules with connections are related to the topology of the singularity.

Our investigations indicate that it is interesting to consider this notion of connections on modules over singularities, but they also raise many questions, and some of these seem difficult to answer in general. Nevertheless, we are able to give some interesting results and examples. For the simple curve singularities, all MCM modules admit a connection. However, there are other curve singularities of finite CM type with MCM modules that do not admit a connection. In dimension two, all MCM modules over any CM finite singularity admit a connection. In higher dimensions, it seems that very few singularities have the property that all MCM modules admit a connection. For instance, an MCM module over a simple singularity in dimension $d \geq 3$ admits a connection if and only if it is free.

Existence of connections on modules has previously been considered by several authors, including Buchweitz, Christophersen, Ile, Kahn, Källström, Laudal and Maakestad; see for instance [8, 16, 18, 19, 22].

9.2 Preliminaries

We will denote by R a Cohen–Macaulay local $\mathbb{C}$-algebra which is an integral domain, and we will assume that $R \cong \mathscr{O}_{X,x}$ where (X,x) is an isolated singularity of an analytic space X. We can embed $(X,x) \subset (\mathbb{C}^n,0)$ and thus $R \cong \mathbb{C}\{x_1,\dots,x_n\}/(f_1,\dots,f_m)$.

A finitely generated R-module M is maximal Cohen–Macaulay if $\operatorname{depth} M = \dim R$, and we denote by $\operatorname{Der}_{\mathbb{C}}(R) \cong \operatorname{Hom}_R(\Omega^1_R, R)$ the module of derivations on R. Note that $\operatorname{Der}_{\mathbb{C}}(R)$ is a left R-module and a $\mathbb{C}$-Lie-algebra.

9.2.1 Connections

Let M be a finitely generated R-module. An *action of* $\operatorname{Der}_{\mathbb{C}}(R)$ on M is a $\mathbb{C}$-linear map $\nabla : \operatorname{Der}_{\mathbb{C}}(R) \to \operatorname{End}_{\mathbb{C}}(M)$ which for all $a \in R$, $m \in M$ and $D \in \operatorname{Der}_{\mathbb{C}}(R)$ satisfy the *Leibniz rule*

$$\nabla_D(am) = a\nabla_D(m) + D(a)m. \tag{9.1}$$

An action $\nabla : \mathrm{Der}_{\mathbb{C}}(R) \to \mathrm{End}_{\mathbb{C}}(M)$ of $\mathrm{Der}_{\mathbb{C}}(R)$ on M is said to be a *connection* if it is R-linear. We shall see that there are cases where M admits an action of $\mathrm{Der}_{\mathbb{C}}(R)$, but not a connection.

A connection $\nabla : \mathrm{Der}_{\mathbb{C}}(R) \to \mathrm{End}_{\mathbb{C}}(M)$ is *integrable* if it is a Lie-algebra homomorphism, i.e. if $\nabla([D_1,D_2]) = [\nabla(D_1),\nabla(D_2)]$.

We recall that when R is regular, a *connection* on M in the sense of André Weil is defined as a $\mathbb{C}$-linear map $\nabla : M \to M \otimes_R \Omega^1_R$ such that $\nabla(am) = a\nabla(m) + m \otimes d(a)$ for all $a \in R$, $m \in M$, where $d : R \to \Omega^1_R$ is the universal derivation, see for instance Katz [20]. Moreover, the *curvature* of ∇ is defined as the R-linear map $K_\nabla : M \to M \otimes_R \Omega^2_R$ given by $K_\nabla = \nabla^1 \circ \nabla$, where ∇^1 is the natural extension of ∇ to $M \otimes_R \Omega_R$, and ∇ is an *integrable connection* if $K_\nabla = 0$. To avoid confusion, we shall refer to connections in the sense of Andre Weil as Ω-connections in this paper.

Lemma 9.1. *Let R be a regular local analytic $\mathbb{C}$-algebra, and let M be a finitely generated R-module. If there is an Ω-connection on M, then M is free.*

Lemma 9.2. *There is a natural functor $\Omega\mathsf{MC}(R) \to \mathsf{MC}(R)$, from the category of R-modules with Ω-connection to the category of R-modules with connection, and an induced functor $\Omega\mathsf{MIC}(R) \to \mathsf{MIC}(R)$ between categories with modules with integrable connections. If Ω_R and $\mathrm{Der}_{\mathbb{C}}(R)$ are projective R-modules of finite presentation, then these functors are equivalences of categories.*

Proof. Any Ω-connection on M induces a connection on M, and this assignment preserves integrability. Moreover, any connection ∇ on M may be considered as a $\mathbb{C}$-linear map $M \to \mathrm{Hom}_R(\mathrm{Der}_{\mathbb{C}}(R), M)$, given by $m \mapsto \{D \mapsto \nabla_D(m)\}$. It is sufficient to show that the natural map $M \otimes_R \Omega_R \to \mathrm{Hom}_R(\mathrm{Der}_{\mathbb{C}}(R), M)$, given by $m \otimes \omega \mapsto \{D \mapsto \phi_D(\omega)m\}$, is an isomorphism. But this is clearly the case when Ω_R and $\mathrm{Der}_{\mathbb{C}}(R)$ are projective R-modules of finite presentation. □

We see that if R is a regular, then there is a bijective correspondence between (integrable) connections on M and (integrable) Ω-connections on M for any R-module M. In contrast, there are many modules that admit connections but not Ω-connections when R is singular.

9.2.2 Representations of the Local Fundamental Group in Dimension Two

We may always choose a representative X of a normal surface singularity (X,x), such that $X \setminus \{x\}$ is connected and $(X,x) \subset (\mathbb{C}^n,0)$. If $\varepsilon > 0$ is small and B_ε is a ball in $\mathbb{C}^n$ of radius ε, then $L := X \cap \partial B_\varepsilon$ is a smooth, compact, connected and oriented real manifold, called the link of (X,x), see Mumford [23]. The isomorphism class of L is independent of (small) ε. We define the local fundamental group $\pi_1^{\mathrm{loc}}(X,x) := \pi_1(L)$, and we will always assume that the representative X is such that $\pi_1^{\mathrm{loc}}(X,x) = \pi_1(X \setminus \{x\})$.

Assume that M is a maximal Cohen–Macaulay R-module of rank r, represented by a sheaf $\mathscr{M}$ on X. Then $\mathscr{M}$ is locally free on $U = X \setminus \{x\}$. Assume that M admits an integrable (flat) connection ∇. Then it follows by localization that $\mathscr{M}_{|U}$ admits an integrable connection $\nabla_U : \mathscr{M}_{|U} \to \mathscr{M}_{|U} \otimes \Omega^1_U$. The kernel $\operatorname{Ker} \nabla_U$ is a local system on U and corresponds to a representation

$$\rho_{(M,\nabla)} : \pi_1^{\mathrm{loc}}(X,x) \to \mathrm{Gl}(\mathbb{C},r)$$

called the monodromy representation of (M,∇).

9.3 Obstruction Theory

Let R be a $\mathbb{C}$-algebra. For any R-modules M,M', we refer to Weibel [25] for the definition of the *Hochschild cohomology groups* $\mathrm{HH}^n(R,\mathrm{Hom}_{\mathbb{C}}(M,M'))$ of R with values in the bimodule $\mathrm{Hom}_{\mathbb{C}}(M,M')$, and we recall that there is a natural isomorphism of $\mathbb{C}$-linear vector spaces $\mathrm{Ext}^n_R(M,M') \to \mathrm{HH}^n(R,\mathrm{Hom}_{\mathbb{C}}(M,M'))$ for any $n \geq 0$. Note that there is an exact sequence of $\mathbb{C}$-linear vector spaces

$$\begin{aligned} 0 \to \mathrm{HH}^0(R,\mathrm{Hom}_{\mathbb{C}}(M,M')) = \mathrm{Hom}_R(M,M') \to \mathrm{Hom}_{\mathbb{C}}(M,M') \\ \xrightarrow{\mu} \mathrm{Der}_{\mathbb{C}}(R,\mathrm{Hom}_{\mathbb{C}}(M,M')) \to \mathrm{HH}^1(R,\mathrm{Hom}_{\mathbb{C}}(M,M')) \to 0, \end{aligned} \tag{9.2}$$

where μ is the assignment which maps $\phi \in \mathrm{Hom}_{\mathbb{C}}(M,M')$ to the trivial derivation given by $(a,m) \mapsto a\phi(m) - \phi(am)$ for all $a \in R$, $m \in M$.

Proposition 9.1. *There is a canonical map* $g : \mathrm{Der}_{\mathbb{C}}(R) \to \mathrm{Ext}^1_R(M,M)$*, called the Kodaira–Spencer map of* M*, with the following properties:*

1. *The Kodaira–Spencer kernel* $\mathsf{V}(M) = \mathrm{Ker}(g)$ *is a Lie algebra and an* R*-module.*
2. *For any* $D \in \mathrm{Der}_{\mathbb{C}}(R)$*, there exists an operator* $\nabla_D \in \mathrm{End}_{\mathbb{C}}(M)$ *satisfying the Leibniz rule (9.1) if and only if* $D \in \mathsf{V}(M)$.

In particular, there is an action of $\mathrm{Der}_{\mathbb{C}}(R)$ *on* M *if and only if* g *is trivial.*

Proof. Let $\psi_D \in \mathrm{Der}_{\mathbb{C}}(R,\mathrm{End}_{\mathbb{C}}(M))$ be given by $\psi_D(a)(m) = D(a)m$ for any $D \in \mathrm{Der}_{\mathbb{C}}(R)$, and denote by $g(D)$ the class in $\mathrm{Ext}^1_{\mathbb{C}}(M,M)$ corresponding to the class of ψ_D in $\mathrm{HH}^1(R,\mathrm{End}_{\mathbb{C}}(M))$. This defines the Kodaira–Spencer map g of M, which is R-linear by definition. Clearly, its kernel $\mathsf{V}(M)$ is closed under the Lie product. Using the exact sequence (9.2), it easily follows that there exists an operator ∇_D satisfying the Leibniz rule (9.1) with respect to D if and only if $D \in \mathsf{V}(M)$. Hence, if g is trivial, we may choose a $\mathbb{C}$-linear operator ∇_D with this property for each derivation $D \in \mathrm{Der}_k(R)$ in a $\mathbb{C}$-linear basis for $\mathrm{Der}_{\mathbb{C}}(R)$, and we obtain an action $\nabla : \mathrm{Der}_{\mathbb{C}}(R) \to \mathrm{End}_{\mathbb{C}}(M)$ of $\mathrm{Der}_{\mathbb{C}}(R)$ on M. One the other hand, if there is an action of $\mathrm{Der}_{\mathbb{C}}(R)$ on M, we must have $\mathsf{V}(M) = \mathrm{Der}_{\mathbb{C}}(R)$. □

We remark that the Kodaira–Spencer map g is the contraction against the first Atiyah class of the R-module M. See Källström [19], Sect. 2.2 for another proof of Proposition 9.1. Some useful properties of the Kodaira–Spencer kernel may be found in Lemma 5 in Eriksen and Gustavsen [10]. See also Buchweitz and Liu, [4], Lemma 3.4.

Proposition 9.2. *Assume that the Kodaira–Spencer map of M is trivial. Then there is a canonical class* $\mathsf{lc}(M) \in \mathrm{Ext}^1_R(\mathrm{Der}_{\mathbb{C}}(R), \mathrm{End}_R(M))$ *such that* $\mathsf{lc}(M) = 0$ *if and only if there exists a connection on M. Moreover, if* $\mathsf{lc}(M) = 0$*, then there is a transitive and effective action of* $\mathrm{Hom}_R(\mathrm{Der}_{\mathbb{C}}(R), \mathrm{End}_R(M))$ *on the set of connections on M.*

Proof. Choose a $\mathbb{C}$-linear connection $\nabla : \mathrm{Der}_{\mathbb{C}}(R) \to \mathrm{End}_{\mathbb{C}}(M)$ on M, and let $\phi \in \mathrm{Der}_{\mathbb{C}}(R, \mathrm{Hom}_{\mathbb{C}}(\mathrm{Der}_{\mathbb{C}}(R), \mathrm{End}_R(M)))$ be the derivation given by $\phi(a)(D) = a\nabla_D - \nabla_{aD}$. We denote by $\mathsf{lc}(M)$ the class in $\mathrm{Ext}^1_R(\mathrm{Der}_{\mathbb{C}}(R), \mathrm{End}_R(M))$ corresponding to the class of ϕ in $\mathrm{HH}^1(R, \mathrm{Hom}_{\mathbb{C}}(\mathrm{Der}_{\mathbb{C}}(R), \mathrm{End}_R(M)))$, and it is easy to check that this class is independent of ∇. Using the exact sequence (9.2), the proposition follows easily. □

When the Kodaira–Spencer map of M is trivial, there is a natural short exact sequence $0 \to \mathrm{End}_R(M) \to c(M) \to \mathrm{Der}_{\mathbb{C}}(R) \to 0$ of left A-modules, where $c(M) = \{\phi \in \mathrm{End}_{\mathbb{C}}(M) : [\phi, a] \in R \text{ for all } a \in R\}$ is the module of first order differential operators on M with scalar symbol. We remark that $c(M)$ represents $\mathsf{lc}(M)$ as an extension of R-modules, see also Källström [19], Proposition 2.2.10.

9.3.1 Implementation in Singular

We have implemented the obstruction theory in a library `conn.lib` [9] for the computer algebra system Singular 3.0. The implementation is explained in detail in Eriksen and Gustavsen [10] and can be used to compute many interesting examples. Some examples are given in Sect. 9.4.

We have used Hochschild cohomology to define this obstruction theory. However, the description of the obstructions in terms of free resolutions is essential for the implementation; see Sect. 4 of Eriksen and Gustavsen [10].

In the case of simple hypersurface singularities (of type A_n, D_n and E_n) in dimension d, there exists a connection on any MCM module if $d \leq 2$. Using our implementation, we got interesting results in higher dimensions: For $d = 3$, we found that the only MCM modules that admit connections are the free modules if $n \leq 50$, and experimental results indicated that the same result should hold for $d = 4$. Using different techniques, we proved a more general result in [11]: An MCM module over a simple hypersurface singularity of dimension $d \geq 3$ admits a connection if and only if it is free.

9.4 Results and Examples

In this section, we explain some results on existence of connections. The main examples are the so-called CM-finite and CM-tame singularities. We say that (X,x) or $R = \mathcal{O}_{X,x}$ is CM-finite if there are only finitely many indecomposable MCM R-modules. We say that (X,x) or $R = \mathcal{O}_{X,x}$ is CM-tame if it is not CM-finite and if there are at most a finite number of 1-parameter families of indecomposable MCM R-modules, see [6].

Theorem 9.1 (Knörrer [21], Buchweitz–Greuel–Schreyer [5]). *A hypersurface* $(X,x) = (V(f),0) \subseteq (\mathbb{C}^{d+1},0)$ *is CM-finite if and only if it is simple, i.e. f is one of the following:*

$$A_n : f = x^2 + y^{n+1} + z_1^2 + \cdots + z_{d-1}^2 \qquad n \geq 1$$
$$D_n : f = x^2 y + y^{n-1} + z_1^2 + \cdots + z_{d-1}^2 \qquad n \geq 4$$
$$E_6 : f = x^3 + y^4 + z_1^2 + \cdots + z_{d-1}^2$$
$$E_7 : f = x^3 + xy^3 + z_1^2 + \cdots + z_{d-1}^2$$
$$E_8 : f = x^3 + y^5 + z_1^2 + \cdots + z_{d-1}^2$$

9.4.1 Dimension One

In this section we consider curve singularities. We have the following:

Theorem 9.2 (Eriksen [8]). *If R is the local ring of a simple curve singularity and M is an MCM R-module, then there is an integrable connection on M.*

Let R be the local ring of any curve singularity. We say that a local ring S *birationally dominates* S if $R \subseteq S \subseteq R^*$, where R^* is the integral closure of R in its total quotient ring. It is known that R has finite CM representation type if and only if it birationally dominates the complete local ring of a simple curve singularity, see Greuel and Knörrer [13].

This result leads to a complete classification of curve singularities of finite CM representation type. The only Gorenstein curve singularities of finite CM type are the simple singularities, and the non-Gorenstein curve singularities of finite CM representation type are of the following form:

$$D_n^s : R = \mathbb{C}\{x,y,z\}/(x^2 - y^n, xz, yz) \text{ for } n \geq 2$$
$$E_6^s : R = \mathbb{C}\{t^3,t^4,t^5\} \subseteq \mathbb{C}\{t\}$$
$$E_7^s : R = \mathbb{C}\{x,y,z\}/(x^3 - y^4, xz - y^2, y^2 z - x^2, yz^2 - xy)$$
$$E_8^s : R = \mathbb{C}\{t^3,t^5,t^7\} \subseteq \mathbb{C}\{t\}$$

Using SINGULAR 3.0 [14] and our library CONN.LIB [9], we show that not all MCM R-modules admit connections in these cases. In fact, the canonical module ω_R does not admit a connection when R is the local ring of the singularities E_6^s, E_7^s, E_8^s or D_n^s for $n \leq 100$.

Theorem 9.3. *Let R be the local ring of a monomial curve singularity. Then all formally gradable MCM R-modules of rank one admits a connection if and only if R is Gorenstein.*

Proof. See Theorem 13 in Eriksen and Gustavsen [11]. □

Let us consider the non-Gorenstein monomial curve singularities E_6^s and E_8^s of finite CM representation type. By Yoshino [26], Theorem 15.14, all MCM R-modules are formally gradable in these cases. For E_6^s, we have three non-free rank one MCM R-modules M_1, M_2, M_{12}. One can show that M_2 and M_{12} admit connections, while M_1 does not, and that M_1 is the canonical module in this case. A similar consideration for E_8^s shows that M_{14}, M_2, M_4, M_{24} and M_{124} are the non-free rank one MCM A-modules, and that the canonical module M_2 is the only rank one MCM A-module that does not admit connections.

Finally, we remark that any connection on an MCM R-module is integrable when R is a monomial curve singularity.

9.4.2 Dimension Two

In dimension two, the situation is somewhat different. Notice that a Cohen–Macaulay isolated surface singularity is normal.

Theorem 9.4 (Herzog [17], Auslander [1], Esnault [12]). *Let (X,x) be a normal surface singularity. Then (X,x) is CM-finite if and only if it is a quotient singularity with local ring $R = \mathbb{C}\{u,v\}^G$ for a finite group $G = \pi_1^{\mathrm{loc}}(X,x)$, and in this case there is a one-to-one correspondence between representations of $\pi_1^{\mathrm{loc}}(X,x)$ and MCM R-modules.*

Proposition 9.3. *Any MCM module on a quotient surface singularity admits an integrable connection. In particular; any MCM module on a CM-finite surface singularity admits an integrable connection.*

Proof. Let $S = \mathbb{C}\{u,v\}$, and $M = (S \otimes_{\mathbb{C}} V)^G$. There is a canonical integrable connection $\nabla' : \mathrm{Der}_k(S) \to \mathrm{End}_k(S \otimes_k V)$ on the free S-module $S \otimes_k V$, given by

$$\nabla'_D(\sum s_i \otimes v_i) = \sum D(s_i) \otimes v_i$$

for any $D \in \mathrm{Der}_k(S)$, $s_i \in S$, $v_i \in V$. But the natural map $\mathrm{Der}_k(S)^G \to \mathrm{Der}_k(S^G)$ is an isomorphism, see Schlessinger [24], hence ∇' induces an integrable connection ∇ on M. □

Theorem 9.5 (Gustavsen and Ile [16]). *Let (X,x) be a normal surface singularity with $R = \mathcal{O}_{X,x}$. Then the monodromy representation gives an equivalence of categories between the category of MCM R-modules with integrable connection and horizontal (i.e. compatible) morphisms, and the category of finite-dimensional representations of the local fundamental group $\pi_1^{\mathrm{loc}}(X,x)$.*

We now consider the tame case. Any MCM module over a simple elliptic surface singularity admits an integrable connection, see Kahn [18], and this result has been generalized to quotients of simple elliptic surface singularities in Gustavsen and Ile [15]. Kurt Behnke has pointed out that it might be true for cusp singularities as well, see Behnke [3]. More generally, it is probable that any MCM module over a log canonical surface singularity admits an integrable connection.

When R is a surface singularity, we have not found any examples of an MCM R-module that does not admit an integrable connection.

We consider the following example. Let $Y = \mathrm{Specan}(S)$ where $S = \mathbb{C}[x,y,z]/(x^3 + y^3 + z^3)$, and let $E = \mathrm{Projan}(S)$. Let $G = \mathbb{Z}/(3)$, and let ω be a fixed primitive third root of unity. We fix the action of G on S given by $(x,y,z) \mapsto (\omega x, \omega^2 y, \omega z)$ and we let X be the quotient of Y under this action. (Note that the action on S is compatible with the grading. In particular we have an action on E.)

The quotient $X = Y/G$ is an elliptic quotient and Y is its canonical covering. One finds that there are 27 rank one and 54 rank two MCM modules on X. There is a one parameter family of non-isomorphic rank three MCM modules on X.

Using Mumford [23], one can show that the local fundamental group of the quotient (X,x) is the group given by two generators a_1 and a_2 and the relations $a_1^3 = a_2^3 = (a_1^2 a_2^2)^3$. The representations of rank one and two are found in an appendix of Gustavsen and Ile [15]. There are 27 rank one and 54 rank two indecomposable modules. Thus in rank one and two, there is a one-to-one correspondence between indecomposable representations of the local fundamental group and indecomposable MCM modules. In particular; rank one and rank two MCM modules admit *unique integrable connections*. On (Y,y), in contrast, there is a positive dimensional family of connections on each indecomposable MCM module, see Kahn [18, Theorem 6.30].

9.4.3 Connections on MCM Modules in Dimension Three

In larger dimensions things are different:

Theorem 9.6 (Eriksen and Gustavsen [11]). *Let R be the analytic local ring of a simple singularity of dimension $d \geq 3$. Then an MCM R-module M admits a connection if and only if it is free.*

Note that if R is Gorenstein and of finite CM representation type, then R is a simple singularity, see Herzog [17]. The classification of non-Gorenstein singularities of finite CM representation type is not known in dimension $d \geq 3$. However, partial results are given in Eisenbud and Herzog [7], Auslander and Reiten [2] and Yoshino [26].

In Auslander and Reiten [2], it was shown that there is only one quotient singularity of dimension $d \geq 3$ with finite CM representation type, the cyclic threefold quotient singularity of type $\frac{1}{2}(1,1,1)$. Its local ring $R = S^G$ is the invariant ring of the action of the group $G = \mathbf{Z}_2$ on $S = \mathbb{C}\{x_1, x_2, x_3\}$ given by $\sigma x_i = -x_i$ for $i = 1, 2, 3$, where $\sigma \in G$ is the non-trivial element. There are exactly two non-free indecomposable MCM R-modules M_1 and M_2. The module M_1 has rank one, and is induced by the non-trivial representation of G of dimension one, see Auslander and Reiten [2]. It follows that M_1 admits an integrable connection, and one may show that M_1 is the canonical module of A.

It is also known that the threefold scroll of type $(2,1)$, with local ring

$$R = \mathbb{C}\{x,y,z,u,v\}/(xz - y^2, xv - yu, yv - zu),$$

has finite CM representation type, see Auslander and Reiten [2]. There are four non-free indecomposable MCM R-modules, and none of these admit connections. In particular, the canonical module ω_R does not admit a connection.

To the best of our knowledge, no other examples of singularities of finite CM representation type are known in dimension $d \geq 3$.

References

1. M. Auslander, *Rational singularities and almost split sequences*, Trans. Am. Math. Soc. **293**, no. 2, 511–531 (1986).
2. M. Auslander and I. Reiten, *The Cohen-Macaulay type of Cohen-Macaulay rings*, Adv. Math. **73**, no. 1, 1–23 (1989).
3. Kurt Behnke, *On Auslander modules of normal surface singularities*, Manuscripta Math. **66**, no. 2, 205–223 (1989).
4. R.-O. Buchweitz and S. Liu, *Hochschild cohomology and representation-finite algebras*, Proc. London Math. Soc. (3) **88**, no. 2, 355–380 (2004).
5. R.-O. Buchweitz, G.-M. Greuel, and F.-O. Schreyer, *Cohen–Macaulay modules on hypersurface singularities. II*, Invent. Math. **88**, no. 1, 165–182 (1987).
6. Yu. A. Drozd and G.-M. Greuel, *Tame-wild dichotomy for Cohen-Macaulay modules*, Math. Ann. **294**, no. 3, 387–394 (1992).
7. D. Eisenbud and J. Herzog, *The classification of homogeneous Cohen-Macaulay rings of finite representation type*, Math. Ann. **280**, no. 2, 347–352 (1988).
8. E. Eriksen, *Connections on Modules Over Quasi-Homogeneous Plane Curves*, Comm. Algebra 36, no. 8, 3032–3041 (2008)
9. E. Eriksen and T. S. Gustavsen, CONN.LIB, *A* SINGULAR *library to compute obstructions for existence of connections on modules*, 2006, Available at http://home.hio.no/~eeriksen/connections.html.
10. E. Eriksen and T. S. Gustavsen, *Computing obstructions for existence of connections on modules*, J. Symb. Comput. **42**, no. 3, 313–323 (2007).
11. E. Eriksen and T. S. Gustavsen, *Connections on modules over singularities of finite CM representation type*, J. Pure Appl. Algebra **212**, no. 7, 1561–1574 (2008).
12. Hélène Esnault, *Reflexive modules on quotient surface singularities*, J. Reine Angew. Math. **362**, 63–71 (1985).

13. G.-M. Greuel and H. Knörrer, *Einfache Kurvensingularitäten und torsionsfreie Moduln*, Math. Ann. **270**, 417–425 (1985).
14. G.-M. Greuel, G. Pfister, and H. Schönemann, SINGULAR 3.0, A Computer Algebra System for Polynomial Computations, Center for Computer Algebra, University of Kaiserslautern, (2005), http://www.singular.uni-kl.de/
15. T. S. Gustavsen and R. Ile, *Reflexive modules on log-canonical surface singularities*, Preprint, http://www.math.uio.no/~stolen, 2006.
16. T. S. Gustavsen and R. Ile, *Reflexive modules on normal surface singularities and representations of the local fundamental group*, J. Pure Appl. Algebra **212**, no. 4, 851–862 (2008).
17. Jürgen Herzog, *Ringe mit nur endlich vielen Isomorphieklassen von maximalen, unzerlegbaren Cohen-Macaulay-Moduln*, Math. Ann. **233**, no. 1, 21–34 (1978).
18. Constantin P. M. Kahn, *Reflexive Moduln auf einfach-elliptischen Flächensingularitäten*, Bonner Mathematische Schriften, 188, Universität Bonn Mathematisches Institut, Bonn, 1988, Dissertation.
19. Rolf Källström, *Preservations of defect sub-schemes by the action of the tangent sheaf*, J. Pure Appl. Algebra **203**, 166–188 (2005).
20. Nicholas M. Katz, *Nilpotent connections and the monodromy theorem: Applications of a result of Turrittin*, Inst. Hautes Études Sci. Publ. Math., no. 39, 175–232 (1970).
21. Horst Knörrer, *Cohen-Macaulay modules on hypersurface singularities I*, Invent. Math. **88**, 153–164 (1987).
22. Helge Maakestad, *The Chern character for L ie-Rinehart algebras*, Ann. Inst. Fourier (Grenoble) **55**, no. 7, 2551–2574 (2005).
23. David Mumford, *The topology of normal singularities of an algebraic surface and a criterion for simplicity*, Inst. Hautes Études Sci. Publ. Math., no. 9, 5–22 (1961).
24. Michael Schlessinger, *Rigidity of quotient singularities*, Invent. Math. **14**, 17–26 (1971).
25. Charles A. Weibel, *An introduction to homological algebra*, Cambridge Studies in Advanced Mathematics, vol. 38, Cambridge University Press, Cambridge, 1994.
26. Yuji Yoshino, *Cohen–Macaulay modules over Cohen–Macaulay rings*, London Mathematical Society Lecture Note Series, vol. 146, Cambridge University Press, Cambridge, 1990.

Chapter 10
Computing Noncommutative Global Deformations of D-Modules

Eivind Eriksen

Abstract Let $(X, \mathscr{D})$ be a D-scheme in the sense of Beilinson and Bernstein, given by an algebraic variety X and a morphism $\mathscr{O}_X \to \mathscr{D}$ of sheaves of rings on X. We consider noncommutative deformations of quasi-coherent sheaves of left $\mathscr{D}$-modules on X, and show how to compute their pro-representing hulls. As an application, we compute the noncommutative deformations of the left $\mathscr{D}_X$-module $\mathscr{O}_X$ when X is any elliptic curve.

10.1 Introduction

Let k be an algebraically closed field of characteristic 0, and let X be an algebraic variety over k, i.e. an integral separated scheme of finite type over k. A *D-algebra* in the sense of Beilinson and Bernstein [2] is a sheaf $\mathscr{D}$ of associative rings on X, together with a morphism $i : \mathscr{O}_X \to \mathscr{D}$ of sheaves of rings on X, such that the following conditions hold: (1) $\mathscr{D}$ is quasi-coherent as a left and right $\mathscr{O}_X$-module via i, and (2) for any open subset $U \subseteq X$ and any section $P \in \mathscr{D}(U)$, there is an integer $n \geq 0$ such that

$$[\dots[[P, a_1], a_2], \dots, a_n] = 0$$

for all sections $a_1, \dots, a_n \in \mathscr{O}_X(U)$, where $[P, Q] = PQ - QP$ is the commutator in $\mathscr{D}(U)$. When $\mathscr{D}$ is a D-algebra on X, the ringed space $(X, \mathscr{D})$ is called a *D-scheme*.

Let us denote the sheaf of k-linear on X by $\mathscr{D}_X$, and for any Lie algebroid $\mathfrak{g}$ of X/k, let us denote the enveloping D-algebra of $\mathfrak{g}$ by $\mathrm{U}(\mathfrak{g})$. We see that $\mathscr{D}_X$ and $\mathrm{U}(\mathfrak{g})$ are examples of noncommutative D-algebras on X, and that $\mathscr{O}_X$ is an example of a commutative D-algebra on X.

E. Eriksen
Oslo University College, P.O. Box 4, St Olavs Plass, 0130 Oslo, Norway
e-mail: eeriksen@hio.no

S. Silvestrov et al. (eds.), *Generalized Lie Theory in Mathematics, Physics and Beyond,*

Let us define a *D-module* on a D-scheme $(X, \mathscr{D})$ to be a quasi-coherent sheaf of left $\mathscr{D}$-modules on X. In Eriksen [3], we developed a noncommutative global deformation theory of D-modules that generalizes the usual (commutative) global deformation theory of D-modules, and the noncommutative deformation theory of modules in the affine case, due to Laudal. In Sects. 10.2–10.3, we review the essential parts of this theory, including the global Hochschild cohomology and the global obstruction calculus, all the time with a view towards concrete computations.

The purpose of this paper is to show how to apply the theory in order to compute noncommutative global deformations of interesting D-modules. In Sect. 10.4, we consider the noncommutative deformation functor

$$\mathsf{Def}_{\mathscr{O}_X} : \mathsf{a}_1 \to \mathsf{Sets}$$

of $\mathscr{O}_X$ considered as a left $\mathscr{D}_X$-module when X is any over k. Recall that in this case, a quasi-coherent $\mathscr{D}_X$-module structure on $\mathscr{O}_X$ is the same as an integrable connection on $\mathscr{O}_X$, and according to a theorem due to André Weil, see Weil [7] and also Atiyah [1], a line bundle admits an integrable connection if and only if it has degree zero.

We show that the noncommutative deformation functor $\mathsf{Def}_{\mathscr{O}_X} : \mathsf{a}_1 \to \mathsf{Sets}$ has pro-representing hull $H = k\langle\langle t_1, t_2\rangle\rangle/(t_1t_2 - t_2t_1) \cong k[[t_1, t_2]]$ that is commutative, smooth and of dimension two. We also compute the corresponding versal family in concrete terms, and remark that it does not admit an algebraization.

10.2 Noncommutative Global Deformations of D-Modules

Let $(X, \mathscr{D})$ be a D-scheme, and let $\mathsf{QCoh}(\mathscr{D})$ be the category of quasi-coherent sheaves of left $\mathscr{D}$-modules on X. This is the full subcategory of $\mathsf{Sh}(X, \mathscr{D})$, the category of sheaves of left $\mathscr{D}$-modules on X, consisting of quasi-coherent sheaves. We recall that a sheaf $\mathscr{F}$ of left $\mathscr{D}$-modules on X is quasi-coherent if for every point $x \in X$, there exists an open neighbourhood $U \subseteq X$ of x, free sheaves $\mathscr{L}_0, \mathscr{L}_1$ of left $\mathscr{D}|_U$-modules on U, and an exact sequence $0 \leftarrow \mathscr{F}|_U \leftarrow \mathscr{L}_0 \leftarrow \mathscr{L}_1$ of sheaves of left $\mathscr{D}|_U$-modules on U. We shall refer to the quasi-coherent sheaves of left $\mathscr{D}$-modules on X as *D-modules* on the D-scheme $(X, \mathscr{D})$.

For any D-scheme $(X, \mathscr{D})$, $\mathsf{QCoh}(\mathscr{D})$ is an Abelian k-category, and we consider noncommutative deformations in $\mathsf{QCoh}(\mathscr{D})$. For a finite family $\mathscr{F} = \{\mathscr{F}_1, \dots, \mathscr{F}_p\}$ of quasi-coherent left $\mathscr{D}$-modules on X, there is a noncommutative deformation functor $\mathsf{Def}^{qc}_{\mathscr{F}} : \mathsf{a}_p \to \mathsf{Sets}$ of $\mathscr{F}$ in $\mathsf{QCoh}(\mathscr{D})$, generalizing the noncommutative deformation functor of modules introduced in Laudal [5]. We shall provide a brief description of $\mathsf{Def}^{qh}_{\mathscr{F}}$ below; see Eriksen [3] for further details.

We recall that the objects of the category a_p of p-pointed *noncommutative Artin rings* are Artinian rings R, together with pairs of structural morphisms $f : k^p \to R$ and $g : R \to k^p$, such that $g \circ f = \mathrm{id}$ and the radical $I(R) = \mathrm{Ker}(g)$ is nilpotent. The morphisms are the natural commutative diagrams. For any $R \in \mathsf{a}_p$, there are

p isomorphism classes of simple left R-modules, represented by $\{k_1, k_2, \dots, k_p\}$, where $k_i = 0 \times \cdots \times k \times \cdots \times 0$ is the i'th projection of k^p for $1 \leq i \leq p$.

We remark that any $R \in \mathsf{a}_p$ is a $p \times p$ *matrix ring*, in the sense that there are p indecomposable idempotents $\{e_1, \dots, e_p\}$ in R with $e_1 + \cdots + e_p = 1$ and a decomposition $R = \oplus R_{ij}$, given by $R_{ij} = e_i R e_j$, such that elements of R multiply as matrices. We shall therefore use matrix notation, and write $R = (R_{ij})$ when $R \in \mathsf{a}_p$, and $(V_{ij}) = \oplus V_{ij}$ when $\{V_{ij} : 1 \leq i, j \leq p\}$ is any family of vector spaces.

For any $R \in \mathsf{a}_p$, a lifting of $\mathscr{F}$ to R is a quasi-coherent sheaf $\mathscr{F}_R$ of left $\mathscr{D}$-modules with a compatible right R-module structure, together with isomorphisms $\eta_i : \mathscr{F}_R \otimes_R k_i \to \mathscr{F}_i$ in $\mathsf{QCoh}(\mathscr{D})$ for $1 \leq i \leq p$, such that $\mathscr{F}_R(U) \cong (\mathscr{F}_i(U) \otimes_k R_{ij})$ as right R-modules for all open subsets $U \subseteq X$. We say that two liftings $(\mathscr{F}_R, \eta_i)$ and $(\mathscr{F}'_R, \eta'_i)$ are equivalent if there is an isomorphism $\tau : \mathscr{F}_R \to \mathscr{F}'_R$ of $\mathscr{D}$-R bimodules on X such that $\eta'_i \circ (\tau \otimes_k k_i) = \eta_i$ for $1 \leq i \leq p$, and denote the set of equivalence classes of liftings of $\mathscr{F}$ to R by $\mathsf{Def}^{qc}_{\mathscr{F}}(R)$. This defines the noncommutative deformation functor $\mathsf{Def}^{qc}_{\mathscr{F}} : \mathsf{a}_p \to \mathsf{Sets}$.

10.3 Computing Noncommutative Global Deformations

Let $(X, \mathscr{D})$ be a D-scheme, and let U be an open affine cover of X that is finite and closed under intersections. We shall explain how to compute noncommutative deformations in $\mathsf{QCoh}(\mathscr{D})$ effectively using the open cover U.

We may consider U as a small category, where the objects are the open subsets $U \in \mathsf{U}$, and the morphisms from U to V are the (opposite) inclusions $U \supseteq V$. There is a natural forgetful functor $\mathsf{QCoh}(\mathscr{D}) \to \mathsf{PreSh}(\mathsf{U}, \mathscr{D})$, where $\mathsf{PreSh}(\mathsf{U}, \mathscr{D})$ is the Abelian k-category of (covariant) presheaves of left $\mathscr{D}$-modules on U. For any finite family $\mathscr{F}$ in $\mathsf{QCoh}(\mathscr{D})$, this forgetful functor induces an isomorphism of noncommutative deformation functors $\mathsf{Def}^{qc}_{\mathscr{F}} \to \mathsf{Def}_{\mathscr{F}}$, where $\mathsf{Def}^{qc}_{\mathscr{F}} : \mathsf{a}_p \to \mathsf{Sets}$ is the noncommutative deformation functor of $\mathscr{F}$ in $\mathsf{QCoh}(\mathscr{D})$ defined in Sect. 10.2, and $\mathsf{Def}_{\mathscr{F}} : \mathsf{a}_p \to \mathsf{Sets}$ is the noncommutative deformation functor of $\mathscr{F}$ in $\mathsf{PreSh}(\mathsf{U}, \mathscr{D})$, defined in a similar way; see Eriksen [3] for details.

Theorem 10.1. *Let $(X, \mathscr{D})$ be a D-scheme, and let $\mathscr{F} = \{\mathscr{F}_1, \dots, \mathscr{F}_p\}$ be a finite family in $\mathsf{QCoh}(\mathscr{D})$. If the global Hochschild cohomology*

$$(\mathrm{HH}^n(\mathsf{U}, \mathscr{D}, \mathrm{Hom}_k(\mathscr{F}_j, \mathscr{F}_i)))$$

has finite k-dimension for $n = 1, 2$, then the noncommutative deformation functor $\mathsf{Def}^{qc}_{\mathscr{F}} : \mathsf{a}_p \to \mathsf{Sets}$ has a pro-representing hull $H = H(\mathscr{F})$, completely determined by $(\mathrm{HH}^n(\mathsf{U}, \mathscr{D}, \mathrm{Hom}_k(\mathscr{F}_j, \mathscr{F}_i)))$ for $n = 1, 2$ and their generalized Massey products.

In fact, there is a constructive proof of the fact that $\mathsf{Def}_{\mathscr{F}} : \mathsf{a}_p \to \mathsf{Sets}$ of $\mathscr{F}$ in $\mathsf{PreSh}(\mathsf{U}, \mathscr{D})$ has a pro-representing hull; see Eriksen [3] for details. The construction uses the global Hochschild cohomology $(\mathrm{HH}^n(\mathsf{U}, \mathscr{D}, \mathrm{Hom}_k(\mathscr{F}_j, \mathscr{F}_i)))$ for $n = 1, 2$, and the obstruction calculus of $\mathsf{Def}_{\mathscr{F}}$, which can be expressed in terms of generalized Massey products on these cohomology groups. We give a brief description of the global Hochschild cohomology and the obstruction calculus below.

10.3.1 Cohomology

For any presheaves $\mathscr{F}, \mathscr{G}$ of left $\mathscr{D}$-modules on U, we recall the definition of *global Hochschild cohomology* $\mathrm{HH}^n(\mathsf{U}, \mathscr{D}, \mathrm{Hom}_k(\mathscr{F}, \mathscr{G}))$ of $\mathscr{D}$ with values in the bimodule $\mathrm{Hom}_k(\mathscr{F}, \mathscr{G})$ on U. For any inclusion $U \supseteq V$ in U, we consider the Hochschild complex $\mathsf{HC}^*(\mathscr{D}(U), \mathrm{Hom}_k(\mathscr{F}(U), \mathscr{G}(V))$ of $\mathscr{D}(U)$ with values in the bimodule $\mathrm{Hom}_k(\mathscr{F}(U), \mathscr{G}(V))$. We define the category $\mathsf{Mor}\ \mathsf{U}$ to have (opposite) inclusions $U \supseteq V$ in U as its objects, and nested inclusions $U' \supseteq U \supseteq V \supseteq V'$ in U as its morphisms from $U \supseteq V$ to $U' \supseteq V'$. It follows that we may consider the Hochschild complex

$$\mathsf{HC}^*(\mathscr{D}, \mathrm{Hom}_k(\mathscr{F}, \mathscr{G})) : \mathsf{Mor}\ \mathsf{U} \to \mathsf{Compl}(k)$$

as a functor on $\mathsf{Mor}\ \mathsf{U}$. The global Hochschild complex is the total complex of the double complex $\mathsf{D}^{**} = \mathsf{D}^*(\mathsf{U}, \mathsf{HC}^*(\mathscr{D}, \mathrm{Hom}_k(\mathscr{F}, \mathscr{G})))$, where

$$\mathsf{D}^*(\mathsf{U}, -) : \mathsf{PreSh}(\mathsf{Mor}\ \mathsf{U}, k) \to \mathsf{Compl}(k)$$

is the resolving complex of the projective limit functor; see Laudal [4] for details. We denote the global Hochschild complex by $\mathsf{HC}^*(\mathsf{U}, \mathscr{D}, \mathrm{Hom}_k(\mathscr{F}, \mathscr{G}))$, and define the global Hochschild cohomology $\mathrm{HH}^n(\mathsf{U}, \mathscr{D}, \mathrm{Hom}_k(\mathscr{F}, \mathscr{G}))$ to be its cohomology.

Note that $\mathsf{H}^n(\mathsf{HC}^*(\mathscr{D}(U), \mathrm{Hom}_k(\mathscr{F}(U), \mathscr{G}(V)))) \cong \mathrm{Ext}^n_{\mathscr{D}(U)}(\mathscr{F}(U), \mathscr{G}(V))$ for any $U \supseteq V$ in U since k is a field. Hence there is a spectral sequence converging to the global Hochschild cohomology $\mathrm{HH}^n(\mathsf{U}, \mathscr{D}, \mathrm{Hom}_k(\mathscr{F}, \mathscr{G}))$ with

$$E_2^{pq} = \mathsf{H}^p(\mathsf{U}, \mathrm{Ext}^q_{\mathscr{D}}(\mathscr{F}, \mathscr{G})), \tag{10.1}$$

where $\mathsf{H}^p(\mathsf{U}, -) = \mathsf{H}^p(\mathsf{D}^*(\mathsf{U}, -))$ and $\mathrm{Ext}^q_{\mathscr{D}}(\mathscr{F}, \mathscr{G}) : \mathsf{Mor}\ \mathsf{U} \to \mathsf{Mod}(k)$ is the functor on $\mathsf{Mor}\ \mathsf{U}$ given by $\{U \supseteq V\} \mapsto \mathrm{Ext}^q_{\mathscr{D}(U)}(\mathscr{F}(U), \mathscr{G}(V))$ for all $q \geq 0$.

10.3.2 Obstruction Calculus

Let $R \in \mathsf{a}_p$ and let $I = I(R)$ be the radical of R. For any lifting $\mathscr{F}_R \in \mathsf{Def}_{\mathscr{F}}(R)$ of the family $\mathscr{F}$ in $\mathsf{PreSh}(\mathsf{U}, \mathscr{D})$ to R, we have that $\mathscr{F}_R(U) \cong (\mathscr{F}_i(U) \otimes_k R_{ij})$ as a right R-module for all $U \in \mathsf{U}$. Moreover, the lifting $\mathscr{F}_R$ is completely determined by the left multiplication of $\mathscr{D}(U)$ on $\mathscr{F}_R(U)$ for all $U \in \mathsf{U}$ and the restriction map $\mathscr{F}_R(U) \to \mathscr{F}_R(V)$ for all $U \supseteq V$ in U. Let us write $Q^R(U,V) = (\mathrm{Hom}_k(\mathscr{F}_j(U), \mathscr{F}_i(V) \otimes_k R_{ij}))$ and $Q^R(U) = Q^R(U,U)$ for all $U \supseteq V$ in U. Then $\mathscr{F}_R \in \mathsf{Def}_{\mathscr{F}}(R)$ is completely described by the following data:

1. For all $U \in \mathsf{U}$, a k-algebra homomorphism $L(U) : \mathscr{D}(U) \to Q^R(U)$ satisfying $L(U)(P)(f_j) = Pf_j \otimes e_j + (\mathscr{F}_i(U) \otimes_k I_{ij})$ for all $P \in \mathscr{D}(U)$, $f_j \in \mathscr{F}_j(U)$.
2. For all inclusions $U \supseteq V$ in U, a restriction map $L(U,V) \in Q^R(U,V)$ satisfying $L(U,V)(f_j) = (f_j|_V) \otimes e_j + (\mathscr{F}_i(V) \otimes I_{ij})$ for all $f_j \in \mathscr{F}_j(U)$ and $L(U,V) \circ L(U)(P) = L(V)(P|_V) \circ L(U,V)$ for all $P \in \mathscr{D}(U)$.
3. For all inclusions $U \supseteq V \supseteq W$ in U, $L(V,W)L(U,V) = L(U,W)$ and $L(U,U) = \mathrm{id}$.

A *small surjection* in a_p is a surjective morphism $u : R \to S$ in a_p such that $IK = KI = 0$, where $K = \mathrm{Ker}(u)$ and $I = I(R)$ is the radical of R. To describe the obstruction calculus of $\mathsf{Def}_{\mathscr{F}}$, it is enough to consider the following problem: Given a small surjection $u : R \to S$ and a deformation $\mathscr{F}_S \in \mathsf{Def}_{\mathscr{F}}(S)$, what are the possible liftings of $\mathscr{F}_S$ to R? The answer is given by the following proposition; see Eriksen [3] for details:

Proposition 10.1. *Let $u : R \to S$ be a small surjection in a_p with kernel K, and let $\mathscr{F}_S \in \mathsf{Def}_{\mathscr{F}}(S)$ be a deformation. Then there exists a canonical obstruction*

$$o(u, \mathscr{F}_S) \in (\mathrm{HH}^2(\mathsf{U}, \mathscr{D}, \mathrm{Hom}_k(\mathscr{F}_j, \mathscr{F}_i)) \otimes_k K_{ij})$$

such that $o(u, \mathscr{F}_S) = 0$ if and only if there is a deformation $\mathscr{F}_R \in \mathsf{Def}_{\mathscr{F}}(R)$ lifting $\mathscr{F}_S$ to R. Moreover, if $o(u, \mathscr{F}_S) = 0$, then there is a transitive and effective action of $(\mathrm{HH}^1(\mathsf{U}, \mathscr{D}, \mathrm{Hom}_k(\mathscr{F}_j, \mathscr{F}_i)) \otimes_k K_{ij})$ on the set of liftings of $\mathscr{F}_S$ to R.

In fact, let $\mathscr{F}_S \in \mathsf{Def}_{\mathscr{F}}(S)$ be given by $L^S(U) : \mathscr{D}(U) \to Q^S(U)$ and $L^S(U,V) \in Q^S(U,V)$ for all $U \supseteq V$ in U, and let $\sigma : S \to R$ be a k-linear section of $u : R \to S$ such that $\sigma(e_i) = e_i$ and $\sigma(S_{ij}) \subseteq R_{ij}$ for $1 \le i,j \le p$. We consider $L^R(U) : \mathscr{D}(U) \to Q^R(U)$ given by $L^R(U) = \sigma \circ L^S(U)$ and $L^R(U,V) = \sigma(L^S(U,V))$ for all $U \supseteq V$ in U. The obstruction $o(U, \mathscr{F}_S)$ for lifting $\mathscr{F}_S$ to R is given by

1. $(P,Q) \mapsto L^R(U)(PQ) - L^R(U)(P) \circ L^R(U)(Q)$ for all $U \in \mathsf{U}$ and $P, Q \in \mathscr{D}(U)$
2. $P \mapsto L^R(U,V) \circ L^R(U)(P) - L^R(V)(P|_V) \circ L^R(U,V)$ for all $U \supseteq V$ in U, $P \in \mathscr{D}(U)$
3. $L^R(V,W) \circ L^R(U,V) - L^R(U,W)$ for all $U \supseteq V \supseteq W$ in U

We see that these expressions are exactly the obstructions for $L^R(U)$ and $L^R(U,V)$ to satisfy conditions (1)–(3) in the characterization of $\mathsf{Def}_{\mathscr{F}}(R)$ given above.

10.4 Calculations for D-Modules on Elliptic Curves

Let $X \subseteq \mathbf{P}^2$ be the irreducible projective plane curve given by the homogeneous equation $f = 0$, where $f = y^2z - x^3 - axz^2 - bz^3$ for fixed parameters $(a,b) \in k^2$. We assume that $\Delta = 4a^3 + 27b^2 \neq 0$, so that X is smooth and therefore an elliptic curve over k. We shall compute the noncommutative deformations of $\mathscr{O}_X$, considered as a quasi-coherent left $\mathscr{D}_X$-module via the natural left action of $\mathscr{D}_X$ on $\mathscr{O}_X$.

We choose an open affine cover $\mathsf{U} = \{U_1, U_2, U_3\}$ of X closed under intersections, given by $U_1 = D_+(y)$, $U_2 = D_+(z)$ and $U_3 = U_1 \cap U_2$. We recall that the open subset $D_+(h) \subseteq X$ is given by $D_+(h) = \{p \in X : h(p) \neq 0\}$ for $h = y$ or $h = z$. It follows from the results in Sect. 10.3 that the noncommutative deformation functor $\mathsf{Def}_{\mathscr{O}_X} : \mathsf{a}_1 \to \mathsf{Sets}$ has a pro-representing hull H, completely determined by the global Hochschild cohomology groups and some generalized Massey products on them. We shall therefore compute $\mathrm{HH}^n(\mathsf{U}, \mathscr{D}, \mathrm{End}_k(\mathscr{O}_X))$ for $n = 1, 2$.

It is known that $\mathscr{D}_X(U)$ is a simple Noetherian ring of global dimension one and that $\mathscr{O}_X(U)$ is a simple left $\mathscr{D}_X(U)$-module for any open affine subset $U \subseteq X$; see for instance Smith and Stafford [6]. Hence, the functor $\mathrm{Ext}^q_{\mathscr{D}_X}(\mathscr{O}_X, \mathscr{O}_X) : \mathsf{Mor}\,\mathsf{U} \to \mathsf{Mod}(k)$ satisfies $\mathrm{Ext}^q_{\mathscr{D}_X}(\mathscr{O}_X, \mathscr{O}_X) = 0$ for $q \geq 2$ and $\mathrm{End}_{\mathscr{D}_X}(\mathscr{O}_X) = k$. Since the spectral sequence for global Hochschild cohomology given in Sect. 10.3 degenerates,

$$\mathrm{HH}^n(\mathsf{U}, \mathscr{D}_X, \mathrm{End}_k(\mathscr{O}_X)) \cong \mathsf{H}^{n-1}(\mathsf{U}, \mathrm{Ext}^1_{\mathscr{D}_X}(\mathscr{O}_X, \mathscr{O}_X)) \text{ for } n \geq 1$$
$$\mathrm{HH}^0(\mathsf{U}, \mathscr{D}_X, \mathrm{End}_k(\mathscr{O}_X)) \cong k$$

We compute $\mathrm{Ext}^1_{\mathscr{D}_X}(\mathscr{O}_X, \mathscr{O}_X)$ and use the result to find $\mathsf{H}^{n-1}(\mathsf{U}, \mathrm{Ext}^1_{\mathscr{D}_X}(\mathscr{O}_X, \mathscr{O}_X))$ for $n = 1, 2$.

Let $A_i = \mathscr{O}_X(U_i)$ and $D_i = \mathscr{D}_X(U_i)$ for $i = 1, 2, 3$. We see that $A_1 \cong k[x,z]/(f_1)$ and $A_2 \cong k[x,y]/(f_2)$, where $f_1 = z - x^3 - axz^2 - bz^3$ and $f_2 = y^2 - x^3 - ax - b$. Moreover, we have that $\mathrm{Der}_k(A_i) = A_i \partial_i$ and $D_i = A_i\langle \partial_i \rangle$ for $i = 1, 2$, where

$$\partial_1 = (1 - 2axz - 3bz^2)\, \partial/\partial x + (3x^2 + az^2)\, \partial/\partial z$$
$$\partial_2 = -2y\, \partial/\partial x - (3x^2 + a)\, \partial/\partial y$$

We choose an isomorphism $A_3 \cong k[x, y, y^{-1}]/(f_3)$ with $f_3 = f_2$ on the intersection $U_3 = U_1 \cap U_2$, and see that $\mathrm{Der}_k(A_3) = A_3 \partial_3$ and $D_3 = A_3\langle \partial_3 \rangle$ for $\partial_3 = \partial_2$. The restriction maps of $\mathscr{O}_X$ and $\mathscr{D}_X$, considered as presheaves on U, are given by

$$x \mapsto xy^{-1},\ z \mapsto y^{-1},\ \partial_1 \mapsto \partial_3$$

for the inclusion $U_1 \supseteq U_3$, and the natural localization map for $U_2 \supseteq U_3$. Finally, we find a free resolution of A_i as a left D_i-module for $i = 1, 2, 3$, given by

$$0 \leftarrow A_i \leftarrow D_i \xleftarrow{\cdot \partial_i} D_i \leftarrow 0$$

and use this to compute $\mathrm{Ext}^1_{D_i}(A_i, A_j) \cong \mathrm{coker}(\partial_i|_{U_j} : A_j \to A_j)$ for $U_i \supseteq U_j$ in U. We see that $\mathrm{Ext}^1_{D_i}(A_i, A_3) \cong \mathrm{coker}(\partial_3 : A_3 \to A_3)$ is independent of i, and find the following k-linear bases for $\mathrm{Ext}^1_{D_i}(A_i, A_j)$:

	$a \neq 0$:	$a = 0$:
$U_1 \supseteq U_1$	$1, z, z^2, z^3$	$1, z, x, xz$
$U_2 \supseteq U_2$	$1, y^2$	$1, x$
$U_3 \supseteq U_3$	$x^2y^{-1}, 1, y^{-1}, y^{-2}, y^{-3}$	$x^2y^{-1}, 1, y^{-1}, x, xy^{-1}$

The functor $\mathrm{Ext}^1_{\mathscr{D}_X}(\mathscr{O}_X,\mathscr{O}_X) : \mathsf{Mor}\,\mathsf{U} \to \mathsf{Mod}(k)$ defines the diagram in $\mathsf{Mod}(k)$ below, where the maps are induced by the restriction maps on $\mathscr{O}_X$:

$$\begin{array}{ccccc} \mathrm{Ext}^1_{D_1}(A_1,A_1) & & & & \mathrm{Ext}^1_{D_2}(A_2,A_2) \\ \downarrow & & & & \downarrow \\ \mathrm{Ext}^1_{D_1}(A_1,A_3) & = & \mathrm{Ext}^1_{D_3}(A_3,A_3) & = & \mathrm{Ext}^1_{D_2}(A_2,A_3) \end{array}$$

We use that $15\,y^2 = \Delta\, y^{-2}$ in $\mathrm{Ext}^1_{D_3}(A_3,A_3)$ when $a \neq 0$ and that $-3b\,xy^{-2} = x$ in $\mathrm{Ext}^1_{D_3}(A_3,A_3)$ when $a = 0$ to describe these maps in the given bases. We compute $\mathsf{H}^{n-1}(\mathsf{U},\mathrm{Ext}^1_{\mathscr{D}_X}(\mathscr{O}_X,\mathscr{O}_X))$ for $n = 1,2$ using the resolving complex $\mathsf{D}^*(\mathsf{U},-)$; see Laudal [4] for definitions. In particular, we identify $\mathsf{H}^0(\mathsf{U},\mathrm{Ext}^1_{\mathscr{D}_X}(\mathscr{O}_X,\mathscr{O}_X))$ with the set of all pairs (h_1,h_2) with $h_i \in \mathrm{Ext}^1_{D_i}(A_i,A_i)$ that satisfies $h_1|_{U_3} = h_2|_{U_3}$, and we identify $\mathsf{H}^1(\mathsf{U},\mathrm{Ext}^1_{\mathscr{D}_X}(\mathscr{O}_X,\mathscr{O}_X))$ with the set of all pairs (h_{13},h_{23}) with $h_{ij} \in \mathrm{Ext}^1_{D_i}(A_i,A_j)$, modulo the pairs of the form $(h_1|_{U_3} - h_3, h_2|_{U_3} - h_3)$ for triples (h_1,h_2,h_3) with $h_i \in \mathrm{Ext}^1_{D_i}(A_i,A_i)$. We find the following k-linear bases:

	$a \neq 0:$	$a = 0:$
$n = 1$	$\xi_1 = (1,1,1)$, $\xi_2 = (\Delta z^2, 15y^2, \Delta y^{-2})$	$\xi_1 = (1,1,1)$, $\xi_2 = (-3b\,xz, x, x)$
$n = 2$	$\omega = (0,0,0,0,6ax^2y^{-1})$	$\omega = (0,0,0,0,x^2y^{-1})$

We recall that ξ_1,ξ_2 and ω are represented by cocycles of degree $p = 0$ and $p = 1$ in the resolving complex $\mathsf{D}^*(\mathsf{U},\mathrm{Ext}^1_{\mathscr{D}_X}(\mathscr{O}_X,\mathscr{O}_X))$, where

$$\mathsf{D}^p(\mathsf{U},\mathrm{Ext}^1_{\mathscr{D}_X}(\mathscr{O}_X,\mathscr{O}_X)) = \prod_{U_0 \supseteq \cdots \supseteq U_p} \mathrm{Ext}^1_{\mathscr{D}_X}(\mathscr{O}_X,\mathscr{O}_X)(U_0 \supseteq U_p)$$

and the product is indexed by the set $\{U_1 \supseteq U_1, U_2 \supseteq U_2, U_3 \supseteq U_3\}$ when $p = 0$, and $\{U_1 \supseteq U_1, U_2 \supseteq U_2, U_3 \supseteq U_3, U_1 \supseteq U_3, U_2 \supseteq U_3\}$ when $p = 1$.

This shows that for any elliptic curve X over k, the noncommutative deformation functor $\mathsf{Def}_{\mathscr{O}_X} : \mathsf{a}_1 \to \mathsf{Sets}$ of the left $\mathscr{D}_X$-module $\mathscr{O}_X$ has a two-dimensional tangent space (since $\mathrm{HH}^1(\mathsf{U},\mathscr{D}_X,\mathrm{End}_k(\mathscr{O}_X)) \cong k^2$), and a one-dimensional obstruction space (since $\mathrm{HH}^2(\mathsf{U},\mathscr{D}_X,\mathrm{End}_k(\mathscr{O}_X)) \cong k$), and a pro-representing hull $H = k\langle\langle t_1,t_2\rangle\rangle/(F)$ for some noncommutative power series $F \in k\langle\langle t_1,t_2\rangle\rangle$.

We shall compute the noncommutative power series F and the versal family $\mathscr{F}_H \in \mathsf{Def}_{\mathscr{O}_X}(H)$ using the obstruction calculus for $\mathsf{Def}_{\mathscr{O}_X}$. We choose base vectors t_1^*, t_2^* in $\mathrm{HH}^1(\mathsf{U},\mathscr{D},\mathrm{Hom}_k(\mathscr{O}_X,\mathscr{O}_X))$, and representatives $(\psi_l,\tau_l) \in \mathsf{D}^{01} \oplus \mathsf{D}^{10}$ of t_l^* for $l = 1,2$, where $\mathsf{D}^{pq} = \mathsf{D}^p(\mathsf{U},\mathsf{HC}^q(\mathscr{D},\mathrm{End}_k(\mathscr{O}_X)))$. We may choose $\psi_l(U_i)$ to be the derivation defined by

$$\psi_l(U_i)(P_i) = \begin{cases} 0 & \text{if } P_i \in A_i \\ \xi_l(U_i) \cdot \mathrm{id}_{A_i} & \text{if } P_i = \partial_i \end{cases}$$

for $l = 1,2$ and $i = 1,2,3$, and $\tau_l(U_i \supseteq U_j)$ to be the multiplication operator in $\mathrm{Hom}_{A_i}(A_i, A_j) \cong A_j$ given by $\tau_1 = 0$, $\tau_2(U_i \supseteq U_i) = 0$ for $i = 1,2,3$ and

$a \neq 0$:	$a = 0$:
$\tau_2(U_1 \supseteq U_3) = 0$	$\tau_2(U_1 \supseteq U_3) = x^2y^{-1}$
$\tau_2(U_2 \supseteq U_3) = -4a^2y^{-1} - 3xy + 9bxy^{-1} - 6ax^2y^{-1}$	$\tau_2(U_2 \supseteq U_3) = 0$

Let $\mathsf{a}_1(n)$ be the full subcategory of a_1 consisting of all R such that $I(R)^n = 0$ for $n \geq 2$. The restriction of $\mathsf{Def}_{\mathscr{O}_X} : \mathsf{a}_1 \to \mathsf{Sets}$ to $\mathsf{a}_1(2)$ is represented by $(H_2, \mathscr{F}_{H_2})$, where $H_2 = k\langle t_1, t_2 \rangle/(t_1, t_2)^2$ and the deformation $\mathscr{F}_{H_2} \in \mathsf{Def}_{\mathscr{O}_X}(H_2)$ is defined by $\mathscr{F}_{H_2}(U_i) = A_i \otimes_k H_2$ as a right H_2-module for $i = 1,2,3$, with left D_i-module structure given by

$$P_i(m_i \otimes 1) = P_i(m_i) \otimes 1 + \psi_1(U_i)(P_i)(m_i) \otimes t_1 + \psi_2(U_i)(P_i)(m_i) \otimes t_2$$

for $i = 1,2,3$ and for all $P_i \in D_i$, $m_i \in A_i$, and with restriction map for the inclusion $U_i \supseteq U_j$ given by

$$m_i \otimes 1 \mapsto m_i|_{U_j} \otimes 1 + \tau_2(U_i \supseteq U_j)\, m_i|_{U_j} \otimes t_2$$

for $i = 1,2$, $j = 3$ and for all $m_i \in A_i$.

We try to lift the family $\mathscr{F}_{H_2} \in \mathsf{Def}_{\mathscr{O}_X}(H_2)$ to $R = k\langle\langle t_1, t_2 \rangle\rangle/(t_1, t_2)^3$. We let $\mathscr{F}_R(U_i) = A_i \otimes_k R$ as a right R-module for $i = 1,2,3$, with left D_i-module structure given by

$$P_i(m_i \otimes 1) = P_i(m_i) \otimes 1 + \psi_1(U_i)(P_i)(m_i) \otimes t_1 + \psi_2(U_i)(P_i)(m_i) \otimes t_2$$

for $i = 1,2,3$ and for all $P_i \in D_i$, $m_i \in A_i$, and with restriction map for the inclusion $U_i \supseteq U_j$ given by

$$m_i \otimes 1 \mapsto m_i|_{U_j} \otimes 1 + \tau_2(U_i \supseteq U_j)\, m_i|_{U_j} \otimes t_2 + \frac{\tau_2(U_i \supseteq U_j)^2}{2}\, m_i|_{U_j} \otimes t_2^2$$

for $i = 1,2$, $j = 3$ and for all $m_i \in A_i$. We see that $\mathscr{F}_R(U_i)$ is a left $\mathscr{D}_X(U_i)$-module for $i = 1,2,3$, and that $t_1t_2 - t_2t_1 = 0$ is a necessary and sufficient condition for $\mathscr{D}_X$-linearity of the restriction maps for the inclusions $U_1 \supseteq U_3$ and $U_2 \supseteq U_3$. This implies that $\mathscr{F}_R$ is not a lifting of $\mathscr{F}_{H_2}$ to R. But if we define the quotient $H_3 = R/(t_1t_2 - t_2t_1)$, we see that the family $\mathscr{F}_{H_3} \in \mathsf{Def}_{\mathscr{O}_X}(H_3)$ induced by $\mathscr{F}_R$ is a lifting of $\mathscr{F}_{H_2}$ to H_3.

In fact, we claim that the restriction of $\mathsf{Def}_{\mathscr{O}_X} : \mathsf{a}_1 \to \mathsf{Sets}$ to $\mathsf{a}_1(3)$ is represented by $(H_3, \mathscr{F}_{H_3})$. One way to prove this is to show that it is not possible to find any lifting $\mathscr{F}'_R \in \mathsf{Def}_{\mathscr{O}_X}(R)$ of $\mathscr{F}_{H_2}$ to R. Another approach is to calculate the cup products $< t_i^*, t_j^* >$ in global Hochschild cohomology for $i, j = 1,2$, and this gives

$$< t_1^*, t_2^* >= o^*,\ < t_2^*, t_1^* >= -o^* \qquad \text{for } a \neq 0$$
$$< t_1^*, t_2^* >= o^*,\ < t_2^*, t_1^* >= -o^* \qquad \text{for } a = 0$$

where $o^* \in \mathrm{HH}^2(\mathsf{U}, \mathscr{D}_X, \mathrm{End}_k(\mathscr{O}_X))$ is the base vector corresponding to ω. This implies that $F = t_1t_2 - t_2t_1 + (t_1, t_2)^3$, since all other cup products vanish.

Let $H = k\langle\langle t_1, t_2\rangle\rangle/(t_1t_2 - t_2t_1)$. We shall show that it is possible to find a lifting $\mathscr{F}_H \in \mathsf{Def}_{\mathscr{O}_X}(H)$ of $\mathscr{F}_{H_3}$ to H. We let $\mathscr{F}_H(U_i) = A_i \widehat{\otimes}_k H$ as a right H-module for $i = 1, 2, 3$, with left D_i-module structure given by

$$P_i(m_i \otimes 1) = P_i(m_i) \otimes 1 + \psi_1(U_i)(P_i)(m_i) \otimes t_1 + \psi_2(U_i)(P_i)(m_i) \otimes t_2$$

for $i = 1, 2, 3$ and for all $P_i \in \mathscr{D}_i$, $m_i \in A_i$, and with restriction map for the inclusion $U_i \supseteq U_j$ given by

$$m_i \otimes 1 \mapsto \sum_{n=0}^{\infty} \frac{\tau_2(U_i \supseteq U_j)^n}{n!} m_1|_{U_j} \otimes t_2^n = \exp(\tau_2(U_i \supseteq U_j) \otimes t_2) \cdot (m_1|_{U_j} \otimes 1)$$

for $i = 1, 2$, $j = 3$ and for all $m_i \in A_i$. This implies that $(H, \mathscr{F}_H)$ is the pro-representing hull of $\mathsf{Def}_{\mathscr{O}_X}$, and that $F = t_1t_2 - t_2t_1$. We remark that the versal family $\mathscr{F}_H$ does not admit an algebraization, i.e. an algebra H_{alg} of finite type over k such that H is a completion of H_{alg}, together with a deformation in $\mathsf{Def}_{\mathscr{O}_X}(H_{\mathrm{alg}})$ that induces the versal family $\mathscr{F}_H \in \mathsf{Def}_{\mathscr{O}_X}(H)$.

Finally, we mention that there is an algorithm for computing the pro-representing hull H and the versal family $\mathscr{F}_H$ using the cup products and higher generalized Massey products on global Hochschild cohomology. We shall describe this algorithm in a forthcoming paper. In many situations, it is necessary to use the full power of this machinery to compute noncommutative deformation functors.

References

1. M. F. Atiyah, *Complex analytic connections in fibre bundles*, Trans. Am. Math. Soc. **85** (1957), 181–207.
2. A. Beilinson and J. Bernstein, *A proof of the Jantzen conjectures*, Adv. Soviet Math. **16** (1993), no. 1, 1–50.
3. E. Eriksen, *Noncommutative deformations of sheaves and presheaves of modules*, ArXiv: math.AG/0405234, 2005.
4. O. A. Laudal, *Formal moduli of algebraic structures*, Lecture Notes in Mathematics, vol. 754, Springer, Berlin, 1979.
5. O. A. Laudal, *Noncommutative deformations of modules*, Homology Homotopy Appl. **4** (2002), no. 2, part 2, 357–396.
6. S. Paul Smith and J. T. Stafford, *Differential operators on an affine curve*, Proc. London Math. Soc. (3) **56** (1988), no. 2, 229–259.
7. A. Weil, *Généralisation des fonctions abéliennes*, J. Math. Pure Appl. **17** (1938), 47–87.

Chapter 11
Comparing Small Orthogonal Classes

Gabriella D'Este

Abstract We investigate reasonably large partial tilting or cotilting modules, obtained after the cancellation of suitable direct summands.

11.1 Introduction

The aim of this paper is to describe some "discrete" and "combinatorial" results on modules with very special homological properties (i.e. of the so-called "large partial n-tilting" and "large partial n-cotilting" modules). These modules are "approximations" of the tilting modules (and of their "duals"), introduced by S. Brenner and M. C. R. Butler in their famous paper [6] , entitled "Generalizations of the Bernstein–Gelfand–Ponomarev reflection functors". Hence a word coming from Physics, namely the word "reflection", appears in the title of the paper which is at the origin of Tilting Theory. Both examples and proofs contained in this note come from finite dimensional algebras, the context where "tilting modules were born" (see `www.mathematik.uni-muenchen.de/tilting` for the announcement of the Tilting Tagung "Twenty years of Tilting Theory, an Interdisciplinary Symposium", November 2002,Raueninseln, Germany). As we shall see, small algebras of finite representation type, together with the most important "combinatorial" and "geometrical" objects usually associated to these algebras, that is their Auslander–Reiten quivers, are the basic tools used in the sequel. In some sense, we often apply a kind of "cancellation" strategy. For instance, Theorems 11.3 and 11.4 point out the following unexpected fact: our "approximations" of tilting and cotilting modules do not admit obvious direct summands. Indeed, under suitable hypotheses on certain "orthogonal" classes, the indecomposable projective–injective modules (i.e. the "obvious" indecomposable direct summands of all "n-tilting" or "n-cotilting"

G. D'Este
Dipartimento di Matematica, Università di Milano, Via Saldini 50, 20133 Milan, Italy
e-mail: gabriella.deste@mat.unimi.it

S. Silvestrov et al. (eds.), *Generalized Lie Theory in Mathematics, Physics and Beyond,*

modules) become "inessential" direct summands of our "large partial n-tilting or n-cotilting" modules.

We refer to the Handbook of Tilting Theory [2], for the important role that Tilting Theory plays in various areas of mathematics.

In particular, we refer to [18] and [21] for the connection between Tilting Theory and Representation Theory (and/or Module Theory). We refer to [13, 16, 17, 20] and [19] for other, more or less combinatorial, aspects of several algebraic and/or tilting-type objects.

This paper is organized as follows. In Sect. 11.2 we fix the notation and we recall the more discrete and combinatorial definitions needed in the paper. In Sect. 11.3, we collect proofs and examples.

11.2 Preliminaries

Let R be a ring. We denote by $R-Mod$ the category of all left R-modules. If $M \in R-Mod$, then we write $\mathrm{Add}M$ (resp. $\mathrm{Prod}M$) for the class of all modules isomorphic to direct summands of direct sums (resp. direct products) of copies of M. Next, for every cardinal λ, we denote by $M^{(\lambda)}$ (resp. M^{λ}) the direct sum (resp. direct product) of λ copies of M. Moreover, for any $n \geq 1$, we denote by $\mathrm{Gen}_n(M)$ (resp. $\mathrm{Cogen}_n(M)$) the class of all modules X such that there is an exact sequence of the form $M^{(\alpha_n)} \to \cdots \to M^{(\alpha_2)} \to M^{(\alpha_1)} \to X \to 0$ (resp. $0 \to X \to M^{\alpha_1} \to M^{\alpha_2} \to \cdots \to M^{\alpha_n}$) for some cardinals $\alpha_1, \ldots, \alpha_n$. Finally, we write $M^{\perp_\infty}$ (resp. ${}^{\perp_\infty}M$) for the following orthogonal class:

$$M^{\perp_\infty} = \{X \in R-Mod \,|\, \mathrm{Ext}^i_R(M,X) = 0 \text{ for every } i \geq 1\}$$

$$(\text{resp. } {}^{\perp_\infty}M = \{X \in R-Mod \,|\, \mathrm{Ext}^i_R(X,M) = 0 \text{ for every } i \geq 1\}).$$

We shall say that an R-module T is a *partial n-tilting module* if the following conditions hold:

- The projective dimension of T is at most n.
- $\mathrm{Ext}^i_R(T,T^{(\lambda)}) = 0$ for every $i \geq 1$ and every cardinal λ.

Given a partial n-tilting module T, we shall say that T is an *n-tilting module* if there is a long exact sequence of the form

$$0 \to R \to T_0 \to T_1 \to \cdots \to T_n \to 0,$$

where $T_i \in \mathrm{Add}T$ for every $i = 0, \ldots, n$. From now on, we shall say, for brevity, that a partial n-tilting module is a *large partial n-tilting module* if

$$\mathrm{Ker}\,\mathrm{Hom}(T,-) \cap T^{\perp_\infty} = 0.$$

Dually, we say that an R-module C is a *partial n-cotilting module* if the following conditions hold:

- The injective dimension of C is at most n.
- $\mathrm{Ext}^i_R(C^\lambda, C) = 0$ for every $i \geq 1$ and every cardinal λ.

Given a partial n-cotilting module C, we shall say that C is an *n-cotilting module* if there is a long exact sequence of the form

$$0 \to C_n \to \cdots \to C_1 \to C_0 \to E \to 0,$$

where E is an injective cogenerator of $R-Mod$ and $C_i \in \mathrm{Prod}\, C$ for every $i = 0,\ldots,n$. In the sequel, we shall say, for brevity, that a partial n-cotilting module is a *large partial n-cotilting module* if

$$\mathrm{Ker}\,\mathrm{Hom}(-,C) \cap {}^{\perp_\infty}C = 0.$$

Maintaining the terminology introduced above, we recall some properties of large partial n-tilting and n-cotilting modules, both in the "classical case" (that is where $n = 1$), and in the "non-classical case" (that is when $n > 1$).

- For every $n \geq 1$, every n-tilting module T (resp. n-cotilting module C) is a large partial n-tilting (resp. n-cotilting) module [5, p. 371].
- A finitely generated module T is a 1-tilting module if and only if T is a large partial 1-tilting module [8, Theorem 1] (also see [7, Theorem 2.3.1 and Sect. 3.1]).
- A module C is a 1-cotilting module if and only if C is a large partial 1-cotilting module ([3, Proposition 2.3], [9, Theorem 1.7], [10] and [7, Sect. 2.5]).
- Every $n \geq 2$ is the projective (resp. injective) dimension of a non faithful large partial n-tilting (resp. n-cotilting) module [12, Example 4].

It is easy to see that faithful large partial 2-tilting (resp. large partial 2-cotilting) modules are not necessarily 2-tilting (resp. 2-cotilting) modules [11, Corollary 6]. On the other hand, for every large partial n-tilting module T, we clearly have:

- $\mathrm{Hom}(T,I) \neq 0$ for every non-zero injective module I.

Dually, for every large partial n-cotilting module C, we obviously have:

- $\mathrm{Hom}(P,C) \neq 0$ for every non-zero projective module P.

Consequently, if there are only finitely many simple modules, then every large partial tilting (resp. cotilting) module M of finite length is *sincere* [4], that is every simple module appears as a composition factor of M.

The first examples of properly large partial n-tilting and n-cotilting modules (constructed in [11] and [12], and defined over algebras of finite representation type) justify the feeling that these modules exist in nature. We also note that the construction of these modules makes use only of their "discrete" definitions, recalled at the beginning of this section. Moreover, the direct construction of these modules leads to a "theoretic" explanation (Theorems 11.3 and 11.4) of their existence in many different cases (Examples 11.5 and 11.6). More precisely, both "discrete" and "continuous" properties (concerning projective–injective direct summands, and left or right orthogonal classes) show up in some sufficient (but not necessary) conditions, which guarantee the presence of as many as possible large partial n-tilting or n-cotilting direct summands. (compare Theorems 11.3 and 11.4 with Example 11.7).

We also show that a large partial 2-tilting (resp. 2-cotilting) module, defined over an algebra of infinite representation type, may be a faithful module over an algebra of finite representation type (Proposition 11.8).

In the sequel K always denotes an algebraically closed field, and we always identify modules with their isomorphism classes. Moreover, if Λ is a K-algebra given by a quiver and relations, according to [15], then we often replace indecomposable finite dimensional modules by some obvious pictures, describing their composition factor. Over a representation-finite algebra given by a quiver, we often denote by x the simple module $S(x)$, corresponding to the vertex x, while we denote by $P(x)$ and $I(x)$ the projective cover and the injective envelope of $S(x)$, respectively. Finally, given an uniserial module M of finite length, the symbols $\operatorname{soc} M$ and $\operatorname{rad} M$ denote the simple socle of M and its maximal submodule respectively.

For unexplained terminology, we refer to [1] and [4].

11.3 Proofs and Examples

In the sequel we shall often use the following preliminary result.

Lemma 11.1. *Let A be a representation-finite K-algebra of finite global dimension m. The following facts hold:*

(a) The regular module ${}_AA$ is an n-cotilting module for some $n \le m$.
(b) The injective cogenerator ${}_AD = \operatorname{Hom}_K(A_A, K)$ is an n-tilting module for some $n \le m$.

Proof. See [11, Lemmas 1 and 2]. □

Given a finite dimensional K-algebra A, a simple module S and a module M of finite length, it is well-known [15, p. 68] that the multiplicity of S (as a composition factor of M) coincides with the K-dimension of the vector spaces $\operatorname{Hom}_A(P,M)$ and $\operatorname{Hom}_A(M,I)$, where P and I are the projective cover and the injective envelope of S respectively. In the next lemma we collect the relationship between some Hom groups used in the following.

Lemma 11.2. *Let R be a semiperfect ring, let S be a simple R-module, and let P (resp. I) be the projective cover (resp. injective envelope) of S. The following facts hold for every module M:*

(a) $\operatorname{Hom}_R(P,M) \neq 0$ *iff* $\operatorname{Hom}_R(M,I) \neq 0$.
(b) $\operatorname{Hom}_R(P,M) \neq 0$ *and* $\operatorname{Hom}_R(M,I) \neq 0$ *for every module M such that* $\operatorname{Ker}\operatorname{Hom}_R(M,-) \cap M^{\perp_\infty} = 0$ *(resp.* $\operatorname{Ker}\operatorname{Hom}_R(-,M) \cap {}^{\perp_\infty}M = 0$*).*

Proof. (a) Given non-zero morphism: $f : P \longrightarrow M$ (resp. $g : M \longrightarrow I$), it suffices to consider a commutative diagram of the following form

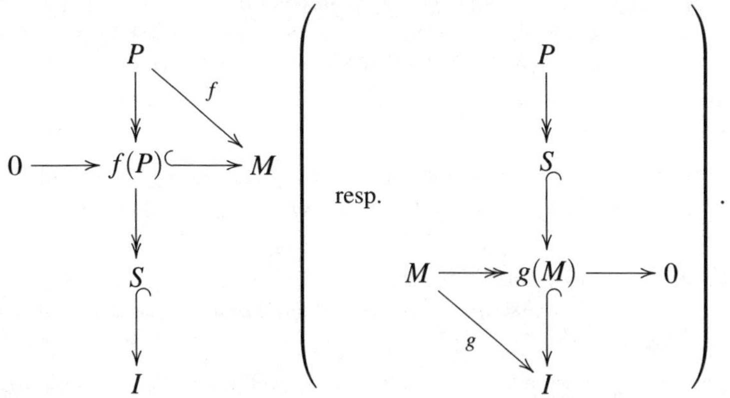

(b) Since $I \in M^{\perp_\infty}$ and $P \in {}^{\perp_\infty}M$, we have $\mathrm{Hom}_R(M,I) \neq 0$ (resp. $\mathrm{Hom}_R(P,M) \neq 0$). Hence (b) follows from (a). □

The next theorems justify the strategy, used in [12], to construct large partial n-tilting or n-cotilting modules by taking big direct summands of very special n-tilting or n-cotilting modules. We begin with a result on projective modules over arbitrary rings.

Theorem 11.3. *Let M be a projective large partial n-cotilting module of the form $M = C \oplus Q$ with the following properties:*

(a) C is sincere.
(b) Q is injective.
(c) ${}^{\perp_\infty}M$ is the class of all projective modules.

Then C is a large partial n-cotilting module.

Proof. Since C is a direct summand of M, it is enough to check that $\mathrm{Ker}\,\mathrm{Hom}(-,C) \cap {}^{\perp_\infty}C = 0$. To this end, let X be a non-zero module such that $X \in {}^{\perp_\infty}C$. Then (b) tells us that $X \in {}^{\perp_\infty}M$. By (c), this implies that X is projective. This observation and (a) guarantee that $\mathrm{Hom}(X,C) \neq 0$. The proof is complete. □

We now prove a partial dual result.

Theorem 11.4. *Let R be a noetherian and semiperfect ring such that any indecomposable injective module has a simple socle. Let M be an injective large partial n-tilting module of the form $M = T \oplus Q$ with the following properties:*

(a) T is sincere.
(b) Q is projective.
(c) $M^{\perp_\infty}$ is the class of all injective modules.

Then T is a large partial n-tilting module.

Proof. Since T is a direct summand of M, it suffices to show that $\operatorname{Ker}\operatorname{Hom}_R(T,-)\cap T^{\perp_\infty}=0$. To see this, take a non-zero module $X\in T^{\perp_\infty}$. Then we deduce from (b) that $X\in M^{\perp_\infty}$. This remark and (c) imply that X is injective. Consequently, our assumptions on R guarantee that X has an indecomposable summand I with simple socle S. Let P be the projective cover of S. Then (a) tells us that $\operatorname{Hom}_R(P,T)\neq 0$. Hence, by Lemma 11.2(a), we have $\operatorname{Hom}_R(T,I)\neq 0$. Therefore we have $\operatorname{Hom}_R(T,X)\neq 0$, and the theorem is proved. □

It is easy to see that both decomposable and indecomposable finite dimensional modules (of minimal length among all sincere modules) admit a minimal orthogonal class.

Example 11.5. *Any natural number $n\geq 2$ occurs as the projective (resp. injective) dimension of a decomposable and non faithful module T (resp. C) with the following properties:*

(a) $T^{\perp_\infty}$ *(resp.* ${}^{\perp_\infty}C$*) is the class of all injective (resp. projective) modules.*
(b) $\operatorname{Gen}_n(T)=\operatorname{Add}(T)$ *and* $\operatorname{Cogen}_n(C)=Add(C)$.
(c) T (resp. C) is a large partial n-tilting (resp. n-cotilting) module of minimal length.

Construction. Let A be the K-algebra given by the quiver

$$\underset{1}{\bullet}\xrightarrow{\alpha_1}\underset{2}{\bullet}\xrightarrow{\alpha_2}\underset{3}{\bullet}\cdots\cdots\underset{n}{\bullet}\xrightarrow{\alpha_n}\underset{n+1}{\bullet}$$

with relations $\alpha_{i+1}\alpha_i=0$ for $i=1,\dots,n-1$. Next, let T denote the injective module

$$T=\begin{cases}\begin{smallmatrix}n\\n+1\end{smallmatrix}\oplus\begin{smallmatrix}n-2\\n-1\end{smallmatrix}\oplus\cdots\oplus\begin{smallmatrix}2\\3\end{smallmatrix}\oplus 1 & \text{if } n \text{ is even}\\ \begin{smallmatrix}n\\n+1\end{smallmatrix}\oplus\begin{smallmatrix}n-1\\n\end{smallmatrix}\oplus\cdots\oplus\begin{smallmatrix}2\\3\end{smallmatrix}\oplus 1 & \text{if } n \text{ is odd}\end{cases}$$

Finally, let C denote the projective module

$$C=\begin{cases}n+1\oplus\begin{smallmatrix}n-1\\n\end{smallmatrix}\oplus\cdots\oplus\begin{smallmatrix}3\\4\end{smallmatrix}\oplus\begin{smallmatrix}1\\2\end{smallmatrix} & \text{if } n \text{ is even}\\ n+1\oplus\begin{smallmatrix}n\\n+1\end{smallmatrix}\oplus\cdots\oplus\begin{smallmatrix}3\\4\end{smallmatrix}\oplus\begin{smallmatrix}1\\2\end{smallmatrix} & \text{if } n \text{ is odd}\end{cases}$$

Then (a) and (b) clearly hold. Hence, by Theorems 11.3 and 11.4, T (resp. C) is a large partial n-tilting (resp. n-cotilting) module. On the other hand, it is easy to check [12, Example 4(iv)] that $\dim_K(T)=\dim_K(C)$ is the smallest possible dimension of any large partial n-tilting or n-cotilting module. Hence also (c) holds. □

Example 11.6. *Every even natural number $m\geq 2$ occurs as the projective (resp. injective) dimension of a uniserial non faithful module T (resp. C) with the following properties:*

(a) $T^{\perp_\infty}$ *(resp.* ${}^{\perp_\infty}C$*) is the class of all injective (resp. projective) modules.*
(b) $\mathrm{Gen}_2(T) = \mathrm{Add}(T)$ *and* $\mathrm{Cogen}_2(C) = \mathrm{Add}(C)$.
(c) T *(resp.* C*) is a large partial n-tilting (resp. n-cotilting) module of minimal length.*

Construction. Let Λ denote the Nakajama algebra, considered in [14, Example 3.2], given by the quiver

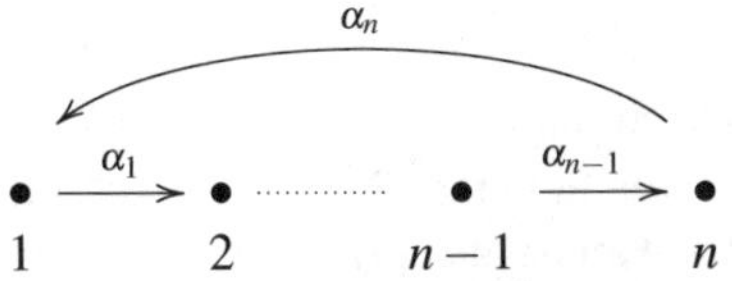

with relation $\alpha_n \cdot \cdots \cdot \alpha_1 = 0$, where $n = \frac{m+2}{2}$. Next let T and C denote the modules $I(1)$ and $P(1)$ respectively. Then (b) clearly holds, while (c) follows from [12, Example 7]. By the definition of Λ, the indecomposable projective modules are of the form

$$\begin{matrix} 1 \\ 2 \\ \vdots \\ n-1 \\ n \end{matrix}, \quad \begin{matrix} 2 \\ 3 \\ \vdots \\ 1 \\ 2 \end{matrix}, \quad \ldots, \quad \begin{matrix} n-1 \\ n \\ \vdots \\ n-2 \\ n-1 \end{matrix}, \quad \begin{matrix} n \\ 1 \\ \vdots \\ n-1 \\ n \end{matrix}.$$

Consequently, as observed in [14, Example 3.2], we have a long exact sequence (i.e., the sequence of complements of the almost complete tilting module $P(2) \oplus \cdots \oplus P(n)$) of the following form:

$$(1) \qquad \begin{aligned} 0 \to C = P(1) \to P(n) \to P(n) \to \ldots \\ \cdots \to P(2) \to P(2) \to T = I(1) \to 0. \end{aligned}$$

Therefore, by dimension shifting, we obtain

(2) $\mathrm{Ext}^1_\Lambda(S(i), -) \simeq \mathrm{Ext}^{2i-2}_\Lambda(T, -)$ for every $i = 2, \ldots, n$;
(3) $\mathrm{Ext}^1_\Lambda(-, S(i)) \simeq \mathrm{Ext}^{2n+2-2i}_\Lambda(-, C)$ for every $i = 2, \ldots, n$;
(4) $\mathrm{Ext}^1_\Lambda(P(i)/S(i), -) \simeq \mathrm{Ext}^{2i-3}_\Lambda(T, -)$ for every $i = 2, \ldots, n$;
(5) $\mathrm{Ext}^1_\Lambda(-, \mathrm{rad}\, P(i)) \simeq \mathrm{Ext}^{2n+1-2i}_\Lambda(-, C)$ for every $i = 2, \ldots, n$.

Let X be an indecomposable module of the form $\begin{matrix} a \\ \vdots \\ b \end{matrix}$. Then we obviously have

(6) $\mathrm{Ext}^1_\Lambda(S(n), X) \neq 0$ if $a = 1$, and $\mathrm{Ext}^1_\Lambda(S(a-1), X) \neq 0$ if X is not injective and $a > 2$.
(7) $\mathrm{Ext}^1_\Lambda(X, S(b+1)) \neq 0$ if X is not projective and $b \neq n$.

Assume now X is not injective and $a = 2$. Then we have $b \neq 1$, and there is a non splitting exact sequence of the form

(8) $0 \to X \to P(b) \oplus X' \to P(b)/S(b) \to 0$,

where $X' = 0$ if X is simple and $X' = X/\operatorname{soc} X$ otherwise. Finally, suppose X is not projective and $b = n$. Then we have $a \neq 1$, and there is a non splitting exact sequence of the form

(9) $0 \to \operatorname{rad} P(a) \to P(a) \oplus X' \to X \to 0$,

where $X' = 0$ if X is simple and $X' = \operatorname{rad} X$ otherwise. Putting (6), ..., (9) and (2), ..., (5) together, we conclude that $T^{\perp_\infty}$ (resp. ${}^{\perp_\infty}C$) is the class of all injective (resp. projective) modules. Hence also (a) holds. □

As the next example shows, large partial n-cotilting (resp. n-tilting) modules, such that many direct summands have the same property, are not necessarily projective (resp. injective). Consequently, they do not necessarily satisfy condition (c) of Theorem 11.3 (resp. 11.4).

Example 11.7. *There exist a 2-cotilting module M and a 2-tilting module N with the following properties:*

(a) M is not projective and N is not injective.
(b) Every sincere direct summand of M (resp. N) with an injective (resp. a projective) complement is a large partial 2-cotilting (resp. 2-tilting) module.

Construction. Let Λ be the K-algebra given by a quiver of the form

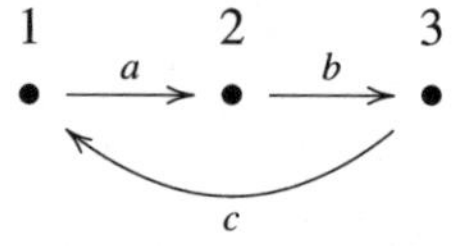

with relations $ac = 0$ and $cb = 0$. Next, let M and N denote the following modules:

$$M = \begin{smallmatrix} 1 \\ 2 \\ 3 \end{smallmatrix} \oplus \begin{smallmatrix} 3 \\ 1 \end{smallmatrix} \oplus 1\,, \qquad N = \begin{smallmatrix} 1 \\ 2 \\ 3 \end{smallmatrix} \oplus \begin{smallmatrix} 3 \\ 1 \end{smallmatrix} \oplus 3\,.$$

Then M is a 2-cotilting module of projective dimension 1, while N is a 2-tilting module of injective dimension 1. Hence (a) holds. Finally, let C and T denote the modules $\begin{smallmatrix} 1 \\ 2 \\ 3 \end{smallmatrix} \oplus 1$ and $\begin{smallmatrix} 1 \\ 2 \\ 3 \end{smallmatrix} \oplus 3$ respectively. Then C (resp. T) is the unique sincere summand of M (resp. N) with a non-zero injective (resp. projective) complement. On the other hand, the simple module 2 is the unique indecomposable module belonging to $\operatorname{Ker}\operatorname{Hom}_\Lambda(-,C)$ (resp. $\operatorname{Ker}\operatorname{Hom}_\Lambda(T,-)$). Since $\operatorname{Ext}^2_\Lambda(2,1) \simeq \operatorname{Ext}^1_\Lambda(2,3) \neq 0$ and $\operatorname{Ext}^2_\Lambda(3,2) \simeq \operatorname{Ext}^1_\Lambda(1,2) \neq 0$, it follows that also (b) holds. □

As the next statement shows, the annihilator of large partial n-tilting or n-cotilting modules may be extremely big.

Proposition 11.8. *Let A be a finite dimensional algebra, let I be the annihilator of a large partial n-tilting or n-cotilting module. Then A and A/I may have different representation types.*

Proof. Assume first Λ is the K-algebra given by the quiver

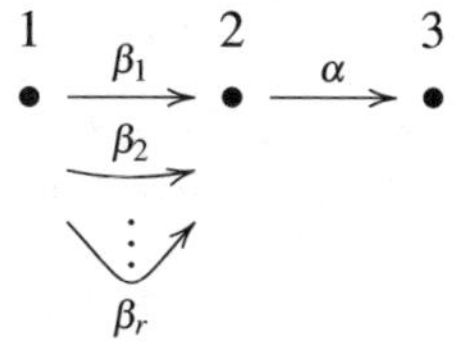

where $r \geq 2$ and the arrows β_i satisfy the relations $\alpha\beta_i =$ for $i = 1,\ldots,r$. Next, let T denote the injective module $= 1 \oplus \begin{smallmatrix}2\\3\end{smallmatrix}$. Then $\mathrm{Ker}\,\mathrm{Hom}_\Lambda(T,-)$ consists of semisimple projective modules. On the other hand, we have $\mathrm{Ext}^2_\Lambda(1,-) \simeq \mathrm{Ext}^1_\Lambda(\underbrace{2 \oplus \cdots \oplus 2}_{r},-)$.

It follows that

(1) $T = 1 \oplus \begin{smallmatrix}2\\3\end{smallmatrix}$ is a large partial 2-tilting module.

Dually, let Γ be the algebra given by the quiver

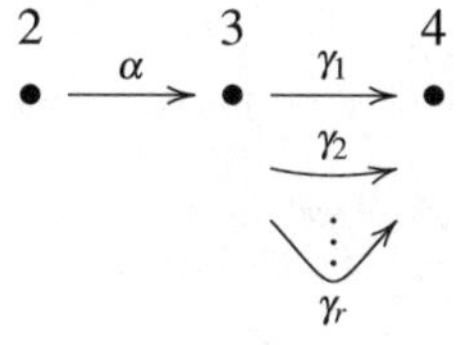

where $r \geq 2$ and the arrows γ_i satisfy the relations $\gamma_i\alpha = 0$ for $i = 1,\ldots,r$. Now, let C denote the projective module $C = \begin{smallmatrix}2\\3\end{smallmatrix} \oplus 4$. Then $\mathrm{Ker}\,\mathrm{Hom}_\Lambda(-,C)$ consists of semisimple injective modules. Moreover, we clearly have

$$\mathrm{Ext}^2_\Gamma(-,4) \simeq \mathrm{Ext}^1_\Gamma(-,\underbrace{3 \oplus \cdots \oplus 3}_{r}).$$

This implies that

(2) $C = \begin{smallmatrix}2\\3\end{smallmatrix} \oplus 4$ is a large partial 2-cotilting module.

Putting (1) and (2) together, we conclude that Λ and $\mathrm{ann}_\Lambda(T)$ (resp. Γ and $\mathrm{ann}_\Gamma(C)$) have all the desired properties. □

Acknowledgements This work is dedicated to the memory of Professor Adalberto Orsatti.

References

1. F. W. Anderson and K. R. Fuller, Rings and categories of modules, GTM 13, Springer, Berlin, 1992.
2. L. Angeleri–Hügel, D. Happel and H. Krause, Handbook of Tilting Theory, London Mathematical Society Lecture Note Series 332, Cambridge University Press, Cambridge 2007.
3. L. Angeleri Hügel, A. Tonolo and J.Trlifaj, Tilting preenvelopes and cotilting precovers, Algebra and Representation Theory 4 (2001), 155–170.
4. M. Auslander, I. Reiten and S. O. Smalø, Representation theory of artin algebras, Cambridge University Press, Cambridge, 1995.
5. S. Bazzoni, A characterization of n-cotilting and n-tilting modules, J. Algebra 273 (2004), 359–372.
6. S. Brenner and M. C. R. Butler, Generalizations of the Bernstein–Gelfand–Ponomarev reflection functors, Representation Theory II (Ottawa, 1979), V.Dlab and P. Gabriel (eds.), Lecture Notes in Mathematics, vol. 832, Springer, Berlin, 1980, 103–169.
7. R. R. Colby and K. R. Fuller, Equivalence and Duality for Module Categories, Cambridge University Press, Cambridge, 2004.
8. R. Colpi, Tilting modules and $*$-modules, Comm. Algebra 21(4) (1993), 1095–1102.
9. R. Colpi, G. D'Este and A. Tonolo, Quasi tilting modules and counter equivalences, J. Algebra 191 (1997), 461–494.
10. R. Colpi, G. D'Este and A. Tonolo, Corrigendum, J. Algebra 206 (1998), 370–370.
11. G. D'Este, On tilting and cotilting-type modules, Comment. Math. Univ. Carolinae 46(2) (2005), 281–291.
12. G. D'Este, On large selforthogonal modules, Comment. Math. Univ. Carolinae 47(4) (2006), 549–560.
13. D. Happel, I. Reiten and S. O. Smalø, Tilting in abelian categories and quasitilted algebras, Memoirs in AMS, vol. 575 (1996).
14. F. Mantese, Complements of projective almost complete tilting modules, Comm. Algebra 33(4) (2005), 2921–2940.
15. C. M. Ringel, Tame algebras and integral quadratic forms, Springer LMN 1099 (1984)
16. C. M. Ringel, Algebra at the Turn of the Century, South East Asian Bull. of Math. 25 (2001), 147–160.
17. C. M. Ringel, Combinatorial Representation Theory: History and Future, Representations of Algebras, Vol. 1 (ed. D. Happel and Y. B. Zhang) BNU Press, Beijing, 2002, 122–144.
18. C. M. Ringel, Some Remarks Concerning Tilting Modules and Tilting Algebras. Origin. Relevance. Future. An appendix to the Handbook of Tilting Theory, London Math. Soc., Cambridge University Press, vol. 332 (2007), 413–472.
19. J. Rickard, Morita theory for derived categories, J. London Math. Soc. (2) 39 (1989), 436–456.
20. M. Schaps and E. Zakay-Illouz, Combinatorial partial tilting complexes for the Brauer star algebras, Lecture Notes in Pure and Applied Mathematics, vol. 224, Marcel Dekker, New York, 2002, 187–207.
21. J. Trlifaj, Infinite dimensional tilting modules and cotorsion pairs, Handbook of Tilting Theory, London Math. Soc., Cambridge University Press, vol. 332 (2007), 279–321.

Part III
Groups and Actions

Chapter 12
How to Compose Lagrangian?

Eugen Paal and Jüri Virkepu

Abstract A method for constructing Lagrangians for the Lie transformation groups is explained. As examples, Lagrangians for real plane rotations and affine transformations of the real line are constructed.

12.1 Introduction

It is a well-known problem in physics and mechanics how to construct Lagrangians for mechanical systems via their equations of motion. This *inverse variational problem* has been investigated for some types of equations of motion in [4].

In [1, 5], the plane rotation group $SO(2)$ was considered as a toy model of the Hamilton–Dirac mechanics with constraints. By introducing a Lagrangian in a particular form, canonical formalism for $SO(2)$ was developed. The crucial idea of this approach is that the Euler–Lagrange and Hamilton canonical equations must in a sense coincide with Lie equations of the Lie transformation group.

In this paper, the method for constructing such a Lagrangian is proposed. It is shown, how it is possible to find the Lagrangian, based on the Lie equations of the Lie transformation group.

By composing Lagrangian, it is possible to describe given Lie transformation group as a mechanical system and to develop corresponding Lagrange and Hamilton formalisms.

E. Paal
Department of Mathematics, Tallinn University of Technology, Ehitajate tee 5,
19086 Tallinn, Estonia
e-mail: eugen.paal@ttu.ee

J. Virkepu
Department of Mathematics, Tallinn University of Technology, Ehitajate tee 5,
19086 Tallinn, Estonia
e-mail: jvirkepu@staff.ttu.ee

S. Silvestrov et al. (eds.), *Generalized Lie Theory in Mathematics, Physics and Beyond,*

12.2 General Method for Constructing Lagrangians

Let G be an r-parametric Lie group with unit $e \in G$ and let g^i $(i = 1,\ldots,r)$ denote the local coordinates of an element $g \in G$ from the vicinity of e. Let $\mathbf{X}$ be an n-dimensional manifold and denote the local coordinates of $X \in \mathbf{X}$ by X^α $(\alpha = 1,\ldots,n)$. Consider a (left) differentiable action of G on $\mathbf{X}$ given by

$$X' = S_g X \quad \in \mathbf{X}\,.$$

Let gh denote the *multiplication* of G. Then

$$S_g S_h = S_{gh}, \qquad \forall g,h \in G\,.$$

By introducing the *auxiliary functions* u^i_j and S^α_j by

$$(gh)^i \doteq h^i + u^i_j(h) g^j + \ldots\,,$$
$$(S_g X)^\alpha \doteq X^\alpha + S^\alpha_j(X) g^j + \ldots\,,$$

the Lie equations read

$$\varphi^\alpha_j(X;g) \doteq u^s_j(g) \frac{\partial (S_g X)^\alpha}{\partial g^s} - S^\alpha_j(S_g X) = 0\,.$$

The expressions φ^α_j are said to be *constraints* for the Lie transformation group $(\mathbf{X}, G)$. Then we search for such a vector Lagrangian $\mathbf{L} \doteq (L_1,\ldots,L_r)$ with components

$$L_k \doteq \sum_{\alpha=1}^{n} \sum_{s=1}^{r} \lambda^s_{k\alpha} \varphi^\alpha_s, \quad k = 1,2,\ldots,r$$

and such *Lagrange multipliers* $\lambda^s_{k\alpha}$ that the Euler–Lagrange equations in a sense coincide with the Lie equations.

The notion of a vector Lagrangian was introduced and developed in [2,6].

Definition 12.1 (weak equality). The functions A and B are called *weakly equal*, if

$$(A - B)\Big|_{\varphi^\alpha_j = 0} = 0, \quad j = 1,2,\ldots,r, \quad \alpha = 1,2,\ldots,n\,.$$

In this case we write $A \approx B$.

By denoting

$$X'^\alpha_i \doteq \frac{\partial X'^\alpha}{\partial g^i}\,,$$

the conditions for the Lagrange multipliers read as the *weak Euler–Lagrange equations*

$$L_{k\alpha} \doteq \frac{\partial L_k}{\partial X'^\alpha} - \sum_{i=1}^{r} \frac{\partial}{\partial g^i} \frac{\partial L_k}{\partial X'^\alpha_i} \approx 0\,.$$

Finally, one must check by direct calculations that the Euler–Lagrange equations $L_{k\alpha} = 0$ imply the Lie equations of the Lie transformation group.

12.3 Lagrangian for *SO*(2)

First consider the 1-parameter Lie transformation group $SO(2)$, *the rotation group of the real two-plane* $\mathbb{R}^2$. In this case $n = 2$ and $r = 1$. Rotation of the plane $\mathbb{R}^2$ by an angle $g \in \mathbb{R}$ is given by the transformation

$$\begin{cases} (S_g X)^1 = X'^1 = X'^1(X^1, X^2, g) \doteq X^1 \cos g - X^2 \sin g \\ (S_g X)^2 = X'^2 = X'^2(X^1, X^2, g) \doteq X^1 \sin g + X^2 \cos g \end{cases} .$$

We consider the rotation angle g as a dynamical variable and the functions X'^1 and X'^2 as *field variables* for the plane rotation group $SO(2)$.

Denote

$$\dot{X}'^\alpha \doteq \frac{\partial X'^\alpha}{\partial g} .$$

The *infinitesimal coefficients* of the transformation are

$$\begin{cases} S^1(X^1, X^2) \doteq \dot{X}'^1(X^1, X^2, e) = -X^2 \\ S^2(X^1, X^2) \doteq \dot{X}'^2(X^1, X^2, e) = X^1 \end{cases} .$$

and the Lie equations read

$$\begin{cases} \dot{X}'^1 = S^1(X'^1, X'^2) = -X'^2 \\ \dot{X}'^2 = S^2(X'^1, X'^2) = X'^1 \end{cases} .$$

Rewrite the Lie equations in implicit form as follows:

$$\begin{cases} \varphi_1^1 \doteq \dot{X}'^1 + X'^2 = 0 \\ \varphi_1^2 \doteq \dot{X}'^2 - X'^1 = 0 \end{cases}$$

We search a Lagrangian of $SO(2)$ in the form

$$L_1 = \sum_{\alpha=1}^{2} \sum_{s=1}^{1} \lambda_{1\alpha}^s \varphi_s^\alpha = \lambda_{11}^1 \varphi_1^1 + \lambda_{12}^1 \varphi_1^2 .$$

It is more convenient to rewrite it as follows:

$$L \doteq \lambda_1 \varphi^1 + \lambda_2 \varphi^2$$

where the Lagrange multipliers λ_1 and λ_2 are to be found from the weak Euler–Lagrange equations

$$\frac{\partial L}{\partial X'^1} - \frac{\partial}{\partial g}\frac{\partial L}{\partial \dot{X}'^1} \approx 0, \qquad \frac{\partial L}{\partial X'^2} - \frac{\partial}{\partial g}\frac{\partial L}{\partial \dot{X}'^2} \approx 0\,.$$

Calculate

$$\begin{aligned}
\frac{\partial L}{\partial X'^1} &= \frac{\partial}{\partial X'^1}\left[\lambda_1(\dot{X}'^1 + X'^2) + \lambda_2(\dot{X}'^2 - X'^1)\right] \\
&= \frac{\partial \lambda_1}{\partial X'^1}\varphi^1 + \frac{\partial \lambda_2}{\partial X'^1}\varphi^2 - \lambda_2 \approx -\lambda_2\,, \\
\frac{\partial L}{\partial \dot{X}'^1} &= \frac{\partial}{\partial \dot{X}'^1}\left[\lambda_1(\dot{X}'^1 + X'^2) + \lambda_2(\dot{X}'^2 - X'^1)\right] \approx \lambda_1\,, \\
\frac{\partial}{\partial g}\frac{\partial L}{\partial \dot{X}'^1} &= \frac{\partial \lambda_1}{\partial g} = \frac{\partial \lambda_1}{\partial X'^1}\dot{X}'^1 + \frac{\partial \lambda_1}{\partial X'^2}\dot{X}'^2 \approx -\frac{\partial \lambda_1}{\partial X'^1}X'^2 + \frac{\partial \lambda_1}{\partial X'^2}X'^1\,,
\end{aligned}$$

from which it follows

$$\frac{\partial L}{\partial X'^1} - \frac{\partial}{\partial g}\frac{\partial L}{\partial \dot{X}'^1} \approx 0 \qquad \Longleftrightarrow \qquad -\lambda_2 + \frac{\partial \lambda_1}{\partial X'^1}X'^2 - \frac{\partial \lambda_1}{\partial X'^2}X'^1 \approx 0\,.$$

Analogously calculate

$$\begin{aligned}
\frac{\partial L}{\partial X'^2} &= \frac{\partial}{\partial X'^2}\left[\lambda_1(\dot{X}'^1 + X'^2) + \lambda_2(\dot{X}'^2 - X'^1)\right] \\
&= \frac{\partial \lambda_1}{\partial X'^2}\varphi^1 + \frac{\partial \lambda_2}{\partial X'^2}\varphi^2 + \lambda_1 \approx \lambda_1\,, \\
\frac{\partial L}{\partial \dot{X}'^2} &= \frac{\partial}{\partial \dot{X}'^2}\left[\lambda_1(\dot{X}'^1 + X'^2) + \lambda_2(\dot{X}'^2 - X'^1)\right] \approx \lambda_2\,, \\
\frac{\partial}{\partial g}\frac{\partial L}{\partial \dot{X}'^2} &= \frac{\partial \lambda_2}{\partial g} = \frac{\partial \lambda_2}{\partial X'^1}\dot{X}'^1 + \frac{\partial \lambda_2}{\partial X'^2}\dot{X}'^2 \approx -\frac{\partial \lambda_2}{\partial X'^1}X'^2 + \frac{\partial \lambda_2}{\partial X'^2}X'^1\,,
\end{aligned}$$

from which it follows

$$\frac{\partial L}{\partial X'^2} - \frac{\partial}{\partial g}\frac{\partial L}{\partial \dot{X}'^2} \approx 0 \qquad \Longleftrightarrow \qquad \lambda_1 + \frac{\partial \lambda_2}{\partial X'^1}X'^2 - \frac{\partial \lambda_2}{\partial X'^2}X'^1 \approx 0\,.$$

So the calculations imply the following system of differential equations for the Lagrange multipliers:

$$\begin{cases} -\frac{\partial \lambda_1}{\partial X'^1}X'^2 + \frac{\partial \lambda_1}{\partial X'^2}X'^1 \approx -\lambda_2 \\ -\frac{\partial \lambda_2}{\partial X'^1}X'^2 + \frac{\partial \lambda_2}{\partial X'^2}X'^1 \approx \lambda_1 \end{cases}.$$

We are not searching for the general solution for this system of partial differential equations, but the Lagrange multipliers are supposed to be a linear combination of the field variables X'^1 and X'^2,

$$\begin{cases} \lambda_1 \doteq \alpha_1 X'^1 + \alpha_2 X'^2 \\ \lambda_2 \doteq \beta_1 X'^1 + \beta_2 X'^2, \end{cases} \qquad \alpha_1, \alpha_2, \beta_1, \beta_2 \in \mathbb{R}.$$

By using these expressions, one has

$$\begin{cases} -\alpha_1 X'^2 + \alpha_2 X'^1 \approx -\beta_1 X'^1 - \beta_2 X'^2 \\ -\beta_1 X'^2 + \beta_2 X'^1 \approx \alpha_1 X'^1 + \alpha_2 X'^2 \end{cases} \Longleftrightarrow \begin{cases} (\alpha_2 + \beta_1) X'^1 + (\beta_2 - \alpha_1) X'^2 \approx 0 \\ (\beta_2 - \alpha_1) X'^1 - (\alpha_2 + \beta_1) X'^2 \approx 0 \end{cases}.$$

This is a homogeneous system of two linear equations of four unknowns $\alpha_1, \alpha_2, \beta_1$, β_2. The system is satisfied, if

$$\begin{cases} \alpha_2 + \beta_1 = 0 \\ \beta_2 - \alpha_1 = 0 \end{cases} \Longleftrightarrow \begin{cases} \beta_1 = -\alpha_2 \\ \beta_2 = \alpha_1 \end{cases}.$$

The parameters α_1, α_2 are free. Thus

$$\begin{cases} \lambda_1 = \alpha_1 X'^1 + \alpha_2 X'^2 \\ \lambda_2 = -\alpha_2 X'^1 + \alpha_1 X'^2 \end{cases}$$

and the desired Lagrangian for $SO(2)$ reads

$$L = \alpha_1 (X'^1 \dot{X}'^1 + X'^2 \dot{X}'^2) + \alpha_2 \left[X'^2 \dot{X}'^1 + (X'^2)^2 - X'^1 \dot{X}'^2 + (X'^1)^2 \right] \tag{12.1}$$

with free real parameters α_1, α_2. Now we can propose

Theorem 12.1. *The Euler–Lagrange equations for the Lagrangian (12.1) coincide with the Lie equations of* $SO(2)$.

Proof. Calculate

$$\begin{aligned} \frac{\partial L}{\partial X'^1} &= \frac{\partial}{\partial X'^1} \left[\alpha_1 (X'^1 \dot{X}'^1 + X'^2 \dot{X}'^2) + \alpha_2 \left(X'^2 \dot{X}'^1 + (X'^2)^2 - X'^1 \dot{X}'^2 + (X'^1)^2 \right) \right] \\ &= \alpha_1 \dot{X}'^1 - \alpha_2 \dot{X}'^2 + 2\alpha_2 X'^1 , \\ \frac{\partial L}{\partial \dot{X}'^1} &= \frac{\partial}{\partial \dot{X}'^1} \left[\alpha_1 (X'^1 \dot{X}'^1 + X'^2 \dot{X}'^2) + \alpha_2 \left(X'^2 \dot{X}'^1 + (X'^2)^2 - X'^1 \dot{X}'^2 + (X'^1)^2 \right) \right] \\ &= \alpha_1 X'^1 + \alpha_2 X'^2 \quad \Longrightarrow \quad \frac{\partial}{\partial g} \frac{\partial L}{\partial \dot{X}'^1} = \alpha_1 \dot{X}'^1 + \alpha_2 \dot{X}'^2 , \end{aligned}$$

from which it follows

$$\frac{\partial L}{\partial X'^1} - \frac{\partial}{\partial g} \frac{\partial L}{\partial \dot{X}'^1} = 0 \quad \Longleftrightarrow \quad 2\alpha_2 X'^1 - 2\alpha_2 \dot{X}'^2 = 0 \quad \Longleftrightarrow \quad \dot{X}'^2 = X'^1 .$$

Analogously calculate

$$\begin{aligned} \frac{\partial L}{\partial X'^2} &= \frac{\partial}{\partial X'^2} \left[\alpha_1 (X'^1 \dot{X}'^1 + X'^2 \dot{X}'^2) + \alpha_2 \left(X'^2 \dot{X}'^1 + (X'^2)^2 - X'^1 \dot{X}'^2 + (X'^1)^2 \right) \right] \\ &= \alpha_2 \dot{X}'^1 + 2\alpha_2 X'^2 + \alpha_1 \dot{X}'^2 , \\ \frac{\partial L}{\partial \dot{X}'^2} &= \frac{\partial}{\partial \dot{X}'^2} \left[\alpha_1 (X'^1 \dot{X}'^1 + X'^2 \dot{X}'^2) + \alpha_2 \left(X'^2 \dot{X}'^1 + (X'^2)^2 - X'^1 \dot{X}'^2 + (X'^1)^2 \right) \right] \\ &= -\alpha_2 X'^1 + \alpha_1 X'^2 \quad \Longrightarrow \quad \frac{\partial}{\partial g} \frac{\partial L}{\partial \dot{X}'^2} = -\alpha_2 \dot{X}'^1 + \alpha_1 \dot{X}'^2 , \end{aligned}$$

from which it follows

$$\frac{\partial L}{\partial X'^2} - \frac{\partial}{\partial g}\frac{\partial L}{\partial \dot{X}'^2} = 0 \quad \Longleftrightarrow \quad 2\alpha_2 \dot{X}'^1 + 2\alpha_2 X'^2 = 0 \quad \Longleftrightarrow \quad \dot{X}'^1 = -X'^2 .$$

□

12.4 Physical Interpretation

While the Lagrangian L of $SO(2)$ contains two free parameters α_1, α_2, particular forms of it can be found taking into account physical considerations. In particular, if $\alpha_1 = 0$ and $\alpha_2 = -1/2$, then the Lagrangian of $SO(2)$ reads

$$L(X'^1, X'^2, \dot{X}'^1, \dot{X}'^2) \doteq \frac{1}{2}(X'^1\dot{X}'^2 - \dot{X}'^1 X'^2) - \frac{1}{2}\left[(X'^1)^2 + (X'^2)^2\right] .$$

By using the Lie equations one can easily check that

$$X'^1\dot{X}'^2 - \dot{X}'^1 X'^2 = (\dot{X}'^1)^2 + (\dot{X}'^2)^2 .$$

The function

$$T \doteq \frac{1}{2}\left[(\dot{X}'^1)^2 + (\dot{X}'^2)^2\right]$$

is the *kinetic energy* of a moving point $(X'^1, X'^2) \in \mathbb{R}^2$, meanwhile

$$l \doteq X'^1\dot{X}'^2 - \dot{X}'^1 X'^2$$

is its *kinetic momentum* with respect to origin $(0,0) \in \mathbb{R}^2$.

This relation has a simple explanation in the kinematics of a rigid body [3]. The kinetic energy of a point can be represented via its kinetic momentum as follows:

$$\frac{1}{2}\left[(\dot{X}^1)^2 + (\dot{X}'^2)^2\right] = T = \frac{l}{2} = \frac{1}{2}\left[X'^1\dot{X}'^2 - \dot{X}^1 X'^2\right] .$$

Thus we can conclude, that for the given Lie equations (that is, on the extremals) of $SO(2)$ the Lagrangian L gives rise to a Lagrangian of a pair of harmonic oscillators.

12.5 Lagrangian for the Affine Transformations of the Line

Now consider the affine transformations of the real line. The latter may be represented by

$$\begin{cases} X'^1 = X'^1(X^1, X^2, g^1, g^2) \doteq g^1 X^1 + g^2 \\ X'^2 = X'^2(X^1, X^2, g^1, g^2) \doteq 1, \end{cases} \qquad 0 \neq g^1, g^2 \in \mathbb{R}.$$

Thus $r = 2$ and $n = 2$. Denote

$$e \doteq (1,0), \quad g^{-1} \doteq \frac{1}{g^1}(1,-g^2)\,.$$

First, find the multiplication rule

$$\begin{aligned}(X'')^1 &\doteq (X'^1)' = S_{gh}X^1 = S_g(S_h X^1) = S_g(h^1X^1 + h^2)\\ &= g^1(h^1X^1 + h^2) + g^2 = (g^1h^1)X^1 + (g^1h^2 + g^2)\,.\end{aligned}$$

Calculate the infinitesimal coefficients

$$\begin{aligned}S_1^1(X^1,X^2) &\doteq X_1'^1\big|_{g=e} = X_1\,,\\ S_2^1(X^1,X^2) &\doteq X_2'^1\big|_{g=e} = 1\,,\\ S_1^2(X^1,X^2) &\doteq X_1'^2\big|_{g=e} = 0\,,\\ S_2^2(X^1,X^2) &\doteq X_2'^2\big|_{g=e} = 0\end{aligned}$$

and the auxiliary functions

$$\begin{aligned}u_1^1(g) &\doteq \left.\frac{\partial(S_{gh}X)^1}{\partial g^1}\right|_{h=g^{-1}} = \frac{1}{g^1}\,,\\ u_2^1(g) &\doteq \left.\frac{\partial(S_{gh}X)^1}{\partial g^2}\right|_{h=g^{-1}} = 0\,,\\ u_1^2(g) &\doteq \left.\frac{\partial(S_{gh}X)^2}{\partial g^1}\right|_{h=g^{-1}} = -\frac{g^2}{g^1}\,,\\ u_2^2(g) &\doteq \left.\frac{\partial(S_{gh}X)^2}{\partial g^2}\right|_{h=g^{-1}} = 1\,.\end{aligned}$$

Next, write Lie equations and find constraints

$$\begin{cases} X_1'^1 = \frac{1}{g^1}X'^1 - \frac{g^2}{g^1}\\ X_2'^1 = 1\\ X_1'^2 = 0\\ X_2'^2 = 0\end{cases} \quad\Longleftrightarrow\quad \begin{cases}\varphi_1^1 \doteq X_1'^1 - \frac{1}{g^1}X'^1 - \frac{g^2}{g^1}\\ \varphi_2^1 \doteq X_2'^1 - 1\\ \varphi_1^2 \doteq X_1'^2\\ \varphi_2^2 \doteq X_2'^2\end{cases}.$$

We search for a vector Lagrangian $\mathbf{L} = (L_1, L_2)$ as follows:

$$\begin{aligned}L_k &= \sum_{\alpha=1}^{2}\sum_{s=1}^{2}\lambda_{k\alpha}^s\varphi_s^\alpha = \lambda_{k1}^1\varphi_1^1 + \lambda_{k1}^2\varphi_2^1 + \lambda_{k2}^1\varphi_1^2 + \lambda_{k2}^2\varphi_2^2\\ &= \lambda_{k1}^1\left(X_1'^1 - \frac{1}{g^1}X'^1 - \frac{g^2}{g^1}\right) + \lambda_{k1}^2\left(X_2'^1 - 1\right) + \lambda_{k2}^1X_1'^2 + \lambda_{k2}^2X_2'^2, \quad k = 1,2\,.\end{aligned}$$

By substituting the Lagrange multipliers $\lambda^s_{k\alpha}$ into the weak Euler–Lagrange equations

$$\frac{\partial L_k}{\partial X'^\alpha} - \sum_{i=1}^{2} \frac{\partial}{\partial g^i} \frac{\partial L_k}{\partial X'^\alpha_i} \approx 0\,,$$

we get the following PDE system

$$\begin{cases} (X'^1 - g^2)\frac{\partial \lambda^1_{k1}}{\partial X'^1} + g^1 \frac{\partial \lambda^2_{k1}}{\partial X'^1} + \lambda^1_{k1} \approx 0 \\ (X'^1 - g^2)\frac{\partial \lambda^1_{k2}}{\partial X'^1} + g^1 \frac{\partial \lambda^2_{k2}}{\partial X'^1} \approx 0, \quad k = 1,2\,. \end{cases}$$

We find some particular solutions for this system. For example,

$$k = 1: \begin{cases} \lambda^1_{11} \doteq 0 \\ \lambda^2_{11} \doteq \psi^2_{11}(X'^2) \\ \lambda^1_{12} \doteq \psi^1_{12}(X'^2) \\ \lambda^2_{12} \doteq \psi^2_{12}(X'^2) \end{cases} \quad \text{and} \quad k = 2: \begin{cases} \lambda^1_{21} \doteq \psi^1_{21}(X'^2) \\ \lambda^2_{21} \doteq -\frac{X'^1}{g^1}\psi^1_{21}(X'^2) \\ \lambda^1_{22} \doteq 0 \\ \lambda^2_{11} \doteq 0 \end{cases}$$

with $\psi^1_{21}(X'^2), \psi^2_{11}(X'^2), \psi^1_{12}(X'^2), \psi^2_{12}(X'^2)$ as arbitrary real-valued functions of X'^2.

Thus we can define the Lagrangian $\mathbf{L} = (L_1, L_2)$ with

$$\begin{cases} L_1 = \psi^2_{11}(X'^2)(X'^1_2 - 1) + \psi^1_{12}(X'^2)X'^2_1 + \psi^2_{12}(X'^2)X'^2_2 \\ L_2 = \psi^1_{21}(X'^2)\left(X'^1_1 - \frac{1}{g^1}X'^1 + \frac{g^2}{g^1}\right) \end{cases} \tag{12.2}$$

and propose

Theorem 12.2. *The Euler–Lagrange equations for the vector Lagrangian* $\mathbf{L} = (L_1, L_2)$ *with components (12.2) coincide with the Lie equations of the affine transformations of the real line.*

Proof. Calculate

$$\begin{aligned} \frac{\partial L_1}{\partial X'^1} &= \frac{\partial}{\partial X'^1}\left[\psi^2_{11}(X'^2)(X'^1_2 - 1) + \psi^1_{12}(X'^2)X'^2_1 + \psi^2_{12}(X'^2)X'^2_2\right] = 0\,, \\ \frac{\partial}{\partial g^1}\frac{\partial L_1}{\partial X'^1_1} &= \frac{\partial}{\partial g^1}0 = 0\,, \\ \frac{\partial}{\partial g^2}\frac{\partial L_1}{\partial X'^1_2} &= \frac{\partial \psi^2_{11}(X'^2)}{\partial g^2} = \frac{\partial \psi^2_{11}(X'^2)}{\partial X'^2}X'^2_2\,, \end{aligned}$$

from which it follows

$$\frac{\partial L_1}{\partial X'^1} - \sum_{i=1}^{2}\frac{\partial}{\partial g^i}\frac{\partial L_1}{\partial X'^1_i} = 0 \iff \frac{\partial \psi^2_{11}(X'^2)}{\partial X'^2}X'^2_2 = 0 \implies X'^2_2 = 0\,.$$

Analogously calculate

$$\begin{aligned}
\frac{\partial L_1}{\partial X'^2} &= \frac{\partial}{\partial X'^2}\left[\psi^2_{11}(X'^2)(X'^1_2-1)+\psi^1_{12}(X'^2)X'^2_1+\psi^2_{12}(X'^2)X'^2_2\right]\\
&= \frac{\partial\psi^2_{11}(X'^2)}{\partial X'^2}(X'^1_2-1)+\frac{\partial\psi^1_{12}(X'^2)}{\partial X'^2}X'^2_1+\frac{\partial\psi^2_{12}(X'^2)}{\partial X'^2}X'^2_2\,,\\
\frac{\partial}{\partial g^1}\frac{\partial L_1}{\partial X'^2_1} &= \frac{\partial\psi^1_{12}(X'^2)}{\partial g^1}=\frac{\partial\psi^1_{12}(X'^2)}{\partial X'^2}X'^2_1\,,\\
\frac{\partial}{\partial g^2}\frac{\partial L_1}{\partial X'^2_2} &= \frac{\partial\psi^2_{12}(X'^2)}{\partial g^2}=\frac{\partial\psi^2_{12}(X'^2)}{\partial X'^2}X'^2_2\,,
\end{aligned}$$

from which it follows

$$\frac{\partial L_1}{\partial X'^2}-\sum_{i=1}^{2}\frac{\partial}{\partial g^i}\frac{\partial L_1}{\partial X'^2_i}=0 \quad\Longleftrightarrow\quad \frac{\partial\psi^2_{11}(X'^2)}{\partial X'^2}(X'^1_2-1)=0 \quad\Longrightarrow\quad X'^1_2-1=0\,.$$

Now we differentiate the second component of the Lagrangian **L**. Calculate

$$\begin{aligned}
\frac{\partial L_2}{\partial X'^1} &= \frac{\partial}{\partial X'^1}\left[\psi^1_{21}(X'^2)\left(X'^1_1-\frac{1}{g^1}X'^1+\frac{g^2}{g^1}\right)-\frac{1}{g^1}X'^1\psi^1_{21}(X'^2)(X'^1_2-1)\right]\\
&= -\frac{1}{g^1}\psi^1_{21}(X'^2)-\frac{1}{g^1}\psi^1_{21}(X'^2)X'^1_2+\frac{1}{g^1}\psi^1_{21}(X'^2)=-\frac{1}{g^1}\psi^1_{21}(X'^2)X'^1_2\,,\\
\frac{\partial}{\partial g^1}\frac{\partial L_2}{\partial X'^1_1} &= \frac{\partial\psi^1_{21}(X'^2)}{\partial g^1}=\frac{\partial\psi^1_{21}(X'^2)}{\partial X'^2}X'^2_1\,,\\
\frac{\partial}{\partial g^2}\frac{\partial L_2}{\partial X'^1_2} &= \frac{\partial}{\partial g^2}\left(-\frac{1}{g^1}X'^1\psi^1_{21}(X'^2)\right)=-\frac{1}{g^1}\left(\psi^1_{21}(X'^2)X'^1_2+X'^1\frac{\partial\psi^1_{21}(X'^2)}{\partial X'^2}X'^2_2\right),
\end{aligned}$$

from which it follows

$$\begin{aligned}
\frac{\partial L_2}{\partial X'^1}-\sum_{i=1}^{2}\frac{\partial}{\partial g^i}\frac{\partial L_2}{\partial X'^1_i}=0 \quad&\Longleftrightarrow\quad \frac{\partial\psi^1_{21}(X'^2)}{\partial X'^2}\left(X'^2_1-\frac{1}{g^1}X'^1X'^2_2\right)=0\\
&\Longrightarrow\quad X'^2_1-\frac{1}{g^1}X'^1X'^2_2=0\,.
\end{aligned}$$

Analogously calculate

$$\begin{aligned}
\frac{\partial L_2}{\partial X'^2} &= \frac{\partial}{\partial X'^2}\left[\psi^1_{21}(X'^2)\left(X'^1_1-\frac{1}{g^1}X'^1+\frac{g^2}{g^1}\right)-\frac{1}{g^1}X'^1\psi^1_{21}(X'^2)(X'^1_2-1)\right]\\
&= \frac{\partial\psi^1_{21}(X'^2)}{\partial X'^2}\left(X'^1_1-\frac{1}{g^1}X'^1+\frac{g^2}{g^1}\right)-\frac{1}{g^1}\frac{\partial\psi^1_{21}(X'^2)}{\partial X'^2}X'^1\left(X'^1_2-1\right),\\
\frac{\partial}{\partial g^1}\frac{\partial L_2}{\partial X'^2_1} &= \frac{\partial}{\partial g^1}0=0\,,\quad \frac{\partial}{\partial g^2}\frac{\partial L_2}{\partial X'^2_2}=\frac{\partial}{\partial g^2}0=0\,,
\end{aligned}$$

from which it follows

$$\frac{\partial L_2}{\partial X'^2} - \sum_{i=1}^{2} \frac{\partial}{\partial g^i} \frac{\partial L_2}{\partial X_i'^2} = 0 \iff \frac{\partial \psi_{21}^1(X'^2)}{\partial X'^2} \left(X_1'^1 - \frac{1}{g^1} X'^1 X_2'^1 + \frac{g^2}{g^1} \right) = 0$$
$$\implies X_1'^1 - \frac{1}{g^1} X'^1 X_2'^1 + \frac{g^2}{g^1} = 0 \,.$$

Thus the Euler–Lagrange equations read

$$\begin{cases} X_2'^2 = 0 \\ X_2'^1 - 1 = 0 \\ X_1'^2 - \frac{1}{g^1} X'^1 X_2'^2 = 0 \\ X_1'^1 - \frac{1}{g^1} X'^1 X_2'^1 + \frac{g^2}{g^1} = 0 \end{cases} .$$

It can be easily verified, that the latter is equivalent to the system of the Lie equations. □

Remark 12.1. While the Lagrangian **L** contains four arbitrary functions, particular forms of it can be fixed by taking into account physical considerations.

Acknowledgements The research was in part supported by the Estonian Science Foundation, Grant 6912.

References

1. Burdik, Č., Paal, E., Virkepu, J.: $SO(2)$ and Hamilton–Dirac mechanics. J. Nonlinear Math. Phys. **13**, 37–43 (2006)
2. Fushchych, W., Krivsky, I., Simulik, V.: On vector and pseudovector Lagrangians for electromagnetic field. W. Fushchych: Scientific Works, ed. by Boyko, V.M., **3**, 199–222 (Russian), 332–336 (English) (2001) http://www.imath.kiev.ua/~fushchych/
3. Goldstein, H.: Classical mechanics. Addison-Wesley, Cambridge (1953)
4. Lopuszanski, J.: The inverse variational problem in classical mechanics. World Scientific, Singapore (1999)
5. Paal, E., Virkepu, J.: Plane rotations and Hamilton–Dirac mechanics. Czech. J. Phys. **55**, 1503–1508 (2005)
6. Sudbery, A.: A vector Lagrangian for the electromagnetic field. J. Phys. A: Math. Gen. **19**, L33–L36 (1986)

Chapter 13
Semidirect Products of Generalized Quaternion Groups by a Cyclic Group

Peeter Puusemp

Abstract All semidirect products $G = Q_n \leftthreetimes C_2$ of generalized quaternion groups Q_n ($n \geq 3$) by the cyclic group C_2 of order two are found and described by their endomorphism semigroups. It follows from this description that each such semidirect product is determined by its endomorphism semigroup in the class of all groups.

13.1 Introduction

It is well known that all endomorphisms of an Abelian group form a ring and many of its properties can be characterized by this ring. All the main fields of the theory of endomorphism rings of groups are described in the monograph [2]. All endomorphisms of an arbitrary group form only a semigroup. The theory of endomorphism semigroups of groups is quite modestly developed. In many of our papers, we have made efforts to describe some properties of groups by the properties of their endomorphism semigroups. For example, it is shown in [3] and [8] that a direct product of groups and some semidirect products of groups can be characterized by the properties of the endomorphism semigroups of these groups. It is also shown that groups of many well-known classes are determined by their endomorphism semigroups in the class of all groups. Some of such groups are: finite Abelian groups [3, Theorem 4.2], non-torsion divisible Abelian groups [5, Theorem 1], generalized quaternion groups [4, Corollary 1]. On the other hand, there exist many examples of groups with very easy structure that are not determined by their endomorphism semigroups: the alternating group A_4 of order 12 [6, Theorem 4.1], some semidirect products of finite cyclic groups [7, Theorem 1].

P. Puusemp
Department of Mathematics, Tallinn University of Technology, Ehitajate tee 5 19086 Tallinn, Estonia
e-mail: puusemp@staff.ttu.ee

In this paper we will describe all nontrivial semidirect products $G = Q_n \leftthreetimes C_2$ of the generalized quaternion group

$$Q_n = \langle a, b \mid a^{2^n} = 1,\ a^{2^{n-1}} = b^2,\ b^{-1}ab = a^{-1} \rangle \qquad (13.1)$$

($n \geq 3$) by the cyclic group C_2 of order two and characterize these semidirect products by their endomorphism semigroups (nontrivial means here that a considered semidirect product is not a direct product). This description will be given in Theorems 13.1 and 13.2. It follows from this characterization that each such semidirect product is determined by its endomorphism semigroup in the class of all groups (Theorem 13.3). Note that the case $n = 2$ gives a group of order 16 and the groups of order 16 are investigated by us in [9].

We shall use the following notations: G – a group; $\mathrm{End}(G)$ – the endomorphism semigroup of a group G; $\langle a, b, \ldots \rangle$ – the subgroup of G, generated by its elements $a, b, \ldots$; $\widehat{g}$ – the inner automorphism, generated by g; C_n – the cyclic group of the order n; $\mathbb{Z}_n$ – the residue-class ring modulo n; $I(G)$ – the set of all idempotents of $\mathrm{End}(G)$; $K(x) = \{y \in \mathrm{End}(G) \mid yx = xy = y\}$; $V(x) = \{y \in \mathrm{Aut}(G) \mid yx = x\}$; $D(x) = \{y \in \mathrm{Aut}(G) \mid xy = yx = x\}$; $P(x) = \{y \in \mathrm{End}(G) \mid xy = yx = x\}$; $H(x) = \{y \in \mathrm{End}(G) \mid xy = y,\ yx = 0\}$; $W(x) = \{y \in \mathrm{End}(G) \mid xy = y\}$.

A map is written in this paper to the right from the element on which it acts. Let G be a fixed group and H an arbitrary group. If the isomorphism of groups G and H always follows from the isomorphism of semigroups $\mathrm{End}(G)$ and $\mathrm{End}(H)$ then we say that the group G is determined by its endomorphism semigroup in the class of all groups.

13.2 Semidirect Products of Q_n by C_2

In this section we will describe all semidirect products of the generalized quaternion group Q_n ($n \geq 3$) by C_2.

First we recall the notion of the semidirect product of groups. Let G be a group. If A and B are subgroups of G such that A is a normal subgroup, $A \cap B = <1>$ and $G = AB$, then we denote $G = A \leftthreetimes B$ and say that G is *a semidirect product* of A by B. In this case, there exists a homomorphism $\varphi : B \longrightarrow \mathrm{Aut}(A)$ such that

$$a(c\varphi) = c^{-1}ac \qquad (13.2)$$

and the composition rule in G is given as follows

$$ba \cdot cd = bc \cdot a(c\varphi)d \qquad (13.3)$$

($b, c \in B;\ a, d \in A$). In other words, if there are given two groups A, B and a homomorphism $\varphi : B \longrightarrow \mathrm{Aut}(A)$, then the semidirect product of A by B is a group of all pairs ba ($b \in B,\ a \in A$) with composition rule (13.3), in which equality (13.2) holds. Denote the obtained group by $A \leftthreetimes_\varphi B$. Of course, if such φ is fixed, the corresponding subscript by $\leftthreetimes$ will be omitted. We get a direct product, i.e., $A \leftthreetimes_\varphi B = A \times B$, if and only if $b\varphi$ is the unit automorphism of A for each $b \in B$. In this case we say that the semidirect product is *trivial.*

There exists a one-to-one correspondence between the set $I(G)$ of all idempotents of $\mathrm{End}(G)$ and the set of all decompositions of G into the semidirect products. Namely, if $x \in I(G)$, then $G = \mathrm{Ker}x \leftthreetimes \mathrm{Im}x$, $\mathrm{Im}x = \{g \in G \mid gx = g\}$ and x is the projection of G onto its subgroup $\mathrm{Im}x$. Conversely, if $G = A \leftthreetimes B$, then $A = \mathrm{Ker}x$ and $B = \mathrm{Im}x$, where x is the projection of G onto its subgroup B.

Lemma 13.1. *Let A and B be groups. Assume that φ, $\psi \in Hom(B, Aut(A))$ and $\psi = \varphi \circ \widehat{z}$ for some $z \in Aut(A)$. Then the map*

$$T : A \leftthreetimes_\varphi B \longrightarrow A \leftthreetimes_\psi B, \;\; (ba)T = b(az) \;\; (b \in B, \; a \in A) \tag{13.4}$$

is an isomorphism between the semidirect products $A \leftthreetimes_\varphi B$ and $A \leftthreetimes_\psi B$.

Indeed, map (13.4) is one-to-one, and direct calculations show that it preserves the multiplication.

Let A and B be groups, $\varphi \in \mathrm{Hom}(B, \mathrm{Aut}(A))$, $z \in \mathrm{Aut}(A)$ and $\psi = \varphi \circ \widehat{z}$. Denote the projection of $G_\varphi = A \leftthreetimes_\varphi B$ onto B by x_φ. Similarly, x_ψ denotes the projection of $G_\psi = A \leftthreetimes_\psi B$ onto B. Under these assumptions, the following lemma holds.

Lemma 13.2. *The isomorphism T, defined in Lemma 13.1, induces a natural isomorphism*

$$Z : End(G_\varphi) \longrightarrow End(G_\psi), \; yZ = T^{-1} \circ y \circ T \;\; (y \in End(G_\varphi)),$$

which maps x_φ to x_ψ and

$$(K(x_\varphi))Z = K(x_\psi), \;\; (V(x_\varphi))Z = V(x_\psi), \;\; (D(x_\varphi))Z = D(x_\psi), \tag{13.5}$$

$$(H(x_\varphi))Z = H(x_\psi), \;\; (P(x_\varphi))Z = P(x_\psi), \;\; (W(x_\varphi))Z = W(x_\psi). \tag{13.6}$$

Proof. Obviously, Z is an isomorphism. Choose $ba \in G_\psi$ $(b \in B, \; a \in A)$ and calculate

$$(ba)(x_\varphi Z) = (ba)(T^{-1} \circ x_\varphi \circ T) = (b(az^{-1}))(x_\varphi \circ T) =$$
$$= ((b(az^{-1}))x_\varphi)T = bT = b = (ba)x_\psi.$$

Therefore, $x_\psi = x_\varphi Z$ and Z maps x_φ to x_ψ. Since Z is an isomorphism and $(K(x_\varphi))Z = K(x_\varphi Z), \ldots, (W(x_\varphi))Z = W(x_\varphi Z)$, equalities (13.5) and (13.6) are true. □

Let us now find all possible non-trivial semidirect products $Q_n \leftthreetimes_\varphi C_2$ where $n \geq 3$, $C_2 = \langle c \rangle$ and Q_n is given by (13.1). It is necessary to find all monomorphisms $\varphi : C_2 \longrightarrow \mathrm{Aut}(Q_n)$ or, equivalently, all automorphisms $y = c\varphi$ of Q_n of order two. In view of (13.1) and (13.2), the corresponding semidirect products are given as follows

$$G_{n,y} = Q_n \leftthreetimes_\varphi C_2 = \langle a, b, c \mid a^{2^n} = 1,\ a^{2^{n-1}} = b^2,\ b^{-1}ab = a^{-1},$$
$$c^2 = 1,\ c^{-1}ac = ay,\ c^{-1}bc = by\rangle. \tag{13.7}$$

Lemma 13.3. *If $y, z \in Aut(Q_n)$ and y is of order two, then $G_{n,y} \cong G_{n,y\widehat{z}}$.*

This is a direct corollary from Lemma 13.1.

Assume that $n \geq 3$ and Q_n is given by (13.1). It is easy to check that $\mathrm{Aut}(Q_n)$ consists of maps

$$y : b \longmapsto ba^j,\ \ y : a \longmapsto a^l, \tag{13.8}$$

where $j, l \in \mathbb{Z}_{2^n}$ and $l \equiv 1 \pmod 2$. Automorphism (13.8) can be identified with a matrix:

$$y = \begin{Vmatrix} 1 & j \\ 0 & l \end{Vmatrix}.$$

Then the multiplication in the group $\mathrm{Aut}(Q_n)$ coincides with the usual multiplication of matrices.

Lemma 13.4. *The elements of order two in $Aut(Q_n)$ $(n \geq 3)$ are*

$$\begin{Vmatrix} 1 & 2^{n-1} \\ 0 & 1 \end{Vmatrix},\ \begin{Vmatrix} 1 & j \\ 0 & -1 \end{Vmatrix},\ \begin{Vmatrix} 1 & 2^{n-1}t \\ 0 & 1+2^{n-1} \end{Vmatrix},\ \begin{Vmatrix} 1 & 2s \\ 0 & -1+2^{n-1} \end{Vmatrix}, \tag{13.9}$$

where $j \in \mathbb{Z}_{2^n}$, $t \in \mathbb{Z}_2$ and $s \in \mathbb{Z}_{2^{n-1}}$.

The statement of Lemma 13.4 can be obtained by the immediate calculations. According to Lemma 13.3, it is necessary to divide automorphisms (13.9) into the conjugacy classes. The direct calculations give the following result.

Lemma 13.5. *The conjugacy classes for elements of order 2 in $Aut(Q_n)$ $(n \geq 3)$ are:*

$$\mathscr{C}_1 = \left\{ \begin{Vmatrix} 1 & 2^{n-1} \\ 0 & 1 \end{Vmatrix} \right\},\ \mathscr{C}_2 = \left\{ \begin{Vmatrix} 1 & 2s \\ 0 & -1 \end{Vmatrix} \mid s \in \mathbb{Z}_{2^{n-1}} \right\},$$

$$\mathscr{C}_3 = \left\{ \begin{Vmatrix} 1 & 2s+1 \\ 0 & -1 \end{Vmatrix} \mid s \in \mathbb{Z}_{2^{n-1}} \right\},\ \mathscr{C}_4 = \left\{ \begin{Vmatrix} 1 & t2^{n-1} \\ 0 & 1+2^{n-1} \end{Vmatrix} \mid t \in \mathbb{Z}_2 \right\},$$

$$\mathscr{C}_5 = \left\{ \begin{Vmatrix} 1 & 2s \\ 0 & -1+2^{n-1} \end{Vmatrix} \mid s \in \mathbb{Z}_{2^{n-1}} \right\}.$$

By Lemma 13.3,

$$u, v \in \mathscr{C}_i \Longrightarrow G_{n,u} \cong G_{n,v}$$

$(i \in \{1, 2, 3, 4, 5\})$. Choose a representative from each conjugacy class:

$$y_1 = \begin{Vmatrix} 1 & 2^{n-1} \\ 0 & 1 \end{Vmatrix} \in \mathscr{C}_1;\ y_2 = \begin{Vmatrix} 1 & 0 \\ 0 & -1 \end{Vmatrix} \in \mathscr{C}_2;\ y_3 = \begin{Vmatrix} 1 & 1 \\ 0 & -1 \end{Vmatrix} \in \mathscr{C}_3;$$

$$y_4 = \begin{Vmatrix} 1 & 0 \\ 0 & 1+2^{n-1} \end{Vmatrix} \in \mathscr{C}_4; \; y_5 = \begin{Vmatrix} 1 & 0 \\ 0 & -1+2^{n-1} \end{Vmatrix} \in \mathscr{C}_5.$$

Next we will investigate groups $G_{n,y_1}, \ldots, G_{n,y_5}$. The generating relations of these groups are given by (13.7). The connections between c and a, b are the following:

$$\begin{aligned} G_{n,y_1} &: \; c^{-1}bc = ba^{2^{n-1}}, \; c^{-1}ac = a; \\ G_{n,y_2} &: \; c^{-1}bc = b, \; c^{-1}ac = a^{-1}; \\ G_{n,y_3} &: \; c^{-1}bc = ba, \; c^{-1}ac = a^{-1}; \\ G_{n,y_4} &: \; c^{-1}bc = b, \; c^{-1}ac = a^{1+2^{n-1}}; \\ G_{n,y_5} &: \; c^{-1}bc = b, \; c^{-1}ac = a^{-1+2^{n-1}}. \end{aligned} \tag{13.10}$$

Lemma 13.6. $G_{n,y_2} \cong G_{n,y_1}$, $G_{n,y_4} \cong G_{n,y_5}$.

Indeed, these isomorphisms are given by maps $T_1 : G_{n,y_2} \longrightarrow G_{n,y_1}$ and $T_2 : G_{n,y_4} \longrightarrow G_{n,y_5}$, where

$$T_1(c) = cba^{2^{n-2}}, \; T_1(b) = b, \; T_1(a) = a,$$

$$T_2(c) = cba^{2^{n-2}}, \; T_2(b) = ba, \; T_2(a) = a.$$

To show that G_{n,y_2}, G_{n,y_3} and G_{n,y_4} are non-isomorphic we have to find the elements of order two in these groups. The direct calculations give the following lemma.

Lemma 13.7. *The elements of order two in groups G_{n,y_2}, G_{n,y_3} and G_{n,y_5} are following*

$G_{n,y_2}: \; b^2, \; ca^j, \; cba^{t2^{n-2}} \;\; (j \in \mathbb{Z}_{2^n}, \; t \in \mathbb{Z}_4, \; t \equiv 1 \pmod 2)$*;*
$G_{n,y_3}: \; b^2, \; ca^j \;\; (j \in \mathbb{Z}_{2^n})$*;*
$G_{n,y_5}: \; b^2, \; ca^{2j}, \; cba^{t2^{n-2}} \;\; (j \in \mathbb{Z}_{2^{n-1}}, \; t \in \mathbb{Z}_4, \; t \equiv 1 \pmod 2)$.

It follows from Lemma 13.7 that the numbers of elements of order two in groups G_{n,y_2}, G_{n,y_3} and G_{n,y_5} are 2^n+3, 2^n+1 and $2^{n-1}+3$, respectively. Therefore, the groups $\mathscr{G}_1 = G_{n,y_2}$, $\mathscr{G}_2 = G_{n,y_3}$ and $\mathscr{G}_3 = G_{n,y_5}$ are non-isomorphic with one another, and we have proved the following theorem.

Theorem 13.1. *For a given integer $n \geq 3$, there exist only three non-isomorphic and non-trivial semidirect products of the generalized quaternion group Q_n by the cyclic group C_2 of order two. They are the following:*

$$\begin{aligned} \mathscr{G}_1 &= \langle a,b,c \mid a^{2^n} = 1, a^{2^{n-1}} = b^2, b^{-1}ab = a^{-1}, c^2 = 1, c^{-1}bc = b, c^{-1}ac = a^{-1} \rangle, \\ \mathscr{G}_2 &= \langle a,b,c \mid a^{2^n} = 1, a^{2^{n-1}} = b^2, b^{-1}ab = a^{-1}, c^2 = 1, c^{-1}bc = ba, \\ &\qquad\qquad\qquad\qquad\qquad\qquad\qquad\qquad\qquad c^{-1}ac = a^{-1} \rangle, \\ \mathscr{G}_3 &= \langle a,b,c \mid a^{2^n} = 1, \; a^{2^{n-1}} = b^2, b^{-1}ab = a^{-1}, c^2 = 1, c^{-1}bc = b, \\ &\qquad\qquad\qquad\qquad\qquad\qquad\qquad\qquad\qquad c^{-1}ac = a^{-1+2^{n-1}} \rangle. \end{aligned}$$

13.3 A Description of $\mathscr{G}_1$, $\mathscr{G}_2$ and $\mathscr{G}_3$ by Their Endomorphisms

Let $G \in \{\mathscr{G}_1, \mathscr{G}_2, \mathscr{G}_3\}$. Then $G = Q_n \rtimes C_2 = \langle a, b\rangle \rtimes \langle c\rangle$ (Theorem 13.1). Denote the projection of G onto $\langle c\rangle$ by x. In this section we shall give a description of $\mathscr{G}_1$, $\mathscr{G}_2$ and $\mathscr{G}_3$ by their endomorphism semigroups. For this we shall use some properties of the idempotent x of $\mathrm{End}(G)$.

Assume now that G is an arbitrary group and $x \in I(G)$. Since G decomposes into the semidirect product $G = \mathrm{Ker}\,x \rtimes \mathrm{Im}\,x$, it is easy to check that

$$K(x) = \{y \in \mathrm{End}(G) \mid (\mathrm{Ker}\,x)y = \langle 1\rangle,\ (\mathrm{Im}\,x)y \subset \mathrm{Im}\,x\} \cong$$
$$\cong \mathrm{End}(\mathrm{Im}\,x), \tag{13.11}$$
$$P(x) = \{y \in \mathrm{End}(G) \mid (\mathrm{Ker}\,x)y \subset \mathrm{Ker}\,x,\ y|_{\mathrm{Im}\,x} = 1_{\mathrm{Im}\,x}\}, \tag{13.12}$$
$$D(x) = \{y \in \mathrm{Aut}(G) \mid (\mathrm{Ker}\,x)y \subset \mathrm{Ker}\,x,\ y|_{\mathrm{Im}\,x} = 1_{\mathrm{Im}\,x}\}, \tag{13.13}$$
$$V(x) = \{y \in \mathrm{Aut}(G) \mid g^{-1}\cdot gy \in \mathrm{Ker}\,x \text{ for each } g \in G\}, \tag{13.14}$$
$$H(x) = \{y \in \mathrm{End}(G) \mid (\mathrm{Ker}\,x)y = \langle 1\rangle,\ (\mathrm{Im}\,x)y \subset \mathrm{Ker}\,x\}, \tag{13.15}$$
$$W(x) = \{y \in \mathrm{End}(G) \mid (\mathrm{Ker}\,x)y = \langle 1\rangle\}. \tag{13.16}$$

Choose $G = G_{n,y_i}$ $(i = 1, 2, 3, 4, 5)$, where G_{n,y_i} is given in Sect. 13.2. Then

$$G = \langle a, b, c\rangle = \langle a, b\rangle \rtimes \langle c\rangle = Q_n \rtimes C_2,$$

where $Q_n = \langle a, b\rangle$, $C_2 = \langle c\rangle$ and the generating relations are given by (13.7) and (13.10). Denote the projection of G onto its subgroup $\langle c\rangle$ by x. Then $x \in I(G)$ and $\mathrm{Im}\,x = \langle c\rangle$, $\mathrm{Ker}\,x = \langle a, b\rangle$. Using equalities (13.11)–(13.15), the immediate calculations give that this x satisfies the properties given in the next lemma.

Lemma 13.8. *Idempotent x satisfies the following six properties:*
1^0 $K(x) \cong \mathrm{End}(C_2)$;
$$2^0\ |V(x)| = \begin{cases} 2\cdot|D(x)| = 4^n, \text{ if } G = G_{n,y_1}, \\ |D(x)|^2 = 4^n, \text{ if } G \in \{G_{n,y_2}, G_{n,y_3}, G_{n,y_5}\}, \\ 2\cdot|D(x)| = 0,5\cdot 4^n, \text{ if } G = G_{n,y_4}; \end{cases}$$
3^0 $|H(x)| = 2$;
4^0 $|P(x)\setminus D(x)| \le 4$;
5^0 *if $y \in I(G)$ and $yx = xy = 0$, then $y = 0$.*

Lemma 13.9. *Assume that x has its previous meaning. Then, depending on G, $|W(x)|$ has the following values:*

$$G = \mathscr{G}_1 = G_{n,y_2} \Longrightarrow |W(x)| = 2^n + 4,$$
$$G = \mathscr{G}_2 = G_{n,y_3} \Longrightarrow |W(x)| = 2^n + 2,$$
$$G = \mathscr{G}_3 = G_{n,y_5} \Longrightarrow |W(x)| = 2^{n-1} + 4.$$

Proof. Assume that $G \in \{\mathscr{G}_1, \mathscr{G}_2, \mathscr{G}_3\}$ and x is the projection of $G = Q_n \leftthreetimes C_2$ onto its subgroup C_2. By (13.16),

$$|W(x)| = |\mathrm{Hom}(\mathrm{Im}x, G)| = |\mathrm{Hom}(C_2, G)| =$$

$$= |\{g \in G \mid g^2 = 1\}| = 1 + |\{g \in G \mid g^2 = 1, \ g \neq 1\}|.$$

Therefore, by Lemma 13.7, the statements of the lemma hold. □

Theorem 13.2. *Let G be a finite group. Then G is isomorphic to a group from the class* $\mathscr{G} = \{\mathscr{G}_1, \mathscr{G}_2, \mathscr{G}_3\}$ *if and only if there exists* $x \in I(G)$ *such that x satisfies the following properties:*
1^0 $K(x) \cong End(C_2)$;
2^0 $|V(x)| = |D(x)|^2 = 4^n$;
3^0 $|H(x)| = 2$;
4^0 $|P(x) \setminus D(x)| \leq 4$;
5^0 *if* $y \in I(G)$ *and* $yx = xy = 0$, *then* $y = 0$.
For each such x

$$|W(x)| \in \{2^n + 4, 2^n + 2, 2^{n-1} + 4\} \tag{13.17}$$

and

$$|W(x)| = 2^n + 4 \Longrightarrow G \cong \mathscr{G}_1, \tag{13.18}$$

$$|W(x)| = 2^n + 2 \Longrightarrow G \cong \mathscr{G}_2, \tag{13.19}$$

$$|W(x)| = 2^{n-1} + 4 \Longrightarrow G \cong \mathscr{G}_3 \tag{13.20}$$

hold.

Proof. Let G be a finite group. Assume that $G \in \mathscr{G}$. Then $G = Q_n \leftthreetimes C_2$. Denote the projection of G onto its subgroup C_2 by x. Then $x \in I(G)$ and, by Lemmas 13.8 and 13.9, x satisfies properties $1^0 - 5^0$ and (13.17)–(13.20).

Conversely, assume now that G is a finite group and there exists $x \in I(G)$ which satisfies properties $1^0 - 5^0$. We will show that $G \in \mathscr{G}$ and x satisfies properties (13.17)–(13.20).

Since x is an idempotent, G decomposes into the semidirect product

$$G = \mathrm{Ker}x \leftthreetimes \mathrm{Im}x. \tag{13.21}$$

By property 1^0 and (13.11), $\mathrm{End}(\mathrm{Im}x) \cong \mathrm{End}(C_2)$. Each finite Abelian group is determined by its endomorphism semigroup in the class of all groups [3, Theorem 4.2]. Therefore, $\mathrm{Im}x \cong C_2$ and $\mathrm{Im}x = < c >$ for some $c \in G$.

Choose $g, h \in G$. Since $\mathrm{Im}x$ is Abelian, we have $g^{-1}h^{-1}gh \in \mathrm{Ker}x$ and

$$g(\widehat{h}x) = (g \cdot g^{-1}h^{-1}gh)x = gx.$$

Therefore, $\widehat{h} \in V(x)$. Hence and by property 2^0, each $2'$-element of G belongs to the centre of G. Thus, $\mathrm{Ker}x = G_1 \times G_2$ and

$$G = G_2 \times (G_1 \leftthreetimes \mathrm{Im}\,x) = (G_2 \times G_1) \leftthreetimes \mathrm{Im}\,x,$$

where G_2 is the Hall $2'$-subgroup of G and G_1 is the Sylow 2-subgroup of $\mathrm{Ker}\,x$. Denote the projection of G onto its subgroup G_2 by y. Then $y \in I(G)$, $yx = xy = 0$ and, by 5^0, $y = 0$. Therefore, $G_2 = \langle 1 \rangle$ and G is a 2-group.

It follows from (13.15), (13.21) and property 3^0 that $\mathrm{Ker}\,x$ has only one element of order two. Hence $\mathrm{Ker}\,x$ is a cyclic 2-group C_{2^m} or a generalized quaternion group Q_m for some integer $m > 0$ [10, Theorem 5.3.6].

Suppose that $\mathrm{Ker}\,x$ is cyclic, i.e., $\mathrm{Ker}\,x = \langle a \rangle \cong C_{2^m}$. In view of (13.12), (13.13) and (13.21), it is clear that $P(x)$ and $D(x)$ consist of maps $c \mapsto c,\ a \mapsto a^i,\ i \in \mathbb{Z}_{2^m}$ and $c \mapsto c,\ a \mapsto a^i,\ i \in \mathbb{Z}_{2^m},\ i \equiv 1 \pmod 2$, respectively. Therefore,

$$|P(x)| = 2^m, \quad |D(x)| = |P(x) \setminus D(x)| = 2^{m-1}. \tag{13.22}$$

Equalities (13.22) and properties 2^0, 4^0 imply $2^{m-1} = 2^n < 4$, i.e., $n < 2$. This contradicts the assumption $n \geq 3$. Hence $\mathrm{Ker}\,x$ is a generalized quaternion group Q_m and $G = G_{m,y}$ for some $m > 0$ and $y \in \mathrm{Aut}(Q_m)$, $y^2 = 1$. If $y = 1$ then $G = \mathrm{Ker}\,x \times \mathrm{Im}\,x$ and $zx = xz = 0$, where z is the projection of G onto $\mathrm{Ker}\,x$. By property 5^0, $z = 0$, i.e., $\mathrm{Ker}\,x = \langle 1 \rangle$. In view of property 3^0, it is impossible. Hence y is an automorphism of order two. By Lemmas 13.1–13.5, we can assume that $y \in \{y_1, \ldots, y_5\}$, where $y_1, \ldots, y_5$ are given in Sect. 13.2 (replace there only n by m). Therefore, $G = G_{m,y_i}$ for some $i \in \{1,2,3,4,5\}$. By property 2^0 and Lemma 13.8, $n = m$ and $G \in \{G_{m,y_2}, G_{m,y_3}, G_{m,y_5}\} = \{\mathcal{G}_1, \mathcal{G}_2, \mathcal{G}_3\} = \mathcal{G}$. Lemma 13.9 implies that x satisfies properties (13.17)–(13.20). □

Theorem 13.3. *If $G \in \{\mathcal{G}_1, \mathcal{G}_2, \mathcal{G}_3\}$, then G is determined by its endomorphism semigroup in the class of all groups.*

Proof. Assume that $G = \mathcal{G}_i$, $i \in \{1,2,3\}$. Let G^* be another group such that the semigroups $\mathrm{End}(G)$ and $\mathrm{End}(G^*)$ are isomorphic:

$$\mathrm{End}(G) \cong \mathrm{End}(G^*). \tag{13.23}$$

Since $\mathrm{End}(G)$ is finite, so is $\mathrm{End}(G^*)$. By [1, Theorem 2], the group G^* is finite, too. In view of Theorem 13.2, there exists $x \in I(G)$ satisfying properties 1^0–5^0 of this theorem. Denote by x^* the idempotent which corresponds to x in the isomorphism (13.23). Due to isomorphism (13.23), x^* satisfies properties, similar to properties 1^0–5^0. By Theorem 13.2, G^* is isomorphic to a group from the class $\{\mathcal{G}_1, \mathcal{G}_2, \mathcal{G}_3\}$. Since $|W(x)| = |W(x^*)|$, it follows from properties (13.18)–(13.20) that $G^* \cong G$. □

Acknowledgements This work was supported in part by the Estonian Science Foundation Research Grant 5900, 2004–2007.

References

1. Alperin, J.L.: Groups with finitely many automorphisms. Pacific J. Math. **12**, no 1, 1–5 (1962).
2. Krylov, P.A., Mikhalev, A.V., Tuganbaev, A.A.: Endomorphism Rings of Abelian Groups. Kluwer, Dordrecht (2003).
3. Puusemp, P.: Idempotents of the endomorphism semigroups of groups. Acta et Comment. Univ. Tartuensis **366**, 76–104 (1975) (in Russian).
4. Puusemp, P.: Endomorphism semigroups of generalized quaternion groups. Acta et Comment. Univ. Tartuensis **390**, 84–103 (1976) (in Russian).
5. Puusemp, P.: A characterization of divisible and torsion Abelian groups by their endomorphism semigroups. Algebras, Groups and Geometries **16**, 183–193 (1999).
6. Puusemp, P.: On endomorphism semigroups of dihedral 2-groups and alternating group A_4. Algebras, Groups and Geometries **16**, 487–500 (1999).
7. Puusemp, P.: Characterization of a semidirect product of cyclic groups by its endomorphism semigroup. Algebras, Groups and Geometries **17**, 479–498 (2000).
8. Puusemp, P.: Characterization of a semidirect product of groups by its endomorphism semigroup. In: Smith, P., Giraldes, E., Martins, P. (eds.) Proc. of the Intern. Conf. on Semigroups, pp. 161–170. World Scientific, Singapore (2000).
9. Puusemp P.: Non-abelian groups of order 16 and their endomorphism semigroups. J. Math. Sci. **131**, no 6, 6098–6111 (2005).
10. Robinson, D.J.S.: A Course in the Group Theory. Springer, New York (1996).

Chapter 14
A Characterization of a Class of 2-Groups by Their Endomorphism Semigroups

Tatjana Gramushnjak and Peeter Puusemp

Abstract In this paper the groups G of order 32 of the form $G = (C_8 \times C_2) \rtimes C_2$ are considered. These groups are described by their endomorphism semigroups. It follows from these descriptions that all these groups are determined by their endomorphism semigroups in the class of all groups.

14.1 Introduction

For any group G, the set of all its endomorphisms $\mathrm{End}(G)$ forms a semigroup with identity. In many of our papers, we have made efforts to describe some properties of groups by the properties of their endomorphism semigroups. For example, it was shown in [4] and [7] that the direct product of groups and some semidirect products of groups can be characterized by the properties of the endomorphism semigroups of these groups. It was also shown that groups of many well-known classes are determined by their endomorphism semigroups in the class of all groups. Some of such groups are: finite abelian groups [4, Theorem 4.2], non-torsion divisible abelian groups [6, Theorem 1], finite symmetric groups [8], etc.

In [9], it was proved that among the finite groups of order less than 32 only the tetrahedral group and the binary tetrahedral group are not determined by their endomorphism semigroups in the class of all groups. In [2], we started to examine the class of finite groups of order 32. Hall and Senior [3] gave a full description of

T. Gramushnjak
Institute of Mathematics and Natural Sciences, Tallinn University, Narva road 25, 10120 Tallinn, Estonia
e-mail: tatjana@tlu.ee

P. Puusemp
Department of Mathematics, Tallinn University of Technology, Ehitajate tee 5, 19086 Tallinn, Estonia
e-mail: puusemp@staff.ttu.ee

S. Silvestrov et al. (eds.), *Generalized Lie Theory in Mathematics, Physics and Beyond*,   151
© Springer-Verlag Berlin Heidelberg 2009

all groups of order 2^n, $n \leq 6$. There exist exactly 51 non-isomorphic groups of order 32. In [3], these groups were numbered by 1, 2,..., 51. We shall mark these groups by $G_1, G_2, \ldots, G_{51}$, respectively. In [2] we investigated the set $\mathscr{G}$ of all groups G of order 32 such that G can be presented in the form $G = (C_4 \times C_4) \leftthreetimes C_2$, where C_4 and C_2 are cyclic groups of orders 4 and 2, respectively. We described all groups of the class $\mathscr{G}$ by their endomorphism semigroups and proved that each group of the class $\mathscr{G}$ is determined by its endomorphism semigroup in the class of all groups.

In this paper we continue the study of the class of groups of order 32. Namely, we consider the groups G of order 32 such that G can be presented in the form $G = (C_8 \times C_2) \leftthreetimes C_2$. We shall characterize these groups by their endomorphism semigroups and show that these groups are determined by their endomorphism semigroups in the class of all groups.

By [3], the groups G of order 32 such that can be presented in the form $G = (C_8 \times C_2) \leftthreetimes C_2$, are G_4, G_{17}, G_{20}, G_{26} and G_{27}, i.e., the groups

$$\begin{aligned}
G_4 &= C_8 \times C_2 \times C_2, \\
G_{17} &= \langle a, b, c \mid a^8 = b^2 = c^2 = 1,\ ab = ba,\ ac = ca,\ c^{-1}bc = ba^4 \rangle, \\
G_{20} &= \langle a, b, c \mid a^8 = b^2 = c^2 = 1,\ ab = ba,\ bc = cb,\ c^{-1}ac = ab \rangle, \\
G_{26} &= \langle a, b, c \mid a^8 = b^2 = c^2 = 1,\ ab = ba,\ c^{-1}ac = a^{-1},\ c^{-1}bc = a^4 b \rangle, \\
G_{27} &= \langle a, b, c \mid a^8 = b^2 = c^2 = 1,\ ab = ba,\ bc = cb,\ c^{-1}ac = a^{-1}b \rangle.
\end{aligned}$$

The group G_4 is abelian and, therefore, it is determined by its endomorphism semigroup in the class of all groups [4, Theorem 4.2]. The group G_{26} can be presented in the form $G_{26} = \langle a, d \rangle \leftthreetimes \langle c \rangle$ and $\langle a, d \rangle \cong Q_3$, where $d = cba^6$ and Q_3 is a generalized quaternion group of order 16. Semidirect products $Q_n \leftthreetimes C_2$, $n \geq 3$, were investigated in [10]. It was proved that they are determined by their endomorphism semigroups in the class of all groups. In this paper, we shall describe the groups G_{17}, G_{20} and G_{27}, by their endomorphism semigroups (Theorems 14.1, 14.3 and 14.5). It follows from these descriptions that these groups are also determined by their endomorphism semigroups in the class of all groups (Theorems 14.2, 14.4 and 14.6).

We shall use the following notations:

G – a group; G' – the derived group of G; $Z(G)$ – the centre of G;
$o(g)$ – the order of an element $g \in G$;
$\mathrm{End}(G)$ – the endomorphism semigroup of G;
$H \leftthreetimes K$ – a semidirect product of a normal subgroup H and a subgroup K;
$\widehat{g}$ – the inner automorphism, generated by g ($h\widehat{g} = g^{-1}hg$);
$[g, h] = g^{-1}h^{-1}gh$ $(g, h \in G)$;
$I(G)$ – the set of all idempotents of $\mathrm{End}(G)$;
$K(x) = \{z \in \mathrm{End}(G) \mid zx = xz = z\}$ $(x \in \mathrm{End}(G))$;
$J(x) = \{z \in \mathrm{End}(G) \mid zx = xz = 0\}$ $(x \in \mathrm{End}(G))$;
$V(x) = \{z \in \mathrm{Aut}(G) \mid zx = x\}$ $(x \in \mathrm{End}(G))$;
$H(x) = \{z \in \mathrm{End}(G) \mid xz = z,\ zx = 0\}$ $(x \in \mathrm{End}(G))$;
$\mathbb{Z}_n$ – the ring of residue classes modulo n.

Assume that $x \in I(G)$. Then G decomposes into the semidirect product $G = \mathrm{Ker}x \leftthreetimes \mathrm{Im}x$ and it is easy to check that

$$K(x) = \{y \in \mathrm{End}(G) \mid (\mathrm{Ker}x)y = \langle 1 \rangle,\ (\mathrm{Im}x)y \subset \mathrm{Im}x\} \cong$$

$$\cong \mathrm{End}(\mathrm{Im}x), \tag{14.1}$$

$$H(x) = \{y \in \mathrm{End}(G) \mid (\mathrm{Ker}x)y = \langle 1 \rangle,\ (\mathrm{Im}x)y \subset \mathrm{Ker}x\}, \tag{14.2}$$

$$J(x) = \{y \in \mathrm{End}(G) \mid (\mathrm{Ker}x)y \subset (\mathrm{Ker}x),\ (\mathrm{Im}x)y = \langle 1 \rangle\}. \tag{14.3}$$

We shall write the mapping right from the element on which it acts. We shall use without any references the following two facts:

(1) If $y, z \in \mathrm{End}(G)$ and $yz = zy$, then $(\mathrm{Im}z)y \subset \mathrm{Im}z$ and $(\mathrm{Ker}z)y \subset \mathrm{Ker}z$.
(2) If $z \in \mathrm{End}(G)$ and $\mathrm{Im}z$ is abelian, then $\widehat{g} \in V(z)$ for each $g \in G$.

Remark that if G is a fixed group and H is an arbitrary group and the isomorphism of the groups G and H follows always from the isomorphism of the semigroups $\mathrm{End}(G)$ and $\mathrm{End}(H)$, then we say that the group G is determined by its endomorphism semigroup in the class of all groups.

14.2 The Group G_{17}

In this section, we shall characterize the group

$$G_{17} = \langle a, b, c \mid a^8 = b^2 = c^2 = 1,\ ab = ba,\ ac = ca,\ c^{-1}bc = ba^4 \rangle =$$

$$= (\langle a \rangle \times \langle b \rangle) \leftthreetimes \langle c \rangle = (\langle a \rangle \times \langle c \rangle) \leftthreetimes \langle b \rangle, \tag{14.4}$$

by its endomorphism semigroup.

Theorem 14.1. *A finite group G is isomorphic to the group G_{17} if and only if there exist $x, y \in I(G)$ which satisfy the following properties:*

1^0 $K(x) \cong K(y) \cong End(C_2)$; 2^0 $xy = yx = 0$; 3^0 $V(x)$ *is a 2-group;*
4^0 $I(G) \cap J(x) \cap J(y) = \{0\}$; 5^0 $|J(x) \cap J(y)| = 4$;
6^0 $|\{z \in End(G) \mid xz = z,\ zx = zy = 0\}| = 2$;
7^0 *there exists* $z \in J(x) \cap J(y)$ *such that* $z^2 \neq 0$ *and* $z^3 = 0$;
8^0 $\{z \in I(G) \mid zx = xz = x,\ zy = yz = y\} = \{1\}$.

Proof. Necessity. Let $G = G_{17}$ be given by (14.4). Denote by x and y the projections of G onto its subgroups $\langle c \rangle$ and $\langle b \rangle$, respectively. Then $x, y \in I(G)$, and direct calculations show that x and y satisfy properties 1^0–8^0.

Sufficiency. Assume that G is a finite group and $x, y \in I(G)$ such that properties 1^0–8^0 hold. By property 2^0, G splits up as follows

$$G = ((\mathrm{Ker}x \cap \mathrm{Ker}y) \leftthreetimes \mathrm{Im}x) \leftthreetimes \mathrm{Im}y = ((\mathrm{Ker}x \cap \mathrm{Ker}y) \leftthreetimes \mathrm{Im}y) \leftthreetimes \mathrm{Im}x.$$

By property 1^0 and condition (14.1), $K(x) \cong \mathrm{End}(\mathrm{Im}\,x) \cong \mathrm{End}(C_2)$. Similarly, $\mathrm{End}(\mathrm{Im}\,y) \cong \mathrm{End}(C_2)$. Since each finite abelian group is determined by its endomorphism semigroup in the class of all groups, we have $\mathrm{Im}\,x \cong \mathrm{Im}\,y \cong C_2$, i.e., $\mathrm{Im}\,x = \langle c \rangle$ and $\mathrm{Im}\,y = \langle b \rangle$ for some elements $c, b \in G$ of order 2. Hence

$$G = (M \leftthreetimes \langle c \rangle) \leftthreetimes \langle b \rangle = (M \leftthreetimes \langle b \rangle) \leftthreetimes \langle c \rangle,$$

where $M = \mathrm{Ker}\,x \cap \mathrm{Ker}\,y$ and $c^2 = b^2 = 1$.

Assume that $g \in G$ is a $2'$-element of G. Then $g \in M$. Since $\mathrm{Im}\,x = \langle c \rangle$ is abelian, $\widehat{g} \in V(x)$ and $\widehat{g}$ is a $2'$-element. Property 3^0 implies that $\widehat{g} = 1$, i.e., $g \in Z(G)$. Therefore, all $2'$-elements of G contain in the centre $Z(G)$ of G and G splits up into the direct product $G = G_{2'} \times G_2$ of its Hall $2'$-subgroup $G_{2'}$ and Sylow 2-subgroup G_2. Denote by z the projection of G onto its subgroup $G_{2'}$. Clearly, $z \in I(G) \cap J(x) \cap J(y)$. Property 4^0 implies $z = 0$, i.e., $G_{2'} = \langle 1 \rangle$ and G is a 2-group.

By property 7^0, the subgroup M of G is non-trivial. Choose an element $d \in M$ of order 2. There exist two endomorphisms z_0 and z_1 of G such that $cz_i = d^i$ and $Mz_i = \langle 1 \rangle$, $bz_i = 1$. These endomorphisms satisfy the equalities $xz_i = z_i$, $z_ix = z_iy = 0$. Therefore, by property 6^0, the subgroup M of G contains only one element of order 2, and hence it is cyclic or a generalized quaternion group [11, Theorem 5.3.6]. Since a product of two proper endomorphisms of a generalized quaternion group is equal to 0 [5, Theorem 14], property 7^0 implies that M is a cyclic subgroup of G, i.e., $M = \langle a \rangle \cong C_{2^n}$, and $n \geq 2$.

Let us show that $n = 3$. Clearly, $G' \subset M = \langle a \rangle$. Assume that $G' = \langle a^{2^m} \rangle$. Choose $z \in J(x) \cap J(y)$. Then $\mathrm{Im}\,z \subset \langle a \rangle$ and $\mathrm{Im}\,z$ is abelian. Hence $G' \subset \mathrm{Ker}\,z$ and z can be presented as a product of following homomorphisms:

$$G \xrightarrow{\varepsilon} G/G' = \langle aG' \rangle \times \langle bG' \rangle \times \langle cG' \rangle \xrightarrow{\pi} \langle aG' \rangle \xrightarrow{\tau} \langle a \rangle,$$

where ε is the natural homomorphism, π is a projection and $(aG')\tau = az$. Conversely, each such product of homomorphisms belongs to $J(x) \cap J(y)$. Hence property 5^0 implies $G' = \langle a^4 \rangle$ and $az = a^{i2^{n-2}}$ for some $i \in \mathbb{Z}_4$. By property 7^0, there exists $i \in \mathbb{Z}_4$ such that $z^2 \neq 0$ and $z^3 = 0$. For such z we have

$$az^2 = a^{i^2 2^{2n-4}} \neq 1, \quad az^3 = a^{i^3 2^{3n-6}} = 1.$$

It follows from here that $i \equiv 1 \pmod 2$, $2n - 4 < n$ and $3n - 6 \geq n$, i.e., $n = 3$.

We have already proved that $M = \langle a \rangle \cong C_8$ and $G' = \langle a^4 \rangle \cong C_2$. Note that $bc \neq cb$. Indeed, if $bc = cb$, then the map $z : G \longrightarrow G$, where $bz = b$, $cz = c$ and $Mz = \langle 1 \rangle$, can be uniquely extended to an endomorphism of G such that $xz = zx = x$, $zy = yz = y$ and $z \neq 1$. This contradicts property 8^0. Therefore, $bc \neq cb$ and $[b, c] = a^4$, i.e.,

$$c^{-1}bc = ba^4.$$

To prove that $G \cong G_{17}$, we consider two possible cases: $ab = ba$ or $ab \neq ba$.

Assume that $ab = ba$. If $ac = ca$, then the elements a, b, c satisfy the generating relations of G_{17} and, therefore, $G \cong G_{17}$. Suppose now that $ac \neq ca$. Then $[a, c] = a^4$, i.e., $c^{-1}ac = a^5$. Denote $\tilde{a} = ab$. Then $\tilde{a}$ is an element of order 8 of G, $G = \langle \tilde{a}, b, c \rangle$, $\tilde{a}b = b\tilde{a}$ and $c^{-1}\tilde{a}c = c^{-1}abc = c^{-1}ac \cdot c^{-1}bc = a^5ba^4 = ab = \tilde{a}$, $\tilde{a}c = c\tilde{a}$. Hence the elements $\tilde{a}, b, c$ satisfy the generating relations of G_{17} and $G \cong G_{17}$.

Assume now that $ab \neq ba$. Then $[a, b] = a^4$ and $b^{-1}ab = a^5$. Denote $\bar{a} = ac$. Then $\bar{a}^2 = acac = a^2[a, c] = a^2a^{4i}$ for some $i \in \mathbb{Z}_2$, $\bar{a}^4 = a^4$ and, therefore, $\bar{a}$ is an element of order 8. In addition, $\bar{a}b = acb = a \cdot c^{-1}bc \cdot c = a \cdot ba^4 \cdot c = b \cdot b^{-1}ab \cdot a^4c = ba^5a^4c = bac = b\bar{a}$. Similarly to the case $ab = ba$ (take now instead a the element $\bar{a}$), we can obtain that $G \cong G_{17}$. □

Theorem 14.2. *The group G_{17} is determined by its endomorphism semigroup in the class of all groups.*

Proof. Let G^* be a group such that the endomorphism semigroups of G^* and G_{17} are isomorphic: $\mathrm{End}(G_{17}) \cong \mathrm{End}(G^*)$. Denote by z^* the image of $z \in \mathrm{End}(G_{17})$ in this isomorphism. Since $\mathrm{End}(G^*)$ is finite, so is G^* [1, Theorem 2]. By Theorem 14.1, there exist $x, y \in I(G_{17})$, satisfying properties 1^0–8^0 of Theorem 14.1. These properties are formulated so that they are preserved in the isomorphism $\mathrm{End}(G_{17}) \cong \mathrm{End}(G^*)$. Therefore, the idempotents x^* and y^* of $\mathrm{End}(G^*)$ satisfy properties similar to properties 1^0–8^0 (it is necessary to change everywhere $z \in \mathrm{End}(G_{17})$ by $z^* \in \mathrm{End}(G^*)$). Using now Theorem 14.1 for G^*, it follows that G^* and G_{17} are isomorphic. □

14.3 The Group G_{20}

In this section, we shall characterize the group

$$G_{20} = \langle a, b, c \mid a^8 = b^2 = c^2 = 1,\ ab = ba,\ bc = cb,\ c^{-1}ac = ab \rangle =$$
$$= (\langle a \rangle \times \langle b \rangle) \leftthreetimes \langle c \rangle = (\langle b \rangle \times \langle c \rangle) \leftthreetimes \langle a \rangle, \tag{14.5}$$

by its endomorphism semigroup.

Theorem 14.3. *A finite group G is isomorphic to the group G_{20} if and only if $Aut(G)$ is a 2-group and there exist $x, y \in I(G)$ which satisfy the following properties:*

1^0 $K(x) \cong End(C_2)$; 2^0 $K(y) \cong End(C_8)$; 3^0 $xy = yx = 0$;
4^0 $J(x) \cap J(y) = \{0\}$; 5^0 $|\{z \in End(G) \mid yz = z,\ zx = zy = 0\}| = 2.$

Proof. Necessity. Let $G = G_{20}$ be given by (14.5). By [3], $|\mathrm{Aut}(G_{20})| = 2^6$. Denote by x and y the projections of G onto its subgroups $\langle c \rangle$ and $\langle a \rangle$, respectively. Then $x, y \in I(G)$, and direct calculations show that x and y satisfy properties 1^0–4^0.

Sufficiency. Assume that G is a finite group, $\mathrm{Aut}(G)$ is a 2-group and $x, y \in I(G)$ such that properties 1^0–4^0 hold. Similarly to the proof of Theorem 14.1, properties 1^0–4^0 imply that G is a 2-group such that $G = (M \leftthreetimes \langle c \rangle) \leftthreetimes \langle a \rangle = (M \leftthreetimes \langle a \rangle) \leftthreetimes \langle c \rangle$, where $M = \mathrm{Ker}\,x \cap \mathrm{Ker}\,y$, $\mathrm{Im}\,x = \langle c\rangle \cong C_2$, $\mathrm{Im}\,y = \langle a\rangle \cong C_8$. Hence $G/M = \langle aM\rangle \times \langle cM\rangle \cong C_8 \times C_2$ and an endomorphism z of G which satisfies equalities $yz = z$ and $zx = zy = 0$ is a product $z = \varepsilon\pi\tau$ of the natural homomorphism $G \xrightarrow{\varepsilon} G/M$, the projection $G/M \xrightarrow{\pi} \langle aM\rangle$ and a homomorphism $\langle aM\rangle \xrightarrow{\tau} M$. Conversely, for each homomorphism $\langle aM\rangle \xrightarrow{\tau} M$ the endomorphism $z = \varepsilon\pi\tau$ of G satisfies equalities $yz = z$ and $zx = zy = 0$. It follows from property 5^0 that the subgroup M of G has only one element of order 2 and does not contain elements of order 4. Therefore, by [11, Theorem 5.3.6], M is a cyclic group of order 2, i.e., $M = \langle b\rangle \cong C_2$ for an element b of G. Since M is an normal subgroup of G, we have $ba = ab$ and $cb = bc$.

The elements a and c do not commute. Indeed, if $ac = ca$, then $G = \langle b\rangle \times \langle a\rangle \times \langle c\rangle$ and the projection z of G onto $\langle b\rangle$ belongs to $J(x) \cap J(y)$ which contradicts property 4^0. Therefore, $ac \neq ca$, $[c,a] = c^{-1}a^{-1}ca \in M = \langle b\rangle \cong C_2$, i.e., $c^{-1}a^{-1}ca = b$ and $c^{-1}ac = ab$. It follows from here that the elements a, b and c of G satisfy the generating relations of the group G_{20}. Consequently, $G \cong G_{20}$. □

Theorem 14.4. *The group G_{20} is determined by its endomorphism semigroup in the class of all groups.*

The proof of Theorem 14.4 is similar to the proof of Theorem 14.2.

14.4 The Group G_{27}

In this section, we shall characterize the group

$$G_{27} = \langle a, b, c \mid a^8 = b^2 = c^2 = 1,\ ab = ba,\ bc = cb,\ c^{-1}ac = a^{-1}b\rangle =$$
$$= (\langle a\rangle \times \langle b\rangle) \leftthreetimes \langle c\rangle \cong (C_8 \times C_2) \leftthreetimes C_2 \qquad (14.6)$$

by its endomorphism semigroup.

Theorem 14.5. *A finite group G is isomorphic to the group G_{27} if and only if there exists $x \in I(G)$ which satisfies the following properties:*

1^0 $K(x) \cong End(C_2)$; 2^0 $J(x) \cap I(G) = \{0\}$; 3^0 $|J(x)| = 8$;
4^0 $|H(x)| = 4$; 5^0 $V(x)$ *is a 2-group;*
6^0 $|\{y \in I(G) \mid xy = y,\ yx = x\}| = 8$;
7^0 *there exists* $z \in H(x)$ *such that* $z \cdot J(x) \neq \{0\}$;
8^0 *if* $z \in V(x)$, $v \in H(x)$ *and* $u \in End(G)$ *such that* $xu = 0$ *and* $ux = u$, *then* $uvz = uv$.

Proof. Necessity. Let $G = G_{27}$ be given by (14.6). Denote by x the projection of G onto its subgroup $\langle c\rangle$. Then $x \in I(G)$ and direct calculations show that x satisfies properties 1^0–8^0.

Sufficiency. Assume that G is a finite group and $x \in I(G)$ such that properties 1^0–8^0 hold. Similarly to the proof of Theorem 14.1, properties 1^0, 2^0 and 5^0 imply that G is a 2-group such that $G = \mathrm{Ker}\,x \leftthreetimes \mathrm{Im}\,x$, $\mathrm{Im}\,x = \langle c\rangle \cong C_2$.

Following (14.2), there exists an one-to-one correspondence between $H(x)$ and Hom(Imx, Ker$x$). Hence by property 4^0, Ker$x$ has three elements of order 2. Since G is a 2-group, one of these belongs to the centre of G. Denote it by b_0. Let a_0 be an element of order 2 in Ker$x$ different from b_0. Then b_0a_0 is the third element of order 2 in Ker$x$.

Since Imx is abelian, we have $G' \subset \mathrm{Ker}x$. Let us prove that G' is a proper subgroup of Ker$x$. By contradiction, assume that $G' = \mathrm{Ker}x$. Choose $y \in J(x),\ y \neq 0$. By (14.3), $\mathrm{Im}x \subset \mathrm{Ker}y$. Therefore,

$$G/\mathrm{Ker}y = \{g \cdot \mathrm{Ker}y \mid g \in \mathrm{Ker}x\} = \{g \cdot \mathrm{Ker}y \mid g \in G'\} = (G/\mathrm{Ker}y)'.$$

This contradicts the well-known fact that a non-trivial finite 2-group does not coincide with its derived subgroup. Hence G' is a proper subgroup of Ker$x$ and the factor-group G/G' splits into direct product $G/G' = \langle a_1G'\rangle \times \ldots \times \langle a_kG'\rangle \times \langle cG'\rangle$ for some $k \geq 1$ ($a_i \in \mathrm{Ker}x \setminus G'$). Then $\varepsilon\pi\tau \in J(x)$, where $G \xrightarrow{\varepsilon} G/G'$ is the natural homomorphism, $G/G' \xrightarrow{\pi} \langle a_1G'\rangle \times \ldots \times \langle a_kG'\rangle$ is the projection and $\tau \in \mathrm{Hom}(\langle a_1G'\rangle \times \ldots \times \langle a_kG'\rangle; \langle a_0\rangle \times \langle b_0\rangle)$. By property 3^0, $k = 1$, i.e., $\mathrm{Ker}x/G' = \langle a_1G'\rangle$. Using again property 3^0 and the elements a_0, b_0, it is easy to check that $\langle a_1G'\rangle \cong C_4$ and

$$z \in J(x) \Longrightarrow G' \subset \mathrm{Ker}z. \tag{14.7}$$

Therefore, $G/G' = \langle aG'\rangle \times \langle cG'\rangle \cong C_4 \times C_2, (a = a_1)$. Remark that $a^4 \neq 1$, because otherwise the product of the natural homomorphism $G \longrightarrow G/G'$, the projection $G/G' \longrightarrow \langle aG'\rangle$ and the isomorphism $\langle aG'\rangle \cong \langle a\rangle$ is a non-zero element in $J(x) \cap I(G)$ which contradicts property 2^0.

It follows from property 7^0, (14.2), (14.3) and condition (14.7) that $\mathrm{Ker}x \setminus G' = \langle a, G'\rangle \setminus G'$ contains an element of order 2. Denote this element by b. Therefore, $b \in \langle a, G'\rangle \setminus G'$, $b^2 = 1$, $b \in \{a_0, b_0, a_0b_0\}$. Since G' has an element of order 2 and Ker$x$ has three elements of order 2, G' has only one element of order 2. Hence G' is cyclic or a generalized quaternion group. If G' is a generalized quaternion group, then it has at least six elements of order 4 and, therefore, we can construct much more than eight endomorphisms which belong to $J(x)$. This contradicts property 3^0. Consequently, G' is a cyclic group and $G' = \langle d\rangle \cong C_{2^m}$, $a^4 \in G'$, $a^4 \neq 1$. Clearly, $a^4 \in \langle d^2\rangle$, because otherwise $\langle a^4\rangle = \langle d\rangle$, $\mathrm{Ker}x = \langle a\rangle$ and this contradicts the fact that Ker$x$ has three elements of order 2. It follows also from here that $o(d) \geq 4$.

Let us prove that each element $h \in \mathrm{Ker}x$ of order 2 belongs to the centre $Z(G)$ of G. The maps u and v, where $cu = 1$, $G'u = \langle 1\rangle$, $au = c$, $(\mathrm{Ker}x)v = \langle 1\rangle$, $cv = h$, can be extended to endomorphisms of G such that $v \in H(x)$, $xu = 0$ and $ux = u$. Denote $z = uv$. Choose an arbitrary element $g \in G$. Since Imx is abelian, $\widehat{g} \in V(x)$. Hence z and $y = \widehat{g}$ satisfy the conditions of property 8^0. Therefore, $zy = z$, $z\widehat{g} = z$. Since $\mathrm{Im}z = \langle h\rangle$, we have $h\widehat{g} = h$, i.e., $hg = gh$ and $h \in Z(G)$. Particularly, $b \in Z(G)$ and $ab = ba$.

In view of $a^4 \neq 1$, we have $b \notin \langle a\rangle$ and hence $\langle a, b\rangle = \langle a\rangle \times \langle b\rangle$. Therefore, all eight elements of $J(x)$ can be obtained as products $\varepsilon\pi\tau_{ij}$, where $G \xrightarrow{\varepsilon} G/G' =$

$\langle aG' \rangle \times \langle cG' \rangle$ is the natural homomorphism, $G/G' \xrightarrow{\pi} \langle aG' \rangle$ is a projection and $(aG')\tau_{ij} = a^{i \cdot o(a)/4} b^j$ $(i \in \mathbb{Z}_4,\ j \in \mathbb{Z}_2)$.

Denote $N = \langle d^2 \rangle$ and $M = \langle a^2, N \rangle$. Then N and M are normal subgroups of G. Since $a^4 \in \langle d^2 \rangle$, we have $\mathrm{Ker}x/N = \langle aN \rangle \times \langle dN \rangle \cong C_4 \times C_2$,

$$\mathrm{Ker}x/M = \langle aM \rangle \times \langle dM \rangle \cong C_2 \times C_2,$$

$$G/M = (\langle aM \rangle \times \langle dM \rangle) \leftthreetimes \langle cM \rangle \cong (C_2 \times C_2) \leftthreetimes C_2.$$

If $cM \cdot aM = aM \cdot cM$, then we have $G/M = \langle aM \rangle \times \langle dM \rangle \times \langle cM \rangle$ and the endomorphism z of G defined by $Mz = \langle 1 \rangle$, $az = cz = 1$, $dz = b$, is an element of $J(x)$ which does not have the form shown above. Therefore, $cM \cdot aM \neq aM \cdot cM$, $[a, c] \notin M = \langle a^2, d^2 \rangle$, $\langle [a, c] \rangle = \langle d \rangle = G'$ and we can assume that $[a, c] = d$, i.e., $c^{-1}ac = ad$ and $a^{-1}ca = [a, c]c = dc$, $dc \cdot dc = a^{-1}c^2a = 1$, $c^{-1}dc = d^{-1}$. Hence for each integer i, the element cd^i of G is an element of order 2: $cd^i \cdot cd^i = c^{-1}d^ic \cdot d^i = d^{-i}d^i = 1$.

Let us define for each integer i an endomorphism z_i of G as

$$(\mathrm{Ker}x)z_i = \langle 1 \rangle, \quad cz_i = cd^i.$$

Define still $z \in \mathrm{End}(G)$ by $(\mathrm{Ker}x)z = \langle 1 \rangle$, $cz = cb$. Then $z \neq z_i$ and

$$z, z_i \in \{y \in I(G) \mid xy = y,\ yx = x\}. \tag{14.8}$$

By property 6^0, $o(d) \geq 4$ and (14.8), we have $o(d) = 4$. Therefore, $|\mathrm{Ker}x| = 16$, $o(a) = 8$, $|\langle a \rangle \times \langle b \rangle| = 16$ and $\mathrm{Ker}x = \langle a \rangle \times \langle b \rangle \cong C_8 \times C_2$.

Let us prove now that $G \cong G_{27}$. By the construction, $b = a^2d^i$ for some $i \in \mathbb{Z}_4$. Since $1 = b^2 = a^4d^{2i} = d^2d^{2i} = d^{2(i+1)}$, we have $i \equiv 1 \pmod 2$, i.e., $i \in \{1, 3\}$. If $i = 1$, then $d = ba^{-2}$, $c^{-1}ac = ad = a^{-1}b$ and a, b, c satisfy the generating relations of the group G_{27}, i.e., $G \cong G_{27}$. If $i = 3$, then $b = a^2d^3 = a^2d^{-1}$, $d = a^2b$, $c^{-1}ac = ad = a^{-1} \cdot a^4b$ and the map $u : G_{27} = \langle a, b, c \rangle \longrightarrow G = \langle a, b, c \rangle$ defined by $cu = c$, $au = a$, $bu = a^4b$, is an isomorphism, i.e., $G \cong G_{27}$. □

Theorem 14.6. *The group G_{27} is determined by its endomorphism semigroup in the class of all groups.*

The proof of Theorem 14.6 is similar to the proof of Theorem 14.2.

Acknowledgements This work was supported in part by the Estonian Science Foundation Research Grant 5900, 2004–2007.

References

1. Alperin, J.L.: Groups with finitely many automorphisms. Pacific J. Math.**12**, no 1, 1–5 (1962).
2. Gramushnjak, T., Puusemp P.: A characterization of a class of groups of order 32 by their endomorphism semigroups. Algebras, Groups and Geometries **22**, no 4, 387–412 (2005).

3. Hall, M., Jr., Senior, J.K.: The groups of order 2^n, $n \leq 6$. Macmillan, New York; Collier-Macmillan, London (1964).
4. Puusemp, P.: Idempotents of the endomorphism semigroups of groups. Acta et Comment. Univ. Tartuensis **366**, 76–104 (1975) (in Russian).
5. Puusemp, P.: Endomorphism semigroups of the generalized quaternion groups. Acta et Comment. Univ. Tartuensis **390**, 84–103 (1976) (in Russian).
6. Puusemp, P.: A characterization of divisible and torsion abelian groups by their endomorphism semigroups. Algebras, Groups and Geometries **16**, 183–193 (1999).
7. Puusemp, P.: Characterization of a semidirect product of groups by its endomorphism semigroup. In: Smith, P., Giraldes, E., Martins, P. (eds.) Proc. of the Int. Conf. on Semigroups, pp. 161–170. World Scientific, Singapore (2000).
8. Puusemp, P.: On endomorphism semigroups of symmetric groups. Acta et Comment. Univ. Tartuensis **700**, 42–49 (1985) (in Russian).
9. Puusemp, P.: Groups of Order Less Than 32 and Their Endomorphism Semigroups. J. Nonlinear Math. Phys. **13**, Supplement, 93–101 (2006).
10. Puusemp, P.: Semidirect products of generalized quaternion groups by a cyclic group. This proceedings.
11. Robinson, D.J.S.: A Course in the Theory of Groups. Springer, New York (1996).

Chapter 15
Adjoint Representations and Movements

Maido Rahula and Vitali Retšnoi

Abstract The aim of this paper is to develop the theory of higher order movements by using the structure of multiple tangent bundles. In this context the meaning of invariants of polynomials and of matrices is explained. The connection with central and raw moments appearing in probability theory is established

15.1 Introduction

Let M be a smooth manifold and $b : M \to M$ be a smooth map. When a transformation (diffeomorphism) $a : M \to M$ is given, b is transformed to $\widetilde{b} = aba^{-1}$ so that the following diagram commutes:

$$\begin{array}{ccc} M & \xrightarrow{\quad b \quad} & M \\ {\scriptstyle a}\Big\downarrow & & \Big\downarrow{\scriptstyle a} \\ M & \xrightarrow{\quad \widetilde{b} \quad} & M \end{array}$$

Suppose G is a group of transformations of M and $b \in G$. Then one can speak about the interior automorphism and adjoint representation of G. In particular, given a vector field X and its flow a_t, b is continuously transformed as follows: $b \rightsquigarrow b_t = a_t b a_t^{-1}$. Any other vector field Y is transformed to $\widetilde{Y} = Ta_t Y$, and the flow of Y is transformed according to the rule $b_s \rightsquigarrow a_t b_s a_t^{-1}$. Thus one can speak about a *movement of movement* and about *movements of higher orders*, or as we say movements in higher floors, see [4,5,7,8].

M. Rahula and V. Retšnoi
Institute of Mathematics, University of Tartu, J. Liivi Str., 2-613, 50409 Tartu, Estonia
e-mail: maido.rahula@ut.ee, vitali@ut.ee

S. Silvestrov et al. (eds.), *Generalized Lie Theory in Mathematics, Physics and Beyond*,

15.2 Generalized Leibnitz Rule

Let M_1, M_2 and M be smooth manifolds. The tangent functor T assigns to a Cartesian product $M_1 \times M_2$ a vector space

$$T(M_1 \times M_2) = (TM_1 \times M_2) \otimes (M_1 \times TM_2);$$

and to a smooth map $\lambda : M_1 \times M_2 \to M, (u,v) \mapsto w = u \cdot v$, its tangent map

$$T\lambda : T(M_1 \times M_2) \to TM, \quad (u_1, v_1) \mapsto w_1 = u_1 \cdot v + u \cdot v_1.$$

There are two smooth maps

$$\begin{aligned} \lambda_v &: M_1 \to M, \quad u \mapsto w = u \cdot v, \quad \forall v \in M_2, \\ \lambda_u &: M_2 \to M, \quad v \mapsto w = u \cdot v, \quad \forall u \in M_1, \end{aligned}$$

associated with λ. Namely, two vectors $u_1 \in T_u M_1$ and $v_1 \in T_v M_2$ at $u \in M_1$ and $v \in M_2$, respectively, are mapped into vectors $T\lambda_v(u_1) = u_1 \cdot v$ and $T\lambda_u(v_1) = u \cdot v_1$ at $w \in M$. The sum of these images is equal to the image of (u_1, v_1) under $T\lambda$:

$$w = u \cdot v \quad \Longrightarrow \quad w_1 = u_1 \cdot v + u \cdot v_1. \tag{15.1}$$

This formula generalizes the ordinary *Leibnitz rule*.

Locally, (15.1) can be presented as the system

$$w^\rho \circ \lambda = \lambda^\rho(u^i, v^\alpha) \quad \Longrightarrow \quad w_1^\rho = \frac{\partial \lambda^\rho}{\partial u^i} u_1^i + \frac{\partial \lambda^\rho}{\partial v^\alpha} v_1^\alpha,$$

where u^i, v^α, w^ρ are coordinates of u, v, w, and $u_1^i, v_1^\alpha, w_1^\rho$ are components of u_1, v_1, w_1, respectively, on corresponding coordinate neighborhoods $U_1 \subset M_1, U_2 \subset M_2$ and $U \subset M$, $i = 1, \ldots, \dim M_1$, $\alpha = 1, \ldots, \dim M_2$, $\rho = 1, \ldots, \dim M$.

15.3 Tangent Group

Let G be a Lie group with the composition law $\gamma : (a,b) \mapsto c = ab$. This group acts on itself by right and left translations:

$$\begin{aligned} r_b &: G \to G, \quad a \mapsto ab, \quad \forall b \in G, \\ l_a &: G \to G, \quad b \mapsto ab, \quad \forall a \in G. \end{aligned}$$

The tangent bundle (*first floor*) TG becomes automatically a Lie group with the composition law

$$T\gamma : TG \times TG \to TG, \quad (a_1, b_1) \mapsto c_1 = a_1 b + a b_1.$$

Two vectors $a_1 \in T_a G$ and $b_1 \in T_b G$ are mapped by $T\gamma$ to the vector $a_1 b + a b_1 = Tr_b(a_1) + Tl_a(b_1) \in T_c G$:

$$c = ab \quad \Longrightarrow \quad c_1 = a_1 b + a b_1. \tag{15.2}$$

The group TG is called a *tangent group* of G. The unit of TG is precisely the zero vector at the unit $e \in G$. The inverse element for $a_1 \in T_aG$ is defined by

$$a_1^{-1} = -a^{-1}a_1a^{-1} \in T_{a^{-1}}G.$$

Systematically applying left and right translations l_a and r_a to any vector $e_1 \in T_eG$ one can obtain left-invariant and right-invariant vector fields ae_1 and e_1a, respectively. The interior automorphism $A_a = l_a \circ r_a^{-1}$ yields the transformation $T_eA_a : e_1 \mapsto ae_1a^{-1}$ in T_eG. Let a basis (frame) be given in T_eG. Then there exists a homomorphism $a \mapsto A(a)$ from G to the general linear group $GL(\dim G, \mathbb{R})$ with center of G $Z \subset G$ as a kernel. An adjoint representation of TG is defined as follows:

$$c = aba^{-1} \quad \Longrightarrow \quad c_1 = (a_1a^{-1})c - c(a_1a^{-1}) + ab_1a^{-1}. \tag{15.3}$$

For $b_1 = 0$ this formula determines an action of TG on G, and for $a_1 = 0$ – an action of G on the group TG.

With any one-parameter subgroup a_t of G we associate one-parameter groups of transformations r_{a_t}, l_{a_t} and A_{a_t} of G that in their turn are generated by a left-invariant vector field X, a right-invariant vector field $\widetilde{X}$ and the field $\widetilde{X} - X$, respectively, i.e.

$$r_{a_t} = \exp tX, \quad l_{a_t} = \exp t\widetilde{X}, \quad A_{a_t} = \exp t(\widetilde{X} - X).$$

This follows from the formulas

$$Xf = (f \circ r_{a_t})'_{t=0}, \quad \widetilde{X}f = (f \circ l_{a_t})'_{t=0}, \quad (\widetilde{X} - X)f = (f \circ A_{a_t})'_{t=0},$$

where f is an arbitrary function on G, and from the fact that left and right translations commute.

15.4 Linear Group $GL(2, \mathbb{R})$

Consider the linear group $GL = GL(2, \mathbb{R})$ and its Lie algebra $gl = gl(2, \mathbb{R})$. Let $A = \begin{pmatrix} a_1 & a_2 \\ a_3 & a_4 \end{pmatrix} \in GL$ and $C = \begin{pmatrix} c_1 & c_2 \\ c_3 & c_4 \end{pmatrix} \in gl$. The exponential map $\kappa : gl \to GL$ takes each additive subgroup $Ct, t \in \mathbb{R}$, to the one-parameter subgroup e^{Ct}, called the exponential of Ct. The left-invariant basis (X_i, ω^i) and right-invariant basis $(\widetilde{X}_i, \omega^i)$, i.e. the corresponding fields of frames and coframes in GL, can be defined as follows:

$$\begin{pmatrix} X_1 & X_2 \\ X_3 & X_4 \end{pmatrix} = \begin{pmatrix} a_1 & a_3 \\ a_2 & a_4 \end{pmatrix} \begin{pmatrix} \partial_1 & \partial_2 \\ \partial_3 & \partial_4 \end{pmatrix}, \quad \begin{pmatrix} \omega^1 & \omega^2 \\ \omega^3 & \omega^4 \end{pmatrix} = \begin{pmatrix} a_1 & a_2 \\ a_3 & a_4 \end{pmatrix}^{-1} \begin{pmatrix} da_1 & da_2 \\ da_3 & da_4 \end{pmatrix},$$

$$\begin{pmatrix} \widetilde{X}_1 & \widetilde{X}_2 \\ \widetilde{X}_3 & \widetilde{X}_4 \end{pmatrix} = \begin{pmatrix} \partial_1 & \partial_2 \\ \partial_3 & \partial_4 \end{pmatrix} \begin{pmatrix} a_1 & a_3 \\ a_2 & a_4 \end{pmatrix}, \quad \begin{pmatrix} \widetilde{\omega}^1 & \widetilde{\omega}^2 \\ \widetilde{\omega}^3 & \widetilde{\omega}^4 \end{pmatrix} = \begin{pmatrix} da_1 & da_2 \\ da_3 & da_4 \end{pmatrix} \begin{pmatrix} a_1 & a_2 \\ a_3 & a_4 \end{pmatrix}^{-1},$$

where $\partial_i = \frac{\partial}{\partial a_i}$, $i = 1, 2, 3, 4$.

Let GL acts on itself by interior automorphisms. Then there exist four basic operators $Y_i = \widetilde{X}_i - X_i$, $i = 1,2,3,4$, that are corresponded to the adjoint representation of GL. Actually, they are linearly dependent: $Y_1 + Y_4 = 0$, $(a_1 - a_4)Y_1 + a_2Y_2 + a_3Y_3 = 0$. Note that the functions

$$\det A = a_1a_4 - a_2a_3, \quad \operatorname{tr} A = a_1 + a_2$$

are common invariants of Y_i, $i = 1,2,3,4$. Thus the integral distribution spanned by Y_i is two-dimensional and its integral surfaces are precisely two-dimensional quadrics (the family of hyperboloids). The operator in general form

$$Y = c_1Y_1 + c_2Y_2 + c_3Y_3 + c_4Y_4,$$

as a linear vector field in GL, determines the system of ODEs and the flow (see [5, p. 49], [6]):

$$A' = CA - AC \quad \Longrightarrow \quad A_t = e^{Ct}Ae^{-Ct}.$$

The corresponding table of commutators looks as follows:

↱	Y_1	Y_2	Y_3	Y_4
Y_1	0	Y_2	$-Y_3$	0
Y_2	$-Y_2$	0	$Y_1 - Y_4$	Y_2
Y_3	Y_3	$Y_4 - Y_1$	0	Y_3
Y_4	0	$-Y_2$	Y_3	0

It allows to describe the dragging of Y_i in the flow of Y (primes here denote the Lie derivative with respect to Y):

$$\begin{aligned}
\begin{pmatrix} Y_1 & Y_2 \\ Y_3 & Y_4 \end{pmatrix}' &= \begin{pmatrix} Y_1 & Y_2 \\ Y_3 & Y_4 \end{pmatrix}\begin{pmatrix} c_1 & c_2 \\ c_3 & c_4 \end{pmatrix} - \begin{pmatrix} c_1 & c_2 \\ c_3 & c_4 \end{pmatrix}\begin{pmatrix} Y_1 & Y_2 \\ Y_3 & Y_4 \end{pmatrix}, \\
\begin{pmatrix} Y_1 & Y_2 \\ Y_3 & Y_4 \end{pmatrix}'' &= \operatorname{tr} C \begin{pmatrix} Y_1 & Y_2 \\ Y_3 & Y_4 \end{pmatrix}', \\
\operatorname{tr} C \neq 0 \Longrightarrow \begin{pmatrix} Y_1 & Y_2 \\ Y_3 & Y_4 \end{pmatrix}_t &= \frac{e^{t \operatorname{tr} C} - 1}{\operatorname{tr} C}\begin{pmatrix} Y_1 & Y_2 \\ Y_3 & Y_4 \end{pmatrix}' + \begin{pmatrix} Y_1 & Y_2 \\ Y_3 & Y_4 \end{pmatrix}, \\
\operatorname{tr} C = 0 \Longrightarrow \begin{pmatrix} Y_1 & Y_2 \\ Y_3 & Y_4 \end{pmatrix}_t &= t\begin{pmatrix} Y_1 & Y_2 \\ Y_3 & Y_4 \end{pmatrix}' + \begin{pmatrix} Y_1 & Y_2 \\ Y_3 & Y_4 \end{pmatrix}.
\end{aligned}$$

Proposition 15.1. *For each operator* $\frac{\partial}{\partial c_i}$, $i = 1,2,3,4$, *in gl there is defined a one-parameter subgroup of GL such that there are operators* X_i, $\widetilde{X}_i$ *and* Y_i, $i = 1,2,3,4$, *associated with it and induced by the corresponding left and right translations, and interior automorphisms, respectively:*

$$\frac{\partial}{\partial c_1} \rightsquigarrow X_1 = a_1\partial_1 + a_3\partial_3, \quad \widetilde{X}_1 = a_1\partial_1 + a_2\partial_2, \quad Y_1 = a_2\partial_2 - a_3\partial_3;$$

$$\frac{\partial}{\partial c_2} \rightsquigarrow X_2 = a_1\partial_2 + a_3\partial_4, \quad \widetilde{X}_2 = a_3\partial_1 + a_4\partial_2, \quad Y_2 = a_3(\partial_1 - \partial_4) - (a_1 - a_4)\partial_2;$$

$$\frac{\partial}{\partial c_3} \rightsquigarrow X_3 = a_2\partial_1 + a_4\partial_3, \quad \widetilde{X}_3 = a_1\partial_3 + a_2\partial_4, \quad Y_3 = -a_2(\partial_1 - \partial_4) + (a_1 - a_4)\partial_3;$$

$$\frac{\partial}{\partial c_4} \rightsquigarrow X_4 = a_2\partial_2 + a_4\partial_4, \quad \widetilde{X}_4 = a_3\partial_3 + a_4\partial_4, \quad Y_4 = -a_2\partial_2 + a_3\partial_3.$$

The proof of this proposition is based on the exponential map κ.

15.5 The Operator of Center

Consider the vector field

$$P = \frac{\partial}{\partial c_1} + \frac{\partial}{\partial c_4},$$

which corresponds to the center of the Lie group gl. The image of P under $T\kappa$ is the homothety operator

$$T\kappa P = a_1\partial_1 + a_2\partial_2 + a_3\partial_3 + a_4\partial_4$$

in GL. Let $s = \frac{1}{2}$ tr C be the canonical parameter of P. Then using the implication $U' = CU \Rightarrow U_t = e^{Ct}U \Rightarrow I = e^{-Cs}U$, see [5], p.49, [6], we obtain its flow and three basic invariants of P:

$$C' = E \quad \Longrightarrow \quad C_t = C + tE \quad \Longrightarrow \quad I = C - sE,$$

where E is the unit matrix. Define the projection $\pi : gl \to \mathbb{R}^3$ by

$$\begin{cases} x \circ \pi = \frac{1}{2}(c_1 - c_4), \\ y \circ \pi = c_2, \\ z \circ \pi = c_3. \end{cases}$$

Thus gl is projected by π onto the xyz space of invariants of P. Using the fact that tr $I = 0$ together with $\det e^C = e^{\operatorname{tr} C}$ we obtain $\det e^I = 1$. It means that the xyz space is the tangent space to the subgroup of matrices with determinant 1. From

$$P(\det C) = \operatorname{tr} C, \quad P^2(\det C) = P(\operatorname{tr} C) = 2$$

it follows that $\det C$ behaves in the flow of P as a quadratic polynomial on t and tr C as a linear one, i.e.

$$(\det C)_t = \det C + \operatorname{tr} C \cdot t + t^2, \quad (\operatorname{tr} C)_t = \operatorname{tr} C + 2t.$$

The substitution of $t = -s$ into $(\det C)_t$ gives us the important invariant

$$\Delta = \frac{1}{4}(\operatorname{tr}^2 C - 4\det C).$$

After regrouping of terms we get $\Delta = (x^2 + yz) \circ \pi$. Note that 4Δ is the discriminant of the quadratic polynomial $(\det C)_t$. The eigenvalues of C are equal up to a sign to the roots of the polynomial $(\det C)_t$. Consider the dragging of a hyper-quadric $\det C = k$, $k \in \mathbb{R}$, along the flow of P in $gl \equiv \mathbb{R}^4$. Then we obtain a one-parameter family of hyper-quadrics defined by $(\det C)_t = k$. The intersection of $\det C = k$ with the plane tr $C = 0$ corresponds to the characteristic which consists of points at that the trajectories of P are tangent to $\det C = k$. The image of this intersection in xyz space is a hyperboloid $x^2 + yz = -k$, which is either hyperboloid of one sheet $(k < 0)$ or two sheets $(k > 0)$, or the light cone $(k = 0)$. The envelope of the family of surfaces is defined by $\Delta = -k$, which is the projecting cylinder with the trajectories of P as generating lines tangent to the characteristic. One can imagine an illumination of the surface $\det C = k$ along the trajectories of P and a shadow on the xyz-screen, the border of which is precisely the hyperboloid $x^2 + yz = -k$. In this case we deal with the cusp singularity of the first type called *fold*, see [1], p.157.

The hyperboloids $x^2 + yz = \text{const}$ in xyz space are precisely the orbits of a generalized group of rotations (see [5], p.77) with operators

$$\widetilde{Y}_1 = y\frac{\partial}{\partial y} - z\frac{\partial}{\partial z}, \quad \widetilde{Y}_2 = z\frac{\partial}{\partial x} - 2x\frac{\partial}{\partial y}, \quad \widetilde{Y}_3 = -y\frac{\partial}{\partial x} + 2x\frac{\partial}{\partial z}.$$

Here $\widetilde{Y}_2$ and $\widetilde{Y}_3$ have parabolic flows and $\widetilde{Y}_1$ has a hyperbolic one. The field $\widetilde{Y}_1$ is an *infinitesimal symmetry* for $\widetilde{Y}_2$ and $\widetilde{Y}_3$. This way we have described the structure of the quotient group $G_1 = G/Z$.

15.6 Discriminant Parabola

Define the map $\zeta : gl \to \mathbb{R}^2$ by $\begin{cases} u \circ \zeta &= \frac{1}{2}\det C, \\ u' \circ \zeta &= \frac{1}{2}\text{tr}\, C, \end{cases}$ where (u, u') denote the coordinates in $\mathbb{R}^2$. The operator P is projected by $T\zeta$ to the vector field $T\zeta P = u'\frac{\partial}{\partial u} + \frac{\partial}{\partial u'}$. Thus the flow $C_t = C + tE$ of P is mapped by ζ to the flow of $T\zeta P$ determined by the system

$$\begin{cases} u_t &= u + u't + \frac{t^2}{2}, \\ u'_t &= u' + t. \end{cases}$$

From $((u')^2 - 2u) \circ \zeta = \Delta$ it follows that the discriminant of the quadratic function u_t is ζ-related to the invariant Δ. Being mapped by ζ, the generating lines of the cylinder $\Delta = 0$ lay down onto the *discriminant parabola* $(u')^2 - 2u = 0$ in $\mathbb{R}^2$.

The classification of linear flows in dimension 2 takes place in the uu' plane with respect to the discriminant parabola, see [2, p. 86], [5, p. 73]. From $U' = CU \Rightarrow U_t = e^{Ct}U$ it follows that a linear flow is determined by the exponential e^{Ct}. The eigenvalues of C satisfy the quadratic equation

$$\lambda^2 - \text{tr}\, C \cdot \lambda + \det C = 0.$$

Depending on the sign of Δ, the eigenvalues may be real: $\lambda_{1,2} = \alpha \pm \beta$, or complex conjugate: $\lambda_{1,2} = \alpha \pm i\beta$, or equal: $\lambda_1 = \lambda_2 = \alpha$, where $\alpha, \beta \in \mathbb{R}$ are given by

$$\operatorname{tr} C = \lambda_1 + \lambda_2 = 2\alpha, \quad \det C = \lambda_1 \lambda_2 = \alpha^2 \pm \beta^2.$$

Here the plus sign corresponds to the case of complex roots and minus to the case of real ones. Thus

$$\Delta = \operatorname{tr}^2 C - 4\det C = (\lambda_1 - \lambda_2)^2 = \mp 4\beta^2$$

and e^{Ct} depends on the sign of Δ as follows:

$$e^{Ct} = \begin{cases} e^{\alpha t}\left[E\cos\beta t + (C-\alpha E)\frac{\sin\beta t}{\beta}\right], & \text{if } \Delta < 0, \\ e^{\alpha t}\left[E\cosh\beta t + (C-\alpha E)\frac{\sinh\beta t}{\beta}\right], & \text{if } \Delta > 0, \\ e^{\alpha t}\left[E + (C-\alpha E)t\right], & \text{if } \Delta = 0. \end{cases}$$

The various possible flow patterns can be summarized as follows:
$\Delta < 0, \det C < 0$ – elliptic flow with focuses;
$\Delta < 0, \det C > 0$ – hyperbolic flow with saddles;
$\Delta > 0$ – hyperbolic knots;
$\Delta = 0$ – parabolic knots.
Depending on the sign of α the knots and focuses may be stable $(\alpha < 0)$ or unstable $(\alpha > 0)$.

15.7 Relations to Moments in Probability Theory

Consider the raw and central moments

$$\nu_k = EX^k, \quad \mu_k = E(X - EX)^k,$$

where X is a random variable and E denotes the expectation value, see [3]. The central moments μ_k, $k = 1, 2, \ldots$, can be expressed in terms of ν_k. The first few cases are given by

$$\begin{aligned} \mu_1 &= 0, \\ \mu_2 &= \nu_2 - \nu_1^2, \\ \mu_3 &= \nu_3 - 3\nu_2\nu_1 + 2\nu_1^3 \\ \mu_4 &= \nu_4 - 4\nu_3\nu_1 + 6\nu_2\nu_1^2 - 3\nu_1^4, \\ &\vdots \end{aligned}$$

Let us show for $k = 2, 3, 4$ that if we identify (up to a constant multiplier) the coefficients in u_t with raw moments, then the invariants of u_t agree with the central ones.

The case $k = 2$. Denote

$$u_t = \frac{1}{2}E(X+t)^2, \qquad u = \frac{1}{2}EX^2 = \frac{1}{2}\nu_2,$$
$$u'_t = E(X+t), \qquad u' = EX = \nu_1.$$

Then

$$u_t = u + u't + \frac{t^2}{2},$$
$$u'_t = u' + t.$$

The substitution $t = -u'$ gives us the fiber invariant that is equal (up to the coefficient) to the central moment μ_2:

$$\Delta = u - \frac{1}{2}(u')^2, \quad \Delta = \frac{1}{2}\mu_2.$$

The case $k = 3$. Denote

$$u_t = \frac{1}{3!}E(X+t)^3, \qquad u = \frac{1}{3!}EX^3 = \frac{1}{3!}\nu_3,$$
$$u'_t = \frac{1}{2}E(X+t)^2, \qquad u' = \frac{1}{2}EX^2 = \frac{1}{2}\nu_2,$$
$$u''_t = E(X+t), \qquad u'' = EX = \nu_1.$$

Then

$$u_t = u + u't + u''\frac{t^2}{2} + \frac{t^3}{3!},$$
$$u'_t = u' + u''t + \frac{t^2}{2},$$
$$u''_t = u'' + t.$$

The substitution $t = -u''$ gives us two fiber invariants equal (up to the coefficients) to the central moments μ_3 and μ_2, respectively:

$$i_0 = u - u'u'' + \frac{1}{3}(u'')^3, \qquad i_0 = \frac{1}{3!}\mu_3,$$
$$i_1 = u' - \frac{1}{2}(u'')^2, \qquad i_1 = \frac{1}{2}\mu_2.$$

Note that the cubic discriminant of u_t expressed in terms of i_0 and i_1 is

$$I = (3i_0)^2 + (2i_1)^3.$$

The case $k = 4$. Denote

$$u_t = \frac{1}{4!}E(X+t)^4, \qquad u = \frac{1}{4!}EX^4 = \frac{1}{4!}\nu_4,$$

$$u'_t = \frac{1}{3!}E(X+t)^3, \qquad u' = \frac{1}{3!}EX^3 = \frac{1}{3!}\nu_3,$$
$$u''_t = \frac{1}{2}E(X+t)^2, \qquad u'' = \frac{1}{2}EX^2 = \frac{1}{2}\nu_2,$$
$$u'''_t = E(X+t), \qquad u''' = EX = \nu_1.$$

Then

$$u_t = u + u't + u''\frac{t^2}{2} + u'''\frac{t^3}{3!} + \frac{t^4}{4!},$$
$$u'_t = u' + u''t + u'''\frac{t^2}{2} + \frac{t^3}{3!},$$
$$u''_t = u'' + u'''t + \frac{t^2}{2},$$
$$u'''_t = u''' + t.$$

The substitution $t = -u'''$ gives us three fiber invariants I_0, I_1, I_2 equal (up to the coefficients) to the central moments μ_4, μ_3 and μ_2, respectively:

$$I_0 = u - u'u''' + \frac{1}{2}u''(u''')^2 - \frac{1}{8}(u''')^4, \qquad I_0 = \frac{1}{4!}\mu_4,$$
$$I_1 = u' - u''u''' + \frac{1}{3}(u''')^3, \qquad I_1 = \frac{1}{3!}\mu_3,$$
$$I_2 = u'' - \frac{1}{2}(u''')^2, \qquad I_2 = \frac{1}{2}\mu_2.$$

These three cases are united by the common scheme (*exponential law*)

$$U' = CU \quad \Longrightarrow \quad U_t = e^{Ct}U \quad \Longrightarrow \quad I = e^{-Ct}U$$

for calculating invariants of total differentiation operator

$$D = \frac{\partial}{\partial t} + u'\frac{\partial}{\partial u} + u''\frac{\partial}{\partial u'} + u'''\frac{\partial}{\partial u''} + \cdots,$$

taking into account $u' = 1$, $u'' = 1$, $u''' = 1, \ldots$, respectively, see [5, p. 49], [6].

15.8 Conclusion

In the previous section we have shown that the central moments from probability theory are invariants of polynomials given by the raw moments. The analogous situation can be observed with moments appearing in mechanics. For instance, the statistic moment in mechanics corresponds to the raw moment $\nu_1 = EX$, the moment of inertia to the central moment $\mu_2 = E(X - EX)^2$ (dispersion) and so on. Rewrite the discriminant of the quadratic function u_t in the form $\Delta = uu'' - \frac{1}{2}(u')^2$, and suppose that u, u' and u'' denote initial path, velocity and acceleration, respectively. Then the first summand in Δ is understood as a potential energy and the second one as a kinetic energy. Thus the equality $\Delta' = 0$ presents the conservation law of energy.

Any matrix C is understood as a linear element of movement. Any perturbation of a movement leads to an interior automorphism of C. The coefficients appearing in Hamilton–Cayley formula for C gives us the matrix invariants under interior automorphisms, including $\det C$ and $\operatorname{tr} C$. The invariants of the next perturbation caused by the vector field P (or D after the map ζ) are precisely the discriminant Δ in dimension 2 ($k = 2$) and the discriminant I in dimension 3 ($k = 3$), and so on.

In general, a tangent vector as an element of a tangent bundle TM of a manifold M is understood as a stop-frame of a movement of order 1, and an element of a k-th order tangent bundle T^kM as a stop-frame of a movement of order k.

References

1. Th. Bröker, L. Lander, Differentiable Germs and Catastrophes, Cambridge University Press, Cambridge, 1975.
2. B.J. Cantwell, Introduction to Symmetry Analysis, Cambridge University Press, Cambridge, 2002.
3. M.G. Kendall, A. Stuart, The Advanced Theory of Statistics, 1, Distribution Theory, Griffin, London, 1966.
4. M. Rahula, New Problems in Differential Geometry, World Scientific, Singapore, 1993.
5. M. Rahula, Vector Fields and Symmetries, University of Tartu Press, Tartu, 2004 (in Russian).
6. M. Rahula, V. Retšnoi, Total Differentiation Under Jet Composition, Proc. of AGMF, *Jour. of Nonlin. Math. Ph.*, 2006, pp.102–109.
7. M. Rahula, Les invariants des monvements, Proc. of the 4th Int. Colloquium of Math. in Engenering and Numerical Physics, Geometry Balkan Press, Bucharest, Pomania, 2007, 145–153.
8. J.E. White, The Methot of Iterated Tangents with Applications in Local Reimannian Geometry, Pitman, Boston, 1982.

Chapter 16
Applications of Hypocontinuous Bilinear Maps in Infinite-Dimensional Differential Calculus

Helge Glöckner

Abstract Paradigms of bilinear maps $\beta\colon E_1 \times E_2 \to F$ between locally convex spaces (like evaluation or composition) are not continuous, but merely hypocontinuous. We describe situations where, nonetheless, compositions of β with Keller C^n_c-maps (on suitable domains) are C^n_c. Our main applications concern holomorphic families of operators, and the foundations of locally convex Poisson vector spaces.

16.1 Introduction

If $\beta\colon E_1 \times E_2 \to F$ is a continuous bilinear map between locally convex spaces, then β is smooth and hence also $\beta \circ f\colon U \to F$ is smooth for each smooth map $f\colon U \to E_1 \times E_2$ on an open subset U of a locally convex space.

Unfortunately, many relevant bilinear maps are discontinuous. For example, the evaluation map $E' \times E \to \mathbb{R}$, $(\lambda, x) \mapsto \lambda(x)$ is discontinuous for each locally convex vector topology on E', if E is a non-normable locally convex space [18, 22]. Hence also the composition map $L(F,G) \times L(E,F) \to L(E,G)$, $(A,B) \mapsto A \circ B$ is discontinuous, for any non-normable locally convex space F, locally convex spaces $E, G \neq \{0\}$, and any locally convex vector topologies on $L(F,G)$, $L(E,F)$ and $L(E,G)$ such that the maps $F \to L(E,F)$, $y \mapsto y \otimes \lambda$ and $L(E,G) \to G$, $A \mapsto A(x)$ are continuous for some $\lambda \in E'$ and some $x \in E$ with $\lambda(x) \neq 0$, where $(y \otimes \lambda)(z) := \lambda(z)y$ (see Remark 16.21; cf. [22] for related results).

Nonetheless, evaluation and composition exhibit a certain weakened continuity property, namely *hypocontinuity*. So far, hypocontinuity arguments have been used in differential calculus on Fréchet spaces in isolated cases (cf. [13] and [30]). In this article, we distill a simple, but useful general principle from these arguments (which is a variant of a result from [28]). Let us say that a Hausdorff topological

H. Glöckner
Institut für Mathematik, Universität Paderborn, Warburger Str. 100, 33098 Paderborn, Germany
e-mail: glockner@math.uni-paderborn.de

S. Silvestrov et al. (eds.), *Generalized Lie Theory in Mathematics, Physics and Beyond,*

space X is a *k^∞-space* if X^n is a k-space for each $n \in \mathbb{N}$. Our observation (recorded in Theorem 16.26) is the following:

If a bilinear map $\beta\colon E_1 \times E_2 \to F$ is hypocontinuous with respect to compact subsets of E_1 or E_2 and $f\colon U \to E_1 \times E_2$ is a C^n-map on an open subset U of a locally convex space X which is a k^∞-space, then $\beta \circ f\colon U \to F$ is a C^n-map.

As a byproduct, we obtain an affirmative solution to an old open problem by Serge Lang (see Corollary 16.27). Our main applications concern two areas.

Application 1: Holomorphic families of operators

In Sect. 16.4, we apply our results to holomorphic families of operators, i.e., holomorphic maps $U \to L(E,F)$ on an open set $U \subseteq \mathbb{C}$ (or $U \subseteq X$ for a locally convex space X). We obtain generalizations (and simpler proofs) for some results from [6].

Application 2: Locally convex Poisson vector spaces

Finite-dimensional Poisson vector spaces arise in finite-dimensional Lie theory as the dual spaces $\mathfrak{g}^*$ of finite-dimensional Lie algebras. The Lie bracket gives rise to a distribution on $\mathfrak{g}^*$ whose maximal integral manifolds are the coadjoint orbits of the corresponding simply connected Lie group G. These are known to play an important role in the representation theory of G (by Kirillov's orbit philosophy).

The study of infinite-dimensional Poisson vector spaces (and manifolds) only began recently with works of A. Odzijewicz and T. S. Ratiu concerning the Banach case [25, 26]. K.-H. Neeb (Darmstadt) asked for a framework of locally convex Poisson vector spaces (which need not be Banach spaces) and the investigation of related topics (like infinite-dimensional versions of the Stefan–Sussmann Theorem). These are currently being explored in joint research with L.R. Lovas (Debrecen) and the author.

In Sect. 16.5, we present a viable setting of locally convex Poisson vector spaces and explain how hypocontinuity arguments can be used to overcome the analytic problems arising beyond the Banach case. In particular, hypocontinuity is the crucial tool needed to define the Poisson bracket and Hamiltonian vector fields.

16.2 Preliminaries and Basic Facts

Throughout this article, $\mathbb{K} \in \{\mathbb{R}, \mathbb{C}\}$ and $\mathbb{D} := \{z \in \mathbb{K}\colon |z| \leq 1\}$. As the default, the letters E, E_1, E_2, F and G denote locally convex topological $\mathbb{K}$-vector spaces. When speaking of linear or bilinear maps, we mean $\mathbb{K}$-linear (resp., $\mathbb{K}$-bilinear) maps. A subset $U \subseteq E$ is called *balanced* if $\mathbb{D}U \subseteq U$. The locally convex spaces considered need not be Hausdorff, but whenever they serve as the domain or range of a differentiable map, we tacitly assume the Hausdorff property. We are working in a setting of infinite-dimensional differential calculus known as Keller's C^n_c-theory (see, e.g., [9] or [15] for streamlined introductions).

Definition 16.1. Let E and F be locally convex spaces over $\mathbb{K} \in \{\mathbb{R}, \mathbb{C}\}$, $U \subseteq E$ be open and $f\colon U \to F$ be a map. We say that f is $C^0_{\mathbb{K}}$ if f is continuous. The map f is called $C^1_{\mathbb{K}}$ if it is continuous, the limit

$$df(x,y) \;=\; \lim_{t \to 0} \frac{f(x+ty)-f(x)}{t}$$

exists for all $x \in U$ and all $y \in E$ (with $0 \neq t \in \mathbb{K}$ sufficiently small), and the map $df\colon U \times E \to F$ is continuous. Given $n \in \mathbb{N}$, we say that f is $C^{n+1}_{\mathbb{K}}$ if f is $C^1_{\mathbb{K}}$ and $df\colon U \times E \to F$ is $C^n_{\mathbb{K}}$. We say that f is $C^\infty_{\mathbb{K}}$ if f is $C^n_{\mathbb{K}}$ for each $n \in \mathbb{N}_0$. If $\mathbb{K}$ is understood, we simply write C^n instead of $C^n_{\mathbb{K}}$, for $n \in \mathbb{N}_0 \cup \{\infty\}$.

If $f\colon E \supseteq U \to F$ is $C^1_{\mathbb{K}}$, then $f'(x) := df(x,\bullet)\colon E \to F$ is a continuous $\mathbb{K}$-linear map, for each $x \in U$. It is known that compositions of composable $C^n_{\mathbb{K}}$-maps are $C^n_{\mathbb{K}}$. Also, continuous (multi)linear maps are $C^\infty_{\mathbb{K}}$ (see [15, Chap. 1] or [9] for all this).

Remark 16.2. Keller's C^n_c-theory is used as the foundation of infinite-dimensional Lie theory by many authors (see [10,11,15,23,24,30]). Others prefer the "convenient differential calculus" by Frölicher, Kriegl and Michor [20].

For some purposes, it is useful to impose certain completeness properties on the locally convex space F involved. These are, in decreasing order of strength: Completeness (every Cauchy net converges); quasi-completeness (every bounded Cauchy net converges); sequential completeness (every Cauchy sequence converges); and Mackey completeness (every Mackey–Cauchy sequence converges, or equivalently: the Riemann integral $\int_0^1 \gamma(t)\,dt$ exists in F, for every smooth curve $\gamma\colon \mathbb{R} \to F$; see [20, Theorem 2.14] for further information).

Remark 16.3. If $\mathbb{K} = \mathbb{C}$, then a mapping $f\colon E \supseteq U \to F$ is $C^\infty_{\mathbb{C}}$ if and only if it is *complex analytic* in the usual sense (as in [4]), i.e., f is continuous and for each $x \in U$, there exists a 0-neighbourhood $Y \subseteq U - x$ and continuous homogeneous polynomials $p_n\colon E \to F$ of degree n such that $f(x+y) = \bigcup_{n=0}^\infty p_n(y)$ for all $y \in Y$. Such maps are also called *holomorphic*. If F is Mackey complete, then f is $C^1_{\mathbb{C}}$ if and only if it is $C^\infty_{\mathbb{C}}$ (see [3, Propositions 7.4 and 7.7] or [15, Chap. 1] for all of this; cf. [9]). For suitable non-Mackey complete F, there are $C^n_{\mathbb{C}}$-maps $\mathbb{C} \to F$ for all $n \in \mathbb{N}$ which are not $C^{n+1}_{\mathbb{C}}$ [12, 16].

Remark 16.4. If $f\colon U \to F$ is a map from an open subset of $\mathbb{K}$ to a locally convex space, then f is $C^1_{\mathbb{K}}$ in the above sense if and only if the (real, respectively, complex) derivative $f^{(1)}(x) = f'(x) = \frac{df}{dx}(x)$ exists for each $x \in U$, and the mapping $f'\colon U \to F$ is continuous. Likewise, f is $C^n_{\mathbb{K}}$ if it has continuous derivatives $f^{(k)}\colon U \to F$ for all $k \in \mathbb{N}_0$ such that $k \leq n$ (where $f^{(k)} := (f^{(k-1)})'$). This is easy to see (and spelled out in [15, Chap. 1]). If $\mathbb{K} = \mathbb{C}$ here, then complex analyticity of f simply means that f can be expressed in the form $f(z) = \sum_{n=0}^\infty (z - z_0)^n a_n$ close to each given point $z_0 \in U$, for suitable elements $a_n \in F$.

Remark 16.5. Consider a map $f\colon U \to F$ from an open set $U \subseteq \mathbb{C}$ to a Mackey complete locally convex space F. Replacing sequential completeness with Mackey completeness in [4, Theorem 3.1] and its proof, one finds that also each of the following conditions is equivalent to f being a $C^\infty_{\mathbb{C}}$-map (see also [16, Chap. II], notably Theorems 2.2, 2.3 and 5.5):[1]

(a) f is weakly holomorphic, i.e., $\lambda \circ f \colon U \to \mathbb{C}$ is holomorphic for every $\lambda \in F'$.

(b) $\int_{\partial\Delta} f(\zeta)\,d\zeta = 0$ for each triangle $\Delta \subseteq U$.

(c) $f(z) = \frac{1}{2\pi i}\int_{|\zeta - z_0| = r} \frac{f(\zeta)}{\zeta - z}\,d\zeta$ for each $z_0 \in U$, $r > 0$ such that $z_0 + r\mathbb{D} \subseteq U$, and each z in the interior of the disk $z_0 + r\mathbb{D}$.

Definition 16.6. Given locally convex spaces E and F, let $L(E,F)$ be the vector space of all continuous linear maps $A\colon E \to F$. If $\mathscr{S}$ is a set of bounded subsets of E, we write $L(E,F)_{\mathscr{S}}$ for $L(E,F)$, equipped with the topology of uniform convergence on the sets $M \in \mathscr{S}$. Finite intersections of sets of the form

$$\lfloor M,U \rfloor := \{A \in L(E,F)\colon A(M) \subseteq U\}$$

(for $M \in \mathscr{S}$ and $U \subseteq F$ a 0-neighbourhood) form a basis for the filter of 0-neighbourhoods of this vector topology. See [5, Chap. III, Sect. 3] for further information. Given $M \subseteq E$ and $N \subseteq E'$, we write $M^\circ := \lfloor M, \mathbb{D} \rfloor \subseteq E'$ and ${}^\circ N := \{x \in E\colon (\forall \lambda \in N)\ \lambda(x) \in \mathbb{D}\}$ for the polar in E' (resp., in E).

Remark 16.7. If F is Hausdorff and $F \neq \{0\}$, then $L(E,F)_{\mathscr{S}}$ is Hausdorff if and only if $\bigcup_{M \in \mathscr{S}} M$ is total in E, i.e., it spans a dense vector subspace.

In fact, totality of $\bigcup \mathscr{S}$ is sufficient for the Hausdorff property by Proposition 3 in [5, Chap. III, Sect. 3, no. 2]. If $V := \operatorname{span}_{\mathbb{K}}(\bigcup \mathscr{S})$ is not dense in E, the Hahn–Banach Theorem provides a linear functional $0 \neq \lambda \in E'$ such that $\lambda|_V = 0$. We pick $0 \neq y \in F$. Then $y \otimes \lambda \in \lfloor M, U \rfloor$ for each $M \in \mathscr{S}$ and 0-neighbourhood $U \subseteq F$, whence $y \otimes \lambda \in W$ for each 0-neighbourhood $W \subseteq L(E,F)_{\mathscr{S}}$. Since $y \otimes \lambda \neq 0$, $L(E,F)_{\mathscr{S}}$ is not Hausdorff.

Proposition 16.8. *Given a separately continuous bilinear map $\beta\colon E_1 \times E_2 \to F$ and a set $\mathscr{S}$ of bounded subsets of E_2, consider the following conditions:*

(a) *For each $M \in \mathscr{S}$ and each 0-neighbourhood $W \subseteq F$, there exists a 0-neighbourhood $V \subseteq E_1$ such that $\beta(V \times M) \subseteq W$.*

(b) *The mapping $\beta^\vee\colon E_1 \to L(E_2,F)_{\mathscr{S}}$, $x \mapsto \beta(x,\bullet)$ is continuous.*

(c) *$\beta|_{E_1 \times M}\colon E_1 \times M \to F$ is continuous, for each $M \in \mathscr{S}$.*

Then (a) *and* (b) *are equivalent, and* (a) *implies* (c). *If*

$$(\forall M \in \mathscr{S})\ (\exists N \in \mathscr{S})\ \ \mathbb{D}M \subseteq N, \tag{16.1}$$

then all of (a)–(c) *are equivalent.*

[1] Considering f as a map into the completion of F, we see that (a), (b) and (c) remain equivalent if F is not Mackey complete. But (a)–(c) do not imply that f is $C^1_{\mathbb{C}}$ (see [16, Example II.2.3] and [12, Theorem 1.1]).

Definition 16.9. A bilinear map β which is separately continuous and satisfies the equivalent conditions (a) and (b) of Proposition 16.8 is called *$\mathscr{S}$-hypocontinuous* (in the second argument), or simply *hypocontinuous* if $\mathscr{S}$ is clear from the context. Hypocontinuity in the first argument with respect to a set of bounded subsets of E_1 is defined analogously.

Proof (of Proposition 16.8). For the equivalence of (a) and (b) and the implication (b)⇒(c), see Proposition 3 and 4 in [5, Chap. III, Sect. 5, no. 3], respectively.

We now show that (c)⇒(a) if (16.1) is satisfied. Given $M \in \mathscr{S}$ and 0-neighbourhood $W \subseteq F$, by hypothesis we find $N \in \mathscr{S}$ such that $\mathbb{D}M \subseteq N$. By continuity of $\beta|_{E_1 \times N}$, there exist 0-neighbourhoods V in E_1 and U in E_2 such that $\beta(V \times (N \cap U)) \subseteq W$. Since M is bounded, $M \subseteq nU$ for some $n \in \mathbb{N}$. Then $\frac{1}{n}M \subseteq N \cap U$. Using that β is bilinear, we obtain $\beta((\frac{1}{n}V) \times M) = \beta(V \times (\frac{1}{n}M)) \subseteq \beta(V \times (N \cap U)) \subseteq W$. □

Remark 16.10. By Proposition 16.8 (b), $\mathscr{S}$-hypocontinuity of $\beta\colon E_1 \times E_2 \to F$ only depends on the topology on $L(E_2,F)_{\mathscr{S}}$, not on $\mathscr{S}$ itself. Given $\mathscr{S}$, define $\mathscr{S}' := \{\mathbb{D}M\colon M \in \mathscr{S}\}$. Then the topologies on $L(E_2,F)_{\mathscr{S}}$ and $L(E_2,F)_{\mathscr{S}'}$ coincide (as is clear), and hence β is $\mathscr{S}$-hypocontinuous if and only if β is $\mathscr{S}'$-hypocontinuous. After replacing $\mathscr{S}$ with $\mathscr{S}'$, we can therefore always assume that (16.1) is satisfied, whenever this is convenient.

Each continuous bilinear map is hypocontinuous (as (a) in Proposition 16.8 is easy to check), but the converse is false. The next proposition compiles useful facts.

Proposition 16.11. *Let $\beta\colon E_1 \times E_2 \to F$ be an $\mathscr{S}$-hypocontinuous bilinear map, for some set $\mathscr{S}$ of bounded subsets of E_2. Then the following holds.*

(a) *$\beta(B \times M)$ is bounded in F, for each bounded subset $B \subseteq E_1$ and each $M \in \mathscr{S}$.*
(b) *Assume that, for each convergent sequence $(y_n)_{n \in \mathbb{N}}$ in E_2, with limit y, there is $M \in \mathscr{S}$ such that $\{y_n\colon n \in \mathbb{N}\} \cup \{y\} \subseteq M$. Then β is sequentially continuous.*

The condition described in (b) *is satisfied, for example, if $\mathscr{S}$ is the set of all bounded subsets of E_2, or the set of all compact subsets of E_2.*

Proof. (a) See Proposition 4 in [5, Chap. III, Sect. 5, no. 3].
(b) See [19, p. 157, Remark following Sect. 40, 1., (5)]. □

Frequently, separately continuous bilinear maps are automatically hypocontinuous. To make this precise, we recall that a subset B of a locally convex space E is called a *barrel* if it is closed, convex, balanced and absorbing. The space E is called *barrelled* if every barrel is a 0-neighbourhood. See Proposition 6 in [5, Chap. III, Sect. 5, no. 3] for the following fact:

Proposition 16.12. *If $\beta\colon E_1 \times E_2 \to F$ is a separately continuous bilinear map and E_1 is barrelled, then β is hypocontinuous with respect to any set $\mathscr{S}$ of bounded subsets of E_2.* □

Another simple fact will be useful.

Lemma 16.13. *Let M be a topological space, F be a locally convex space, and $BC(M,F)$ be the space of bounded F-valued continuous functions on M, equipped with the topology of uniform convergence. Then the evaluation mapping $\mu\colon BC(M,F) \times M \to F$, $\mu(f,x) := f(x)$ is continuous.*

Proof. Let (f_α, x_α) be a convergent net in $BC(M,F) \times M$, convergent to (f,x). Then $\mu(f_\alpha,x_\alpha) - \mu(f,x) = (f_\alpha(x_\alpha) - f(x_\alpha)) + (f(x_\alpha) - f(x))$, where $f_\alpha(x_\alpha) - f(x_\alpha)$ tends to 0 as $f_\alpha \to f$ uniformly and $f(x_\alpha) - f(x) \to 0$ as f is continuous. □

We now turn to paradigmatic bilinear maps, namely evaluation and composition.

Proposition 16.14. *Let E and F be locally convex spaces and $\mathscr{S}$ be a set of bounded subsets of E which covers E, i.e., $\bigcup_{M \in \mathscr{S}} M = E$. Then the evaluation map*

$$\varepsilon\colon L(E,F)_{\mathscr{S}} \times E \to F\,, \quad \varepsilon(A,x) := A(x)$$

is hypocontinuous in the second argument with respect to $\mathscr{S}$. If E is barrelled, then ε is also hypocontinuous in the first argument, with respect to any locally convex topology $\mathscr{O}$ on $L(E,F)$ which is finer than the topology of pointwise convergence, and any set $\mathscr{T}$ of bounded subsets of $(L(E,F),\mathscr{O})$.

Proof. By Remark 16.10, we may assume that $\mathscr{S}$ satisfies (16.1). Given $A \in L(E,F)$, we have $\varepsilon(A,\bullet) = A$, whence ε is continuous in the second argument. It is also continuous in the first argument, as the topology on $L(E,F)_{\mathscr{S}}$ is finer than the topology of pointwise convergence, by the hypothesis on $\mathscr{S}$. Let $M \in \mathscr{S}$ now. As $L(E,F)$ is equipped with the topology of uniform convergence on the sets in $\mathscr{S}$, the restriction mapping $\rho\colon L(E,F) \to BC(M,F)$, $A \mapsto A|_M$ is continuous. By Lemma 16.13, the evaluation mapping $\mu\colon BC(M,F) \times M \to F$ is continuous. Now $\varepsilon|_{L(E,F)\times M} = \mu \circ (\rho \times \mathrm{id}_M)$ shows that $\varepsilon|_{L(E,F)\times M}$ is continuous. Since we assume (16.1), the implication "(c)⇒(a)" in Proposition 16.8 shows that ε is $\mathscr{S}$-hypocontinuous.

Since $\mathscr{O}$ is finer than the topology of pointwise convergence, the map ε remains separately continuous in the situation described at the end of the proposition. Hence, if E is barrelled, Proposition 16.12 ensures hypocontinuity with respect to $\mathscr{T}$. □

While it was sufficient so far to consider an individual set $\mathscr{S}$ of bounded subsets of a given locally convex space, we now frequently wish to select such a set $\mathscr{S}$ simultaneously for each space. The following definition captures such situations.

Definition 16.15. A *bounded set functor* is a functor $\mathscr{S}$ from the category of locally convex spaces to the category of sets, with the following properties:

(a) $\mathscr{S}(E)$ is a set of bounded subsets of E, for each locally convex space E.
(b) If $A\colon E \to F$ is a continuous linear mapping, then $A(M) \in \mathscr{S}(F)$ for each $M \in \mathscr{S}(E)$, and $\mathscr{S}(A)\colon \mathscr{S}(E) \to \mathscr{S}(F)$ is the map taking $M \in \mathscr{S}(E)$ to its image $A(M)$ under A.

Given a bounded set functor $\mathscr{S}$ and locally convex spaces E and F, we write $L(E,F)_{\mathscr{S}}$ as a shorthand for $L(E,F)_{\mathscr{S}(E)}$. An $\mathscr{S}(E_2)$-hypocontinuous bilinear map $\beta\colon E_1 \times E_2 \to F$ will simply be called $\mathscr{S}$-hypocontinuous in the second argument.

Example 16.16. Bounded set functors are obtained if $\mathscr{S}(E)$ denotes the set of all bounded, (quasi-) compact, or finite subsets of E, respectively. We then write b, c, resp., p for $\mathscr{S}$.

Further examples abound: We can let $\mathscr{S}(E)$ be the set of precompact subsets of E, or the set of metrizable compact subsets (if only Hausdorff spaces are considered).

Remark 16.17. If $\mathscr{S}$ is a bounded set functor, and $A\colon E \to F$ a continuous linear mapping between locally convex spaces, then also its adjoint mapping $A'\colon F'_{\mathscr{S}} \to E'_{\mathscr{S}}$, $\lambda \mapsto \lambda \circ A$ is continuous, because $A'(\lfloor A(M), U \rfloor) \subseteq \lfloor M, U \rfloor$ for each $M \in \mathscr{S}(E)$ and 0-neighbourhood $U \subseteq \mathbb{K}$.

The double use of $f'(x)$ (for differentials) and A' (for adjoints) should not confuse.

Definition 16.18. If $\mathscr{S}(E)$ contains all finite subsets of E, then the mapping $\eta_E(x)\colon E'_{\mathscr{S}} \to \mathbb{K}$, $\lambda \mapsto \lambda(x)$ is continuous for each $x \in E$ and we obtain a linear map $\eta_E\colon E \to (E'_{\mathscr{S}})'$, the *evaluation homomorphism*. We say that E is $\mathscr{S}$*-reflexive* if $\eta_E\colon E \to (E'_{\mathscr{S}})'_{\mathscr{S}}$ is an isomorphism of topological vector spaces. If $\mathscr{S} = b$, we simply speak of a *reflexive* space; if $\mathscr{S} = c$, we speak of a *Pontryagin reflexive* space. Occasionally, we call E'_b the *strong* dual of E.

See Proposition 9 in [5, Chap. III, Sect. 5, no. 5] for the following fact in the three cases described in Example 16.16. It might also be deduced from [19, Sect. 40, 5., (6)].

Proposition 16.19. *Let E, F, and G be locally convex spaces and $\mathscr{S}$ be a bounded set functor such that $\mathscr{S}(E)$ covers E and*

$$\forall M \in \mathscr{S}(L(E,F)_{\mathscr{S}})\ \forall N \in \mathscr{S}(E)\ \exists K \in \mathscr{S}(F)\colon \quad \varepsilon(M \times N) \subseteq K, \tag{16.2}$$

where $\varepsilon\colon L(E,F) \times E \to F$, $(A,x) \mapsto A(x)$. Then the composition map

$$\Gamma\colon L(F,G)_{\mathscr{S}} \times L(E,F)_{\mathscr{S}} \to L(E,G)_{\mathscr{S}}, \quad \Gamma(\alpha,\beta) := \alpha \circ \beta$$

is $\mathscr{S}(L(E,F)_{\mathscr{S}})$-hypocontinuous in the second argument.

Remark 16.20. If $\mathscr{S} = b$, condition (16.2) is satisfied by Proposition 16.11 (a). If $\mathscr{S} = p$, then $\varepsilon(M \times N)$ is finite and thus (16.2) holds. If $\mathscr{S} = c$, then $\varepsilon|_{M \times N}$ is continuous since $\varepsilon\colon L(E,F)_{\mathscr{S}} \times E \to F$ is $\mathscr{S}(E)$-hypocontinuous by Proposition 16.14. Hence $\varepsilon(M \times N)$ is compact (and thus (16.2) is satisfied).

Proof (of Proposition 16.19). Γ is continuous in the second argument: Let $M \in \mathscr{S}(E)$, $U \subseteq G$ be a 0-neighbourhood, and $A \in L(F,G)$. Then $A^{-1}(U)$ is a 0-neighbourhood in F. For $B \in L(E,F)$, we have $A(B(M)) \subseteq U$ if and only if $B(M) \subseteq A^{-1}(U)$, showing that $\Gamma(A, \lfloor M, A^{-1}(U) \rfloor) \subseteq \lfloor M, U \rfloor$. Hence, $\Gamma(A, \bullet)$ being linear, it is continuous.

Continuity in the first argument: Let $U \subseteq G$ be a 0-neighbourhood, $B \in L(E,F)$ and $M \in \mathscr{S}(E)$. Then $B(M) \in \mathscr{S}(F)$ by Definition 16.15 (b) and $\Gamma(\lfloor B(M), U \rfloor, B) \subseteq \lfloor M, U \rfloor$.

To complete the proof, let $M \in \mathscr{S}(L(E,F)_{\mathscr{S}})$, $U \subseteq G$ be a 0-neighbourhood and $N \in \mathscr{S}(E)$. By (16.2), there exists $K \in \mathscr{S}(F)$ such that $\varepsilon(M \times N) \subseteq K$. Note that for all $B \in M$ and $A \in \lfloor K,U \rfloor$, we have $\Gamma(A,B).N = (A \circ B)(N) = A(B(N)) \subseteq A(K) \subseteq U$. Thus $\Gamma(\lfloor K,U \rfloor \times M) \subseteq \lfloor N,U \rfloor$. Since $\lfloor K,U \rfloor$ is a 0-neighbourhood in $L(F,G)_{\mathscr{S}}$, condition (a) of Proposition 16.8 is satisfied. □

Remark 16.21. Despite the hypocontinuity of the composition map Γ, it is discontinuous in the situations specified in the introduction. To see this, pick $\lambda \in E'$ and $x \in E$ as described in the introduction. Let $0 \neq z \in G$ and give F' the topology induced by $F' \to L(F,G)$, $\zeta \mapsto z \otimes \zeta$. There is $\mu \in G'$ such that $\mu(z) \neq 0$. If Γ was continuous, then also the following map would be continuous: $F' \times F \to \mathbb{K}$, $(\zeta, y) \mapsto \mu(\Gamma(z \otimes \zeta, y \otimes \lambda)(x)) = \mu(z)\lambda(x)\zeta(y)$. But this mapping is a non-zero multiple of the evaluation map and hence discontinuous [22].

16.3 Differentiability Properties of Compositions with Hypocontinuous Bilinear Mappings

In this section, we introduce a new class of topological spaces ("k^∞-spaces"). We then discuss compositions of hypocontinuous bilinear maps with C^n-maps on open subsets of locally convex spaces which are k^∞-spaces.

Recall that a Hausdorff topological space X is called a *k-space* if, for every subset $A \subseteq X$, the set A is closed in X if and only if $A \cap K$ is closed in K for each compact set $K \subseteq X$. Equivalently, a subset $U \subseteq X$ is open in X if and only if $U \cap K$ is open in K for each compact set $K \subseteq X$. It is clear that closed subsets, as well as open subsets of k-spaces are k-spaces when equipped with the induced topology. If X is a k-space, then a map $f\colon X \to Y$ to a topological space Y is continuous if and only if $f|_K\colon K \to Y$ is continuous for each compact set $K \subseteq X$, as is easy to see. (Given a closed set $A \subseteq Y$, the intersection $f^{-1}(A) \cap K = (f|_K)^{-1}(A)$ is closed in K in the latter case and thus $f^{-1}(A)$ is closed, whence f is continuous). Thus $X = \varinjlim_K K$ as a topological space. This property is crucial.

Definition 16.22. We say that a topological space X is a *k^∞-space* if it is Hausdorff and its n-fold power $X^n = X \times \cdots \times X$ is a k-space, for each $n \in \mathbb{N}$.

Example 16.23. It is well known (and easy to prove) that every metrizable topological space is a k-space. Finite powers of metrizable spaces being metrizable, we see: *Every metrizable topological space is a k^∞-space.*

Example 16.24. A Hausdorff topological space X is called a *k_ω-space* if it is a k-space and *hemicompact*,[2] i.e., there exists a sequence $K_1 \subseteq K_2 \subseteq \cdots$ of compact subsets of X such that $X = \bigcup_{n \in \mathbb{N}} K_n$ and each compact subset of X is contained in some K_n. Since finite products of k_ω-spaces are k_ω-spaces (see, e.g., [14, Proposition 4.2 (c)]), it follows that each k_ω-space is a k^∞-space. For an introduction to k_ω-spaces, the reader may consult [14].

[2] An equivalent definition reads as follows: A Hausdorff space X is a k_ω-space if and only if $X = \varinjlim K_n$ for an ascending sequence $(K_n)_{n \in \mathbb{N}}$ of compact subsets of X with union X.

Remark 16.25. E'_c is a k_ω-space (and hence a k^∞-space), for each metrizable locally convex space E (see [1, Corollary 4.7 and Proposition 5.5]). In particular, every Silva space E is a k_ω-space (and hence a k^∞-space), i.e., every locally convex direct limit $E = \varinjlim E_n$ of an ascending sequence $E_1 \subseteq E_2 \subseteq \cdots$ of Banach spaces, such that the inclusion maps $E_n \to E_{n+1}$ are compact operators (see [11, Example 9.4]).

Having set up the terminology, let us record a simple, but useful observation.

Theorem 16.26. *Let $n \in \mathbb{N}_0 \cup \{\infty\}$. If $n = 0$, let $U = X$ be a topological space. If $n \geq 1$, let X be a locally convex space and $U \subseteq X$ be open. Let $\beta \colon E_1 \times E_2 \to F$ be a bilinear map and $f \colon U \to E_1 \times E_2$ be C^n. Assume that at least one of* (a) *or* (b) *holds:*

(a) *X is metrizable and β is sequentially continuous.*
(b) *X is a k^∞-space and β is hypocontinuous in the second argument with respect to a set $\mathcal{S}$ of bounded subsets of E_2 which contains all compact subsets of E_2.*

Then $\beta \circ f \colon U \to F$ is C^n.

Proof. It suffices to consider the case where $n < \infty$. The proof is by induction.

We assume (a) first. If $n = 0$, let $(x_k)_{k\in\mathbb{N}}$ be a convergent sequence in U, with limit x. Then $f(x_k) \to f(x)$ by continuity of f and hence $\beta(f(x_k)) \to \beta(f(x))$, because β is sequentially continuous.

Now let $n \geq 1$ and assume that the assertion holds if n is replaced with $n-1$. Given $x \in U$ and $y \in X$, let $(t_k)_{k\in\mathbb{N}}$ be a sequence in $\mathbb{K} \setminus \{0\}$ such that $x + t_k y \in U$ for each $k \in \mathbb{N}$, and $\lim_{k\to\infty} t_k = 0$. Write $f = (f_1, f_2)$ with $f_j \colon U \to E_j$. Then

$$\begin{aligned} &\frac{\beta(f(x+t_k y)) - \beta(f(x))}{t_k} \\ &\quad = \beta\left(\frac{f_1(x+t_k y) - f_1(x)}{t_k}, f_2(x+t_k y)\right) + \beta\left(f_1(x), \frac{f_2(x+t_k y) - f_2(x)}{t_k}\right) \\ &\quad \to \beta(df_1(x,y), f_2(x)) + \beta(f_1(x), df_2(x,y)) \quad \text{as } k \to \infty, \end{aligned}$$

by continuity of f and sequential continuity of β. Hence the limit $d(\beta \circ f)(x,y) = \lim\limits_{t\to 0} \frac{\beta(f(x+ty)) - \beta(f(x))}{t}$ exists, and is given by

$$d(\beta \circ f)(x,y) \;=\; \beta(df_1(x,y), f_2(x)) + \beta(f_1(x), df_2(x,y)) . \tag{16.3}$$

The mappings $g_1, g_2 \colon U \times X \to E_1 \times E_2$ defined via $g_1(x,y) := (df_1(x,y), f_2(x))$ and $g_2(x,y) := (f_1(x), df_2(x,y))$ are $C^{n-1}_{\mathbb{K}}$. Since

$$d(\beta \circ f) \;=\; \beta \circ g_1 + \beta \circ g_2 \tag{16.4}$$

by (16.3), we deduce from the inductive hypotheses that $d(\beta \circ f)$ is $C^{n-1}_{\mathbb{K}}$ and hence continuous. Thus $\beta \circ f$ is $C^1_{\mathbb{K}}$ with $d(\beta \circ f)$ a $C^{n-1}_{\mathbb{K}}$-map and hence $\beta \circ f$ is $C^n_{\mathbb{K}}$, which completes the inductive proof in the situation of (a).

In the situation of (b), let $K \subseteq U$ be compact. Then $f_2(K) \subseteq E_2$ is compact and hence $f_2(K) \in \mathscr{S}$, by hypothesis. Since $\beta|_{E_1 \times f_2(K)}$ is continuous by Proposition 16.8 (c), $(\beta \circ f)|_K = \beta|_{E_1 \times f_2(K)} \circ f|_K$ is continuous. Since X and hence also its open subset U is a k-space, it follows that $\beta \circ f$ is continuous, settling the case $n = 0$.

Now let $n \geq 1$ and assume that the assertion holds if n is replaced with $n-1$. Since β is sequentially continuous by Proposition 16.11 (b), we see as in case (a) that the directional derivative $d(\beta \circ f)(x,y)$ exists for all $(x,y) \in U \times X$, and that $d(\beta \circ f)$ is given by (16.4). Since g_1 and g_2 are $C^{n-1}_{\mathbb{K}}$, the inductive hypothesis can be applied to the summands in (16.4). Thus $d(\beta \circ f)$ is $C^{n-1}_{\mathbb{K}}$, whence $\beta \circ f$ is $C^1_{\mathbb{K}}$ with $d(\beta \circ f)$ a $C^{n-1}_{\mathbb{K}}$-map, and so $\beta \circ f$ is $C^n_{\mathbb{K}}$. □

Combining Proposition 16.19 and Theorem 16.26, as a first application we obtain an affirmative answer to an open question formulated by Serge Lang [21, p. 8, Remark].

Corollary 16.27. *Let U be an open subset of a Fréchet space, E, F and G be Fréchet spaces, and $f\colon U \to L(E,F)_b$, $g\colon U \to L(F,G)_b$ be continuous maps. Then also the mapping $U \to L(E,G)_b$, $x \mapsto g(x) \circ f(x)$ is continuous.* □

We mention that an alternative path leading to Theorem 16.26 is to prove, in a first step, smoothness of hypocontinuous bilinear maps (with respect to compact sets) in an alternative sense, replacing "continuity" by "continuity on each compact set" in the definition of a C^n-map [28, Theorem 4.1].[3] The second step is to observe that such C^n-maps coincide with ordinary C^n-maps if the domain is a k^∞-space.

16.4 Holomorphic Families of Operators

We now describe first implications of the previous results and some useful additional material. Specializing to the case $X := \mathbb{K} := \mathbb{C}$ and $n := \infty$, we obtain results concerning holomorphic families of operators, i.e., holomorphic maps $U \to L(E,F)_{\mathscr{S}}$, where U is an open subset of $\mathbb{C}$. Among other things, such holomorphic families are of interest for representation theory and cohomology [6,7].

Proposition 16.28. *Let X, E, F and G be locally convex spaces over $\mathbb{K}$, such that X is a k^∞-space (e.g., $X = \mathbb{K} = \mathbb{C}$). Let $U \subseteq X$ be an open set, $n \in \mathbb{N}_0 \cup \{\infty\}$ and $f\colon U \to L(E,F)_{\mathscr{S}}$ as well as $g\colon U \to L(F,G)_{\mathscr{S}}$ be $C^n_{\mathbb{K}}$-maps, where $\mathscr{S} = b$ or $\mathscr{S} = c$. Then also the map $U \to L(E,G)_{\mathscr{S}}$, $z \mapsto g(z) \circ f(z)$ is $C^n_{\mathbb{K}}$.*

Proof. The composition map $L(F,G)_{\mathscr{S}} \times L(E,F)_{\mathscr{S}} \to L(E,G)_{\mathscr{S}}$ is hypocontinuous with respect to $\mathscr{S}(L(E,F)_{\mathscr{S}})$, by Proposition 16.19. Now use Theorem 16.26 (b). □

The remainder of this section is devoted to the proof of the following result. Here $\mathscr{S}$ is a bounded set functor such that $\mathscr{S}(E)$ covers E, for each locally convex space E.

[3] Compare also [27] for the use of Kelleyfications in differential calculus.

Proposition 16.29. *Let E, F and X be locally convex spaces over $\mathbb{K}$. If $\eta_F\colon F\to(F'_{\mathscr{S}})'_{\mathscr{S}}$ is continuous, then $g\colon U\to L(F'_{\mathscr{S}},E'_{\mathscr{S}})_{\mathscr{S}}$, $z\mapsto f(z)'$ is $C^n_{\mathbb{K}}$, for each $n\in\mathbb{N}_0\cup\{\infty\}$ and $C^n_{\mathbb{K}}$-map $f\colon U\to L(E,F)_{\mathscr{S}}$ on an open set $U\subseteq X$.*

The proof of Proposition 16.29 exploits the continuity of the formation of adjoints.

Proposition 16.30. *Let E and F be locally convex spaces and $\mathscr{S}$ be a bounded set functor such that $\mathscr{S}(F)$ covers F. If $\eta_F\colon F\to(F'_{\mathscr{S}})'_{\mathscr{S}}$ is continuous, then $\Psi\colon L(E,F)_{\mathscr{S}}\to L(F'_{\mathscr{S}},E'_{\mathscr{S}})_{\mathscr{S}}$, $\alpha\mapsto\alpha'$ is a continuous linear map.*

Proof. Replace $\mathscr{S}(V)$ with $\{r_1M_1\cup\cdots\cup r_nM_n\colon r_1,\ldots,r_n\in\mathbb{K},M_1,\ldots,M_n\in\mathscr{S}(V)\}$ for each locally convex space V. This does not change the $\mathscr{S}$-topologies, and allows us to assume that $\mathscr{S}(E)$ is closed under finite unions and multiplication with scalars. Let $M\in\mathscr{S}(F'_{\mathscr{S}})$ and $U\subseteq E'_{\mathscr{S}}$ be a 0-neighbourhood; we have to show that $\Psi^{-1}(\lfloor M,U\rfloor)$ is a 0-neighbourhood in $L(E,F)_{\mathscr{S}}$. After shrinking U, without loss of generality $U=N^\circ$ for some $N\in\mathscr{S}(E)$ (by our hypothesis concerning $\mathscr{S}(E)$). For $\alpha\in L(E,F)$, we have

$$\begin{aligned}\alpha'\in\lfloor M,U\rfloor &\Leftrightarrow (\forall\lambda\in M)\ \lambda\circ\alpha=\alpha'(\lambda)\in U=N^\circ\\ &\Leftrightarrow (\forall\lambda\in M)(\forall x\in N)\ |\lambda(\alpha(x))|\leq 1\\ &\Leftrightarrow \alpha(N)\subseteq{}^\circ M \Leftrightarrow \alpha\in\lfloor N,{}^\circ M\rfloor.\end{aligned}$$

Since $M\in\mathscr{S}(F'_{\mathscr{S}})$ and η_F is continuous, ${}^\circ M=\eta_F^{-1}(M^\circ)$ is a 0-neighbourhood in F. Thus $\lfloor N,{}^\circ M\rfloor=\Psi^{-1}(\lfloor M,U\rfloor)$ is a 0-neighbourhood in $L(E,F)_{\mathscr{S}}$. □

Proof (of Proposition 16.29). Since f is a $C^n_{\mathbb{K}}$-mapping and Ψ in Proposition 16.30 is continuous linear and hence a $C^\infty_{\mathbb{K}}$-map, also $g=\Psi\circ f$ is $C^n_{\mathbb{K}}$. □

The locally convex spaces E such that $\eta_E\colon E\to(E'_b)'_b$ is continuous are known as "quasi-barrelled" spaces. They can characterized easily. Recall that a subset A of a locally convex space E is called *bornivorous* if it absorbs all bounded subsets of E. The space E is called *bornological* if every convex, balanced, bornivorous subset of E is a 0-neighbourhood. See Proposition 2 in [17, Sect. 11.2] (and the lines thereafter) for the following simple fact:

The evaluation homomorphism $\eta_E\colon E\to(E'_b)'_b$ is continuous if and only if each closed, convex, balanced subset $A\subseteq E$ which absorbs all bounded subsets of E (i.e., each bornivorous barrel A) is a 0-neighbourhood in E.

Thus $\eta_E\colon E\to(E'_b)'_b$ is continuous if E is bornological or barrelled. It is also known that $\eta_E\colon E\to(E'_c)'_c$ is continuous if E is a k-space (cf. [1, Corollary 5.12 and Proposition 5.5]).

We mention that Proposition 16.28 generalizes [6, Lemma 2.4], where $\mathscr{S}=b$, $X=\mathbb{K}=\mathbb{C}$, E is assumed to be a Montel space and E, F, G are complete (see last line of [6, p. 637]). The method of proof used in loc. cit. depends on completeness properties of $L(E,G)_b$, because the characterization of holomorphic functions via Cauchy integrals (as in Remark 16.5 (c)) requires Mackey completeness.

Proposition 16.29 generalizes [6, Lemma 2.3], where $X = \mathbb{K} = \mathbb{C}$, $\mathscr{S} = b$ and where continuity of η_F is presumed as well (penultimate sentence of the proof). We remark that the proof of [6, Lemma 2.3] requires that $L(F', E')_b$ is at least Mackey complete (since only weak holomorphicity of g is checked there).

16.5 Locally Convex Poisson Vector Spaces

We now define locally convex Poisson vector spaces and prove fundamental facts concerning such spaces, using hypocontinuity as a tool.

16.31. Throughout this section, we let $\mathscr{S}$ be a bounded set functor such that the following holds for each locally convex space E:

(a) $\mathscr{S}(E)$ contains all compact subsets of E.
(b) For each $M \in \mathscr{S}(E'_{\mathscr{S}})$ and $N \in \mathscr{S}(E)$, the set $\varepsilon(M \times N) \subseteq \mathbb{K}$ is bounded, where $\varepsilon \colon E' \times E \to \mathbb{K}$ is the evaluation map.

Since $\mathscr{S}(\mathbb{K})$ contains all compact sets and each bounded subset of $\mathbb{K}$ is contained in a compact set, condition (b) means that there is $K \in \mathscr{S}(\mathbb{K})$ such that $\varepsilon(M \times N) \subseteq K$.

Definition 16.32. An *$\mathscr{S}$-reflexive locally convex Poisson vector space* is a locally convex space E that is $\mathscr{S}$-reflexive and a k^∞-space, together with an $\mathscr{S}$-hypocontinuous bilinear map $[.,.] \colon E'_{\mathscr{S}} \times E'_{\mathscr{S}} \to E'_{\mathscr{S}}$, $(\lambda, \eta) \mapsto [\lambda, \eta]$ which makes $E'_{\mathscr{S}}$ a Lie algebra.

Of course, we are mostly interested in the case where $[.,.]$ is continuous, but only $\mathscr{S}$-hypocontinuity is required for the basic results described below.

Remark 16.33. We mainly have two choices of $\mathscr{S}$ in mind:

(a) The case $\mathscr{S} = b$. If E is a Hilbert space, a reflexive Banach space, a nuclear Fréchet space, or the strong dual of a nuclear Fréchet space, then reflexivity is satisfied and also the k^∞-property (by Example 16.23 and Remark 16.25).[4]
(b) If $\mathscr{S} = c$, then the scope widens considerably. For example, every Fréchet space E is Pontryagin reflexive [2, Propositions 15.2 and 2.3] and a k^∞-space (see Example 16.23). And the same holds for its dual E'_c (see [1, Proposition 5.9] and Remark 16.25).

Remark 16.34. Dual spaces of topological Lie algebras are paradigms of $\mathscr{S}$-reflexive locally convex Poisson vector spaces. More precisely, let $\mathscr{S} = b$ or $\mathscr{S} = c$, and $(\mathfrak{g}, [.,.]_{\mathfrak{g}})$ be a locally convex topological Lie algebra. If $\mathfrak{g}$ is $\mathscr{S}$-reflexive and

[4] Let E be a nuclear Fréchet space. Then E is Pontryagin reflexive [2, Propositions 15.2 and 2.3]. By [2, Theorem 16.1] and [1, Proposition 5.9], also E'_c is nuclear and Pontryagin reflexive. Since E is metrizable and hence a k-space, E'_c is complete [1, Proposition 4.11]. We now see with [29, Proposition 50.2] that every closed, bounded subset of E (and E'_c) is compact. Hence $E'_b = E'_c$, $(E'_b)'_b = (E'_c)'_c$ and both E and $E'_b = E'_c$ are also reflexive. By Remark 16.25, $E'_b = E'_c$ is a k^∞-space.

$\mathfrak{g}'_{\mathscr{S}}$ happens to be a k^∞-space, then $E := \mathfrak{g}'_{\mathscr{S}}$ is an $\mathscr{S}$-reflexive locally convex Poisson vector space with Lie bracket defined via $[\lambda,\mu] := [\eta_{\mathfrak{g}}^{-1}(\lambda),\eta_{\mathfrak{g}}^{-1}(\mu)]_{\mathfrak{g}}$ for $\lambda,\mu \in E' = (\mathfrak{g}'_{\mathscr{S}})'$, using the isomorphism $\eta_{\mathfrak{g}}\colon \mathfrak{g} \to (\mathfrak{g}'_{\mathscr{S}})'_{\mathscr{S}}$. Here are typical examples:

(a) If $\mathfrak{g}$ is a Banach–Lie algebra whose underlying Banach space is reflexive, then $\mathfrak{g}'_b$ is a *reflexive* locally convex Poisson vector space (i.e., w.r.t. $\mathscr{S} = b$); see Remark 16.33 (a).
(b) If $\mathfrak{g}$ is a Fréchet–Lie algebra (a topological Lie algebra which is a Fréchet space), then $\mathfrak{g}'_c$ is a *Pontryagin reflexive* locally convex Poisson vector space (i.e., with respect to $\mathscr{S} = c$), by Remark 16.33 (b).
(c) If $\mathfrak{g}$ is a Silva–Lie algebra, then $\mathfrak{g}$ is reflexive (thus also Pontryagin reflexive), and $\mathfrak{g}'_b = \mathfrak{g}'_c$ is a Fréchet–Schwartz space (see [8]) and hence a k^∞-space. Therefore $\mathfrak{g}'_b = \mathfrak{g}'_c$ is a reflexive and Pontryagin reflexive locally convex Poisson vector space.

Two well-behaved classes of examples were suggested to the author by K.-H. Neeb:

If a topological group G is a projective limit $\varprojlim G_n$ of a projective sequence $\cdots \to G_2 \to G_1$ of finite-dimensional Lie groups, then $\mathfrak{g} := \varprojlim L(G_n) \cong \mathbb{R}^{\mathbb{N}}$ can be considered as the Lie algebra of G and coadjoint orbits of G in $\mathfrak{g}'$ can be studied, where $\mathfrak{g}'_c$ $(= \mathfrak{g}'_b)$ is a Pontryagin reflexive (and reflexive) locally convex Poisson vector space, by (b).

If a group G is the union $\bigcup_{n\in\mathbb{N}} G_n$ of an ascending sequence $G_1 \subseteq G_2 \subseteq \cdots$ of finite-dimensional Lie groups, then G can be made an infinite-dimensional Lie group with Lie algebra $\mathfrak{g} = \varinjlim L(G_n) \cong \mathbb{R}^{(\mathbb{N})}$ (see [10]), where $\mathfrak{g}'_c$ $(= \mathfrak{g}'_b)$ is a Pontryagin reflexive (and reflexive) locally convex Poisson vector space, by (c). Again coadjoint orbits can be studied. Manifold structures on them do not pose problems, since all homogeneous spaces of G are manifolds [10, Proposition 7.5].

Given a Lie algebra $(\mathfrak{g},[.,.])$ and $x \in \mathfrak{g}$, we write $\mathrm{ad}_x := \mathrm{ad}(x) := [x,.]\colon \mathfrak{g} \to \mathfrak{g}$, $y \mapsto [x,y]$. Definition 16.39 can be adapted to non $\mathscr{S}$-reflexive spaces, mimicking [25] and [26]:

Definition 16.35. A *locally convex Poisson vector space* with respect to $\mathscr{S}$ is a locally convex space E whose evaluation homomorphism $\eta_E\colon E \to (E'_{\mathscr{S}})'_{\mathscr{S}}$ is a topological embedding and which is a k^∞-space, together with an $\mathscr{S}$-hypocontinuous bilinear map $[.,.]\colon E'_{\mathscr{S}} \times E'_{\mathscr{S}} \to E'_{\mathscr{S}}$, $(\lambda,\eta) \mapsto [\lambda,\eta]$ making $E'_{\mathscr{S}}$ a Lie algebra with

$$\eta_E(x) \circ \mathrm{ad}_\lambda \;\in\; \eta_E(E) \quad \text{for all } x \in E \text{ and } \lambda \in E'. \tag{16.5}$$

Remark 16.36. Every $\mathscr{S}$-reflexive Poisson vector space $(E,[.,.])$ in the sense of Definition 16.32 also is a Poisson vector space with respect to $\mathscr{S}$, in the sense of Definition 16.35. In fact, since $[.,.]$ is separately continuous, the linear map $\mathrm{ad}_\lambda = [\lambda,.]\colon E'_{\mathscr{S}} \to E'_{\mathscr{S}}$ is continuous, for each $\lambda \in E'_{\mathscr{S}}$. Hence $\alpha \circ \mathrm{ad}_\lambda \in (E'_{\mathscr{S}})' = \eta_E(E)$ for each $\alpha \in (E'_{\mathscr{S}})'$, and thus (16.5) is satisfied.

Remark 16.37. If $\mathscr{S} = b$, then $\eta_E \colon E \to (E'_b)'_b$ is a topological embedding if and only if η_E is continuous, i.e., if and only if E is quasi-barrelled (see [17, Sect. 11.2]). Most locally convex spaces of practical interest are quasi-barrelled.

If $\mathscr{S} = c$, then η_E is a topological embedding *automatically* in the situation of Definition 16.35 as we assume that E is a k^∞-space (and hence a k-space). In fact, $\eta_E \colon E \to (E'_c)'_c$ is injective (by the Hahn–Banach theorem) for each locally convex space E, and open onto its image (cf. [1, Proposition 6.10] or [2, Lemma 14.3]). Hence $\eta_E \colon E \to (E'_c)'_c$ is an embedding if and only if it is continuous, which holds if E is a k-space (cf. [2, Lemma 14.4]).

Remark 16.38. Since reflexive Banach spaces are rare, the more complicated non-reflexive theory cannot be avoided in the study of Banach–Lie–Poisson vector spaces (as in [25] and [26]). By contrast, typical non-Banach locally convex spaces are reflexive and hence fall within the simple, basic framework of Definition 16.32. And the class of Pontryagin reflexive spaces is even more comprehensive.

Definition 16.39. Let $(E, [.,.])$ be a locally convex Poisson vector space with respect to $\mathscr{S}$, and $U \subseteq E$ be open. If $f, g \in C^\infty_{\mathbb{K}}(U, \mathbb{K})$, define a function $\{f,g\} \colon U \to \mathbb{K}$ via

$$\{f,g\}(x) := \langle [f'(x), g'(x)], x \rangle \quad \text{for } x \in U, \tag{16.6}$$

where $\langle .,. \rangle \colon E' \times E \to \mathbb{K}$, $\langle \lambda, x \rangle := \lambda(x)$ is the evaluation map and $f'(x) = df(x,.)$.

Condition (16.5) in Definition 16.35 enables us to define a map $X_f \colon U \to E$ via

$$X_f(x) := \eta_E^{-1}\big(\eta_E(x) \circ \mathrm{ad}(f'(x))\big) \quad \text{for } x \in U, \tag{16.7}$$

where $\eta_E \colon E \to (E'_{\mathscr{S}})'_{\mathscr{S}}$ is the evaluation homomorphism.

Theorem 16.40. *Let $(E, [.,.])$ be a locally convex Poisson vector space with respect to $\mathscr{S}$ and $U \subseteq E$ be an open subset. Then*

(a) *$\{f,g\} \in C^\infty_{\mathbb{K}}(U, \mathbb{K})$, for all $f, g \in C^\infty_{\mathbb{K}}(U, \mathbb{K})$.*
(b) *For each $f \in C^\infty_{\mathbb{K}}(U, \mathbb{K})$, the map $X_f \colon U \to E$ is $C^\infty_{\mathbb{K}}$.*

The following fact will help us to prove Theorem 16.40.

Lemma 16.41. *Let E and F be locally convex spaces, $U \subseteq E$ be open and $f \colon U \to F$ be a $C^\infty_{\mathbb{K}}$-map. Then also the map $f' \colon U \to L(E,F)_{\mathscr{S}}$, $x \mapsto f'(x) = df(x, \bullet)$ is $C^\infty_{\mathbb{K}}$, for each set $\mathscr{S}$ of bounded subsets of E such that $L(E,F)_{\mathscr{S}}$ is Hausdorff.*

Proof. For $\mathscr{S} = b$, see [13]. The general case follows from the case $\mathscr{S} = b$. □

Proof (of Theorem 16.40). (a) The maps $f' \colon U \to L(E, \mathbb{K})_{\mathscr{S}} = E'_{\mathscr{S}}$ as well as $g' \colon U \to E'_{\mathscr{S}}$ are $C^\infty_{\mathbb{K}}$ by Lemma 16.41, E is a k^∞-space by hypothesis, and $[.,.]$ is $\mathscr{S}$-hypocontinuous. Hence $h := [.,.] \circ (f', g') \colon U \to E'_{\mathscr{S}}, x \mapsto [f'(x), g'(x)]$ is $C^\infty_{\mathbb{K}}$, by Theorem 16.26 (b). The evaluation map $\varepsilon \colon E'_{\mathscr{S}} \times E \to \mathbb{K}$ is $\mathscr{S}$-hypocontinuous in the second argument by Proposition 16.14, and the inclusion map $\iota \colon U \to E$ is $C^\infty_{\mathbb{K}}$. Hence $\{f,g\} = \varepsilon \circ (h, \iota)$ is $C^\infty_{\mathbb{K}}$, by Theorem 16.26 (b).

(b) Since $[.,.]\colon E'_{\mathscr{S}} \times E'_{\mathscr{S}} \to E'_{\mathscr{S}}$ is an $\mathscr{S}$-hypocontinuous bilinear mapping, the linear map $[.,.]^\vee\colon E'_{\mathscr{S}} \to L(E'_{\mathscr{S}}, E'_{\mathscr{S}})_{\mathscr{S}}$, $\lambda \mapsto [\lambda,.] = \mathrm{ad}(\lambda)$ is continuous, by Proposition 16.8 (b). Hence $h\colon U \to L(E'_{\mathscr{S}}, E'_{\mathscr{S}})_{\mathscr{S}}$, $h(x) := \mathrm{ad}(f'(x))$ is $C^\infty_{\mathbb{K}}$, by Lemma 16.41. The composition map $\Gamma\colon (E'_{\mathscr{S}})'_{\mathscr{S}} \times L(E'_{\mathscr{S}}, E'_{\mathscr{S}})_{\mathscr{S}} \to (E'_{\mathscr{S}})'_{\mathscr{S}}$, $(\alpha, A) \mapsto \alpha \circ A$ is $\mathscr{S}$-hypocontinuous in the second argument, because (b) in Sect. 16.31 ensures that Proposition 16.19 can be applied. Then

$$V := \{A \in L(E'_{\mathscr{S}}, E'_{\mathscr{S}})_{\mathscr{S}}\colon (\forall x \in E)\ \eta_E(x) \circ A \in \eta_E(E)\}$$

is a vector subspace of $L(E'_{\mathscr{S}}, E'_{\mathscr{S}})_{\mathscr{S}}$ and the bilinear map

$$\Theta\colon E \times V \to E\,, \quad \Theta(x,A) := \eta_E^{-1}(\Gamma(\eta_E(x), A))$$

is $\mathscr{S}$-hypocontinuous in its second argument, using that η_E is an isomorphism of topological vector spaces onto its image. The inclusion map $\iota\colon U \to E$, $x \mapsto x$ being $C^\infty_{\mathbb{K}}$, Theorem 16.26 shows that $X_f = \Theta \circ (\iota, h)$ is $C^\infty_{\mathbb{K}}$. □

Definition 16.42. Let $(E, [.,.])$ be a locally convex Poisson vector space with respect to $\mathscr{S}$, and $U \subseteq E$ be an open subset.

(a) The map $\{.,.\}\colon C^\infty_{\mathbb{K}}(U,\mathbb{K}) \times C^\infty_{\mathbb{K}}(U,\mathbb{K}) \to C^\infty_{\mathbb{K}}(U,\mathbb{K})$ taking (f,g) to $\{f,g\}$ (as in Definition 16.39) is called the *Poisson bracket*.

(b) Given $f \in C^\infty_{\mathbb{K}}(U,\mathbb{K})$, the map $X_f\colon U \to E$ is a smooth vector field on U, by Theorem 16.40 (b). It is called the *Hamiltonian vector field* associated with f.

Remark 16.43. Basic differentiation rules entail that $\{.,.\}$ makes $C^\infty_{\mathbb{K}}(U,\mathbb{K})$ a Poisson algebra, i.e., $(C^\infty_{\mathbb{K}}(U,\mathbb{K}), \{.,.\})$ is a Lie algebra and the mapping $\{f,.\}\colon C^\infty_{\mathbb{K}}(U,\mathbb{K}) \to C^\infty_{\mathbb{K}}(U,\mathbb{K})$, $g \mapsto \{f,g\}$ is a derivation for the associative $\mathbb{K}$-algebra $(C^\infty_{\mathbb{K}}(U,\mathbb{K}), \cdot)$, for each $f \in C^\infty_{\mathbb{K}}(U,\mathbb{K})$. (The Jacobi identity for $\{.,.\}$ can be shown as in the proof of [25, Theorem 4.2]).

Remark 16.44. The case $\mathscr{S} = c$ is well adapted to the compact-open C^∞-topology on function spaces. Then the Poisson bracket

$$\{.,.\}\colon C^\infty_{\mathbb{K}}(U,\mathbb{K}) \times C^\infty_{\mathbb{K}}(U,\mathbb{K}) \to C^\infty_{\mathbb{K}}(U,\mathbb{K})$$

is hypocontinuous with respect to compact subsets of $C^\infty_{\mathbb{K}}(U,\mathbb{K})$, for each locally convex Poisson vector space $(E,[.,.])$ with respect to $\mathscr{S} = c$, and open set $U \subseteq E$. Moreover, the map $C^\infty_{\mathbb{K}}(U,\mathbb{K}) \to C^\infty_{\mathbb{K}}(U,E)$, $f \mapsto X_f$ is continuous linear in this case. And if $[.,.]$ is continuous, then also the Poisson bracket is continuous (see [13]).

Acknowledgements This article is partially based on an unpublished preprint from 2002 ("Remarks on holomorphic families of operators"), the preparation of which was supported by the German Research Foundation (DFG, FOR 363/1-1). A former referee of that preprint suggested valuable references to the literature. The topic of locally convex Poisson vector spaces was brought up by K.-H. Neeb (Darmstadt), whose questions and suggestions triggered the author's research on this subject. Currently, the author is supported by DFG, GL 357/5-1.

References

1. Außenhofer, L.: Contributions to the duality theory of Abelian topological groups and to the theory of nuclear groups. Diss. Math. **384** (1999)
2. Banaszczyk, W.: Additive Subgroups of Topological Vector Spaces. Springer, Berlin (1991)
3. Bertram, W., Glöckner, H., Neeb, K.-H.: Differential calculus over general base fields and rings. Expo. Math. **22**, 213–282 (2004)
4. Bochnak, J., Siciak, J.: Analytic functions in topological vector spaces. Studia Math. **39**, 77–112 (1971)
5. Bourbaki, N.: Topological Vector Spaces, chaps. 1–5. Springer, Berlin (1987)
6. Bunke, U., Olbrich M.: Group cohomology and the singularities of the Selberg zeta function associated to a Kleinian group. Ann. Math. **149**, 627–689 (1999)
7. Bunke, U., Olbrich M.: The spectrum of Kleinian manifolds. J. Funct. Anal. **172**, 76–164 (2000)
8. Floret, K.: Lokalkonvexe Sequenzen mit kompakten Abbildungen. J. Reine Angew. Math. **247**, 155–195 (1971)
9. Glöckner, H.: Infinite-dimensional Lie groups without completeness restrictions. Strasburger A. et al. (eds.) Geometry and Analysis on Finite- and Infinite-Dimensional Lie Groups, pp. 43–59. Banach Center, Warszawa (2002)
10. Glöckner, H.: Fundamentals of direct limit Lie theory. Compos. Math. **141**, 1551–1577 (2005)
11. Glöckner, H.: Direct limits of infinite-dimensional Lie groups compared to direct limits in related categories. J. Funct. Anal. **245**, 19–61 (2007)
12. Glöckner, H.: Instructive examples of smooth, complex differentiable and complex analytic mappings into locally convex spaces. J. Math. Kyoto Univ. **47**, 631–642 (2007)
13. Glöckner, H.: Aspects of differential calculus related to infinite-dimensional vector bundles and Poisson vector spaces. Manuscript in preparation
14. Glöckner, H., Gramlich, R., Hartnick, T.: Final group topologies, Kac–Moody groups and Pontryagin duality. Preprint, arXiv:math/0603537v3
15. Glöckner, H., Neeb K.-H.: Infinite-Dimensional Lie Groups, vol. I. Book in preparation
16. Große-Erdmann, K.-G.: The Borel–Okada Theorem Revisited. Habilitationsschrift, Fern-Universität Hagen (1992)
17. Jarchow, H.: Locally Convex Spaces. B. G. Teubner, Stuttgart (1981)
18. Keller, H.H.: Räume stetiger multilinearer Abbildungen als Limesräume. Math. Ann. **159**, 259–270 (1965)
19. Köthe, G.: Topological Vector Spaces II. Springer, New York (1979)
20. Kriegl, A., Michor, P.W.: The Convenient Setting of Global Analysis. Am. Math. Soc., Providence (1997)
21. Lang, S.: Fundamentals of Differential Geometry. Springer, New York (1999)
22. Maissen, B.: Über Topologien im Endomorphismenraum eines topologischen Vektorraumes. Math. Ann. **151**, 283–285 (1963)
23. Milnor, J.: Remarks on infinite-dimensional Lie groups. In: DeWitt, B., Stora R. (eds.) Relativité, Groupes et Topologie II, pp. 1007–1057. Elsevier, Amsterdam (1984)
24. Neeb, K.-H.: Towards a Lie theory of locally convex groups. Jap. J. Math. **1**, 291–468 (2006)
25. Odzijewicz, A., Ratiu T.S.: Banach Lie–Poisson spaces and reduction. Comm. Math. Phys. **243**, 1–54 (2003)
26. Odzijewicz, A., Ratiu, T.S.: Extensions of Banach Lie–Poisson spaces. J. Funct. Anal. **217**, 103–125 (2004)
27. Seip, U.: Kompakt erzeugte Vektorräume und Analysis. Springer, Berlin (1972)
28. Thomas, E.G.F.: Calculus on locally convex spaces. Preprint, University of Groningen (1996)
29. Treves, F.: Topological Vector Spaces, Distributions and Kernels. Academic, New York (1967)
30. Wurzbacher, T.: Fermionic second quantization and the geometry of the restricted Grassmannian. In: Huckleberry, A., Wurzbacher, T. (eds.) Infinite-Dimensional Kähler Manifolds, pp. 287–375. Birkhäuser, Basel (2001)

Part IV
Quasi-Lie, Super-Lie, Hom-Hopf and Super-Hopf Structures and Extensions, Deformations and Generalizations of Infinite-Dimensional Lie Algebras

Chapter 17
Hom-Lie Admissible Hom-Coalgebras and Hom-Hopf Algebras

Abdenacer Makhlouf and Sergei Silvestrov

Abstract The aim of this paper is to develop the coalgebra counterpart of the notions introduced by the authors in a previous paper, we introduce the notions of Hom-coalgebra, Hom-coassociative coalgebra and G-Hom-coalgebra for any subgroup G of permutation group $\mathscr{S}_3$. Also we extend the concept of Lie-admissible coalgebra by Goze and Remm to Hom-coalgebras and show that G-Hom-coalgebras are Hom-Lie admissible Hom-coalgebras, and also establish duality correspondence between classes of G-Hom-coalgebras and G-Hom-algebras. In another hand, we provide relevant definitions and basic properties of Hom-Hopf algebras generalizing the classical Hopf algebras and define the module and comodule structure over Hom-associative algebra or Hom-coassociative coalgebra.

17.1 Introduction

In [4, 7, 8], the class of quasi-Lie algebras and subclasses of quasi-hom-Lie algebras and Hom-Lie algebras have been introduced. These classes of algebras are tailored in a way suitable for simultaneous treatment of the Lie algebras, Lie superalgebras, the color Lie algebras and the deformations arising in connection with twisted, discretized or deformed derivatives [5] and corresponding generalizations, discretizations and deformations of vector fields and differential calculus. It has been shown in [4, 7–9] that the class of quasi-Hom-Lie algebras contains as a subclass on the one hand the color Lie algebras and in particular Lie superalgebras and Lie algebras, and

A. Makhlouf
Laboratoire de Mathématiques, Informatique et Applications, Université de Haute Alsace, 4, rue des Frères Lumière, 68093 Mulhouse, France
e-mail: Abdenacer.Makhlouf@uha.fr

S. Silvestrov
Centre for Mathematical Sciences, Lund University, Box 118, 221 00 Lund, Sweden
e-mail: ssilvest@maths.lth.se

S. Silvestrov et al. (eds.), *Generalized Lie Theory in Mathematics, Physics and Beyond,*

on the another hand various known and new single and multi-parameter families of algebras obtained using twisted derivations and constituting deformations and quasi-deformations of universal enveloping algebras of Lie and color Lie algebras and of algebras of vector-fields. The main feature of quasi-Lie algebras, quasi-Hom-Lie algebras and Hom-Lie algebras is that the skew-symmetry and the Jacobi identity are twisted by several deforming twisting maps and also in quasi-Lie and quasi-Hom-Lie algebras the Jacobi identity in general contains six twisted triple bracket terms.

In the paper [12], we provided a different way for constructing Hom-Lie algebras by extending the fundamental construction of Lie algebras from associative algebras via commutator bracket multiplication. To this end we defined the notion of Hom-associative algebras generalizing associative algebras to a situation where associativity law is twisted, and showed that the commutator product defined using the multiplication in a Hom-associative algebra leads naturally to Hom-Lie algebras. We introduced also Hom-Lie-admissible algebras and more general G-Hom-associative algebras with subclasses of Hom-Vinberg and pre-Hom-Lie algebras, generalizing to the twisted situation Lie-admissible algebras, G-associative algebras, Vinberg and pre-Lie algebras respectively, and show that for these classes of algebras the operation of taking commutator leads to Hom-Lie algebras as well. We constructed also all the twistings so that the brackets $[X_1,X_2]=2X_2,\ [X_1,X_3]=-2X_3,\ [X_2,X_3]=X_1$ determine a three-dimensional Hom-Lie algebra. Finally, we provided for a subclass of twistings, the list of all three-dimensional Hom-Lie algebras. This list contains all three-dimensional Lie algebras for some values of structure constants. The families of Hom-Lie algebras in these list can be viewed as deformations of Lie algebras into a class of Hom-Lie algebras. The notion, constructions and properties of the enveloping algebras of Hom-Lie algebras are yet to be properly studied in full generality. An important progress in this direction has been made in the recent work by D. Yau [14].

In the present paper we develop the coalgebra counterpart of the notions and results of [12], extending in particular in the framework of Hom-associative and Hom-Lie algebras and Hom-coalgebras, the notions and results on associative and Lie admissible coalgebras obtained in [2]. In Sect. 17.2 we summarize the relevant definitions of Hom-associative algebra, Hom-Lie algebra, Hom-Leibniz algebra, and define the notions of Hom-coalgebras and Hom-coassociative coalgebras. In Sect. 17.3, we introduce the concept of Hom-Lie admissible Hom-coalgebra, describe some useful relations between coproduct, opposite coproduct, the cocommutator defined as their difference, and their β-twisted coassociators and β-twisted co-Jacobi sums. We also introduce the notion of G-Hom-coalgebra for any subgroup G of permutation group S_3. We show that G-Hom-coalgebras are Hom-Lie admissible Hom-coalgebras, and also establish duality correspondence between classes of G-Hom-coalgebras and G-Hom-algebras. Section 17.4 is dedicated to relevant definitions and basic properties of the Hom-Hopf algebra which generalize the classical Hopf algebra structure. We also define the module and comodule structure over Hom-associative algebra or Hom-coassociative coalgebra.

17.2 Hom-Algebra and Hom-Coalgebra Structures

A Hom-algebra structure is a multiplication on a vector space where the structure is twisted by a homomorphism. The structure of Hom-Lie algebra was introduced by Hartwig et al. [4]. In the following we summarize the definitions of Hom-associative, Hom-Leibniz, and Hom-Lie-admissible algebraic structures introduced in [12] and generalizing the well known associative, Leibniz and Lie-admissible algebras. By dualization of Hom-associative algebra we define the Hom-coassociative coalgebra structure.

17.2.1 Hom-Algebra Structures

Let $\mathbb{K}$ be an algebraically closed field of characteristic 0 and V be a linear space over $\mathbb{K}$.

Definition 17.1. A *Hom-associative algebra* is a triple (V,μ,α) consisting of a linear space V, a linear map $\mu : V\otimes V \to V$ and a homomorphism α satisfying

$$\mu(\alpha(x)\otimes\mu(y\otimes z)) = \mu(\mu(x\otimes y)\otimes\alpha(z)). \tag{17.1}$$

The Hom-associativity condition (17.1) may be expressed by the following commutative diagram.

$$\begin{array}{ccc} V\otimes V\otimes V & \stackrel{\mu\otimes\alpha}{\longrightarrow} & V\otimes V \\ \downarrow{\scriptstyle \alpha\otimes\mu} & & \downarrow{\scriptstyle \mu} \\ V\otimes V & \stackrel{\mu}{\longrightarrow} & V \end{array}$$

The Hom-associative algebra is unital if there exists a homomorphism $\eta : \mathbb{K}\to V$ such that the following diagrams are commutative

$$\begin{array}{ccccc} \mathbb{K}\otimes V & \stackrel{\eta\otimes id}{\longrightarrow} & V\otimes V & \stackrel{id\otimes\eta}{\longleftarrow} & V\otimes\mathbb{K} \\ & \searrow^{\cong} & \downarrow^{\mu} & \swarrow^{\cong} & \\ & & V & & \end{array}$$

In the language of Hopf algebra, a Hom-associative algebra $\mathscr{A}$ is a quadruple (V,μ,α,η) where V is the vector space, μ is the Hom-associative multiplication, α is the twisting homomorphism and η is the unit.

Let (V,μ,α,η) and (V',μ',α',η') be two Hom-associative algebras. A linear map $f : V\to V'$ is a morphism of Hom- associative algebras if

$$\mu'\circ(f\otimes f) = f\circ\mu \quad , \quad f\circ\eta = \eta' \quad \text{and} \quad f\circ\alpha = \alpha'\circ f.$$

In particular, (V,μ,α,η) and (V,μ',α',η') are isomorphic if there exists a bijective linear map f such that

$$\mu = f^{-1}\circ\mu'\circ(f\otimes f) \quad , \quad \eta = f^{-1}\circ\eta' \quad \text{and} \quad \alpha = f^{-1}\circ\alpha'\circ f.$$

The tensor product of two Hom-associative algebras $(V_1, \mu_1, \alpha_1, \eta_1)$ and $(V_2, \mu_2, \alpha_2, \eta_2)$ is defined in an obvious way as the Hom-associative algebra $(V_1 \otimes V_2, \mu_1 \otimes \mu_2, \alpha_1 \otimes \alpha_2, \eta_1 \otimes \eta_2)$.

The Hom-Lie algebras were initially introduced in [4] motivated initially by examples of deformed Lie algebras coming from twisted discretizations of vector fields.

Definition 17.2. A *Hom-Lie algebra* is a triple $(V, [\cdot,\cdot], \alpha)$ consisting of a linear space V, bilinear map $[\cdot,\cdot] : V \times V \to V$ and a linear space homomorphism $\alpha : V \to V$ satisfying

$$[x,y] = -[y,x] \quad \text{(skew-symmetry)}$$
$$\circlearrowleft_{x,y,z} [\alpha(x),[y,z]] = 0 \quad \text{(Hom-Jacobi condition)}$$

for all x, y, z from V, where $\circlearrowleft_{x,y,z}$ denotes summation over the cyclic permutation on x, y, z.

In a similar way we have the following definition of Hom-Leibniz algebra.

Definition 17.3. A *Hom-Leibniz algebra* is a triple $(V, [\cdot,\cdot], \alpha)$ consisting of a linear space V, bilinear map $[\cdot,\cdot] : V \times V \to V$ and a homomorphism $\alpha : V \to V$ satisfying

$$[[x,y], \alpha(z)] = [[x,z], \alpha(y)] + [\alpha(x), [y,z]]. \tag{17.2}$$

Note that if a Hom-Leibniz algebra is skewsymmetric then it is a Hom-Lie algebra.

17.2.2 Hom-Coalgebra Structures

Definition 17.4. A *Hom-coassociative coalgebra* is a quadruple $(V, \Delta, \beta, \varepsilon)$ where V is a $\mathbb{K}$-vector space and

$$\Delta : V \to V \otimes V, \quad \beta : V \to V \text{ and } \varepsilon : V \to \mathbb{K}$$

are linear maps satisfying the following conditions:

(C1) $(\beta \otimes \Delta) \circ \Delta = (\Delta \otimes \beta) \circ \Delta$
(C2) $(id \otimes \varepsilon) \circ \Delta = id$ and $(\varepsilon \otimes id) \circ \Delta = id$.

The condition (C1) expresses the Hom-coassociativity of the comultiplication Δ. Also, it is equivalent to the following commutative diagram:

$$\begin{array}{ccc} V & \xrightarrow{\Delta} & V \otimes V \\ \downarrow{\scriptstyle \Delta} & & \downarrow{\scriptstyle \beta \otimes \Delta} \\ V \otimes V & \xrightarrow{\Delta \otimes \beta} & V \otimes V \otimes V \end{array}$$

The condition (C2) expresses that ε is the counit which is also equivalent to the following commutative diagram:

$$\begin{array}{ccccc} \mathbb{K}\otimes V & \overset{\varepsilon\otimes id_V}{\longleftarrow} & V\otimes V & \overset{id\otimes\varepsilon}{\longrightarrow} & V\otimes\mathbb{K} \\ & \nwarrow^{\cong} & \uparrow^{\Delta} & \nearrow^{\cong} & \\ & & V & & \end{array}$$

Let $(V,\Delta,\beta,\varepsilon)$ and $(V',\Delta',\beta',\varepsilon')$ be two Hom-coassociative coalgebras. A linear map $f : V \rightarrow V'$ is a morphism of Hom-coassociative coalgebras if

$$(f\otimes f)\circ\Delta = \Delta'\circ f \quad , \quad \varepsilon = \varepsilon'\circ f \quad \text{and} \quad f\circ\beta = \beta'\circ f.$$

If $V = V'$, then the previous Hom-coassociative coalgebras are isomorphic if there exists a bijective linear map $f : V \rightarrow V$ such that

$$\Delta' = (f\otimes f)\circ\Delta\circ f^{-1} \quad , \quad \varepsilon' = \varepsilon\circ f^{-1} \quad \text{and} \quad \beta = f^{-1}\circ\beta'\circ f.$$

In the sequel, we call *Hom-coalgebra* a triple (V,Δ,β) where V is a $\mathbb{K}$-vector space, Δ is a comultiplication not necessarily coassociative or Hom-coassociative, that is a linear map $\Delta : V \rightarrow V\otimes V$, and β is a linear map $\beta : V \rightarrow V$.

17.3 Hom-Lie Admissible Hom-Coalgebras

Let $\mathbb{K}$ be an algebraically closed field of characteristic 0 and V be a vector space over $\mathbb{K}$. Let (V,Δ,β) be a Hom-coalgebra where $\Delta : V \rightarrow V\otimes V$ and $\beta : V \rightarrow V$ are linear maps and Δ is not necessarily coassociative or Hom-coassociative.

By a *β-coassociator* of Δ we call a linear map $\mathbf{c}_\beta(\Delta)$ defined by

$$\mathbf{c}_\beta(\Delta) := (\Delta\otimes\beta)\circ\Delta - (\beta\otimes\Delta)\circ\Delta.$$

Let $\mathcal{S}_3$ be the symmetric group of order 3. Given $\sigma\in\mathcal{S}_3$, we define a linear map

$$\Phi_\sigma : V^{\otimes 3} \longrightarrow V^{\otimes 3}$$

by

$$\Phi_\sigma(x_1\otimes x_2\otimes x_3) = x_{\sigma^{-1}(1)}\otimes x_{\sigma^{-1}(2)}\otimes x_{\sigma^{-1}(3)}.$$

Recall that $\Delta^{op} = \tau\circ\Delta$ where τ is the usual flip that is $\tau(x\otimes y) = y\otimes x$.

Definition 17.5. A triple (V,Δ,β) is a *Hom-Lie admissible Hom-coalgebra* if the linear map

$$\Delta_L : V \longrightarrow V\otimes V$$

defined by $\Delta_L = \Delta - \Delta^{op}$, is a Hom-Lie coalgebra multiplication, that is the following condition is satisfied

$$\mathbf{c}_\beta(\Delta_L) + \Phi_{(213)}\circ\mathbf{c}_\beta(\Delta_L) + \Phi_{(231)}\circ\mathbf{c}_\beta(\Delta_L) = 0 \qquad (17.3)$$

where (213) and (231) are the two cyclic permutations of order 3 in $\mathcal{S}_3$.

Remark 17.1. Since $\Delta_L = \Delta - \Delta^{op}$, the equality $\Delta_L^{op} = -\Delta_L$ holds.

Lemma 17.1. *Let (V, Δ, β) be a Hom-coalgebra where $\Delta : V \to V \otimes V$ and $\beta : V \to V$ are linear maps and Δ is not necessarily coassociative or Hom-coassociative, then the following relations are true*

$$\mathbf{c}_\beta(\Delta^{op}) = -\Phi_{(13)} \circ \mathbf{c}_\beta(\Delta) \tag{17.4}$$
$$(\beta \otimes \Delta^{op}) \circ \Delta = \Phi_{(13)} \circ (\Delta \otimes \beta) \circ \Delta^{op} \tag{17.5}$$
$$(\beta \otimes \Delta) \circ \Delta^{op} = \Phi_{(13)} \circ (\Delta^{op} \otimes \beta) \circ \Delta \tag{17.6}$$
$$(\Delta \otimes \beta) \circ \Delta^{op} = \Phi_{(213)} \circ (\beta \otimes \Delta) \circ \Delta \tag{17.7}$$
$$(\Delta^{op} \otimes \beta) \circ \Delta = \Phi_{(12)} \circ (\Delta \otimes \beta) \circ \Delta. \tag{17.8}$$

Lemma 17.2. *The β-coassociator of Δ_L is expressed using Δ and Δ^{op} as follows:*

$$\begin{aligned}\mathbf{c}_\beta(\Delta_L) &= \mathbf{c}_\beta(\Delta) + \mathbf{c}_\beta(\Delta^{op}) && (17.9)\\ &\quad -(\Delta \otimes \beta) \circ \Delta^{op} - (\Delta^{op} \otimes \beta) \circ \Delta + \\ &\quad \Phi_{(13)} \circ (\Delta \otimes \beta) \circ \Delta^{op} + \Phi_{(13)} \circ (\Delta^{op} \otimes \beta) \circ \Delta \\ &= \mathbf{c}_\beta(\Delta) - \Phi_{(13)} \circ \mathbf{c}_\beta(\Delta) && (17.10)\\ &\quad -\Phi_{(213)} \circ (\beta \otimes \Delta) \circ \Delta - \Phi_{(12)} \circ (\Delta \otimes \beta) \circ \Delta \\ &\quad +\Phi_{(23)} \circ (\beta \otimes \Delta) \circ \Delta + \Phi_{(231)} \circ (\Delta \otimes \beta) \circ \Delta.\end{aligned}$$

Proposition 17.1. *Let (V, Δ, β) be a Hom-coalgebra. Then one has*

$$\mathbf{c}_\beta(\Delta_L) + \Phi_{(213)} \circ \mathbf{c}_\beta(\Delta_L) + \Phi_{(231)} \circ \mathbf{c}_\beta(\Delta_L) = 2 \sum_{\sigma \in \mathscr{S}_3} (-1)^{\varepsilon(\sigma)} \Phi_\sigma \circ \mathbf{c}_\beta(\Delta) \tag{17.11}$$

where $(-1)^{\varepsilon(\sigma)}$ is the signature of the permutation σ.

Proof. By (17.10) and multiplication rules in the group $\mathscr{S}_3$, it follows that

$$\begin{aligned}\Phi_{(213)} \circ \mathbf{c}_\beta(\Delta_L) &= \Phi_{(213)} \circ \mathbf{c}_\beta(\Delta) - \Phi_{(213)} \circ \Phi_{(13)} \circ \mathbf{c}_\beta(\Delta) \\ &\quad -\Phi_{(213)} \circ \Phi_{(213)} \circ (\beta \otimes \Delta) \circ \Delta - \Phi_{(213)} \circ \Phi_{(12)} \circ (\Delta \otimes \beta) \circ \Delta \\ &\quad +\Phi_{(213)} \circ \Phi_{(23)} \circ (\beta \otimes \Delta) \circ \Delta + \Phi_{(213)} \circ \Phi_{(231)} \circ (\Delta \otimes \beta) \circ \Delta \\ &= \Phi_{(213)} \circ \mathbf{c}_\beta(\Delta) - \Phi_{(12)} \circ \mathbf{c}_\beta(\Delta) && (17.12)\\ &\quad -\Phi_{(231)} \circ (\beta \otimes \Delta) \circ \Delta - \Phi_{(23)} \circ (\Delta \otimes \beta) \circ \Delta \\ &\quad +\Phi_{(13)} \circ (\beta \otimes \Delta) \circ \Delta + (\Delta \otimes \beta) \circ \Delta, \\ \Phi_{(231)} \circ \mathbf{c}_\beta(\Delta_L) &= \Phi_{(231)} \circ \mathbf{c}_\beta(\Delta) - \Phi_{(231)} \circ \Phi_{(13)} \circ \mathbf{c}_\beta(\Delta) \\ &\quad -\Phi_{(231)} \circ \Phi_{(213)} \circ (\beta \otimes \Delta) \circ \Delta - \Phi_{(231)} \circ \Phi_{(12)} \circ (\Delta \otimes \beta) \circ \Delta \\ &\quad +\Phi_{(231)} \circ \Phi_{(23)} \circ (\beta \otimes \Delta) \circ \Delta + \Phi_{(231)} \circ \Phi_{(231)} \circ (\Delta \otimes \beta) \circ \Delta \\ &= \Phi_{(231)} \circ \mathbf{c}_\beta(\Delta) - \Phi_{(23)} \circ \mathbf{c}_\beta(\Delta) && (17.13)\\ &\quad -(\beta \otimes \Delta) \circ \Delta - \Phi_{(13)} \circ (\Delta \otimes \beta) \circ \Delta \\ &\quad +\Phi_{(12)} \circ (\beta \otimes \Delta) \circ \Delta + \Phi_{(213)} \circ (\Delta \otimes \beta) \circ \Delta.\end{aligned}$$

After summing up the equalities (17.10), (17.12) and (17.13) the terms on the right hand sides may be pairwise combined into the terms of the form $(-1)^{\varepsilon(\sigma)}\Phi_\sigma \circ \mathbf{c}_\beta(\Delta)$ with each one being present in the sum twice for all $\sigma \in \mathscr{S}_3$.

Definition 17.5 together with (17.11) yields the following corollary.

Corollary 17.1. *A triple (V,Δ,β) is a Hom-Lie admissible Hom-coalgebra if and only if*

$$\sum_{\sigma\in\mathscr{S}_3} (-1)^{\varepsilon(\sigma)}\Phi_\sigma \circ \mathbf{c}_\beta(\Delta) = 0$$

where $(-1)^{\varepsilon(\sigma)}$ is the signature of the permutation σ.

17.3.1 G-Hom-Coalgebra Structures

In this section we introduce, as in the multiplication case, the notion of *G-Hom-coalgebra* where G is a subgroup of the symmetric group $\mathscr{S}_3$.

Definition 17.6. Let G be a subgroup of the symmetric group $\mathscr{S}_3$, A Hom-coalgebra (V,Δ,β) is called *G-Hom-coalgebra* if

$$\sum_{\sigma\in G} (-1)^{\varepsilon(\sigma)}\Phi_\sigma \circ \mathbf{c}_\beta(\Delta) = 0 \qquad (17.14)$$

where $(-1)^{\varepsilon(\sigma)}$ is the signature of the permutation σ.

Proposition 17.2. *Let G be a subgroup of the permutations group $\mathscr{S}_3$. Then any G-Hom-Coalgebra (V,Δ,β) is a Hom-Lie admissible Hom-coalgebra.*

Proof. The skew-symmetry follows straightaway from the definition. Take the set of conjugacy classes $\{gG\}_{g\in I}$ where $I \subseteq G$, and for any $\sigma_1,\sigma_2 \in I, \sigma_1 \neq \sigma_2 \Rightarrow \sigma_1 G \bigcap \sigma_1 G = \emptyset$. Then

$$\sum_{\sigma\in\mathscr{S}_3} (-1)^{\varepsilon(\sigma)}\Phi_\sigma \circ \mathbf{c}_\beta(\Delta) = \sum_{\sigma_1\in I}\sum_{\sigma_2\in\sigma_1 G} (-1)^{\varepsilon(\sigma)}\Phi_\sigma \circ \mathbf{c}_\beta(\Delta) = 0.$$

The subgroups of $\mathscr{S}_3$ are

$$G_1 = \{Id\},\ G_2 = \{Id,\tau_{12}\},\ G_3 = \{Id,\tau_{23}\},$$

$$G_4 = \{Id,\tau_{13}\},\ G_5 = A_3,\ G_6 = \mathscr{S}_3,$$

where A_3 is the alternating group and where τ_{ij} is the transposition between i and j.

We obtain the following type of Hom-Lie-admissible Hom-coalgebras:

- The G_1-Hom-coalgebras are the Hom-associative coalgebras defined above.
- The G_2-Hom-coalgebras satisfy the condition

$$\mathbf{c}_\beta(\Delta) + \Phi_{(12)}\mathbf{c}_\beta(\Delta) = 0.$$

- The G_3-Hom-coalgebras satisfy the condition

$$\mathbf{c}_\beta(\Delta)+\Phi_{(23)}\mathbf{c}_\beta(\Delta)=0.$$

- The G_4-Hom-coalgebras satisfy the condition

$$\mathbf{c}_\beta(\Delta)+\Phi_{(13)}\mathbf{c}_\beta(\Delta)=0.$$

- The G_5-Hom-coalgebras satisfy the condition

$$\mathbf{c}_\beta(\Delta)+\Phi_{(213)}\mathbf{c}_\beta(\Delta)+\Phi_{(231)}\mathbf{c}_\beta(\Delta)=0.$$

  If the product μ is skewsymmetric then the previous condition is exactly the Hom-Jacobi identity.
- The G_6-Hom-coalgebras are the Hom-Lie-admissible coalgebras.

The G_2-Hom-coalgebras may be called Vinberg-Hom-coalgebras and G_3-Hom-coalgebras may be called preLie-Hom-coalgebras. The two classes define in fact the same class.

Definition 17.7. A *Vinberg-Hom-coalgebra* is a triple (V,Δ,β) consisting of a linear space V, a linear map $\mu : V \to V \times V$ and a homomorphism β satisfying

$$\mathbf{c}_\beta(\Delta)+\Phi_{(12)}\mathbf{c}_\beta(\Delta)=0.$$

Definition 17.8. A *preLie-Hom-coalgebra* is a triple (V,Δ,β) consisting of a linear space V, a linear map $\mu : V \to V \times V$ and a homomorphism β satisfying

$$\mathbf{c}_\beta(\Delta)+\Phi_{(23)}\mathbf{c}_\beta(\Delta)=0.$$

More generally, by dualization we have a correspondence between G-Hom-associative algebras introduced in [12] and G-Hom-coalgebras for a subgroup G of $\mathscr{S}_3$.

Let G be a subgroup of $\mathscr{S}_3$ and (V,μ,α) be a G-Hom-associative algebra that is $\mu : V\otimes V \to V$ and $\alpha : V \to V$ are linear maps and the following condition is satisfied

$$\sum_{\sigma\in G}(-1)^{\varepsilon(\sigma)}a_{\alpha,\mu}\circ\Phi_\sigma=0. \tag{17.15}$$

where $a_{\alpha,\mu}$ is the α-associator that is $a_{\alpha,\mu}=\mu\circ(\mu\otimes\alpha)-\mu\circ(\alpha\otimes\mu)$.

Setting

$$(\mu\otimes\alpha)_G=\sum_{\sigma\in G}(-1)^{\varepsilon(\sigma)}(\mu\otimes\alpha)\circ\Phi_\sigma \text{ and } (\alpha\otimes\mu)_G=\sum_{\sigma\in G}(-1)^{\varepsilon(\sigma)}(\alpha\otimes\mu)\circ\Phi_\sigma$$

the condition (17.15) is equivalent to the following commutative diagram

$$\begin{array}{ccc} V\otimes V\otimes V & \overset{(\mu\otimes\alpha)_G}{\longrightarrow} & V\otimes V \\ \Big\downarrow (\alpha\otimes\mu)_G & & \downarrow\mu \\ V\otimes V & \overset{\mu}{\longrightarrow} & V \end{array}$$

By the dualization of the square one may obtain the following commutative diagram

$$\begin{array}{ccc} V & \overset{\Delta}{\longrightarrow} & V\otimes V \\ \downarrow\Delta & & \Big\downarrow (\beta\otimes\Delta)_G \\ V\otimes V & \overset{(\Delta\otimes\beta)_G}{\longrightarrow} & V\otimes V\otimes V \end{array}$$

where

$$(\beta\otimes\Delta)_G=\sum_{\sigma\in G}(-1)^{\varepsilon(\sigma)}\Phi_\sigma\circ(\beta\otimes\Delta) \text{ and } (\Delta\otimes\beta)_G=\sum_{\sigma\in G}(-1)^{\varepsilon(\sigma)}\Phi_\sigma\circ(\Delta\otimes\beta).$$

The previous commutative diagram expresses that (V,Δ,β) is a G-Hom-coalgebra. More precisely we have the following connection between G-Hom-coalgebras and G-Hom-associative algebras.

Proposition 17.3. *Let (V,Δ,β) be a G-Hom-coalgebra where G is a subgroup of $\mathscr{S}_3$. Its dual vector space $V^\star$ is provided with a G-Hom-associative algebra $(V^\star,\Delta^\star,\beta^\star)$ where $\Delta^\star,\beta^\star$ are the transpose maps.*

Proof. Let (V,Δ,β) be a G-Hom-coalgebra. Let $V^\star$ be the dual space of V ($V^\star = Hom(V,\mathbb{K})$).

Consider the map

$$\lambda_n : (V^\star)^{\otimes n} \longrightarrow (V^\star)^{\otimes n}$$
$$f_1\otimes\cdots\otimes f_n \longrightarrow \lambda_n(f_1\otimes\cdots\otimes f_n)$$

such that for $v_1\otimes\cdots\otimes v_n\in V^{\otimes n}$

$$\lambda_n(f_1\otimes\cdots\otimes f_n)(v_1\otimes\cdots\otimes v_n)=f_1(v_1)\otimes\cdots\otimes f_n(v_n)$$

and set

$$\mu:=\Delta^\star\circ\lambda_2 \qquad \alpha:=\beta^*$$

where the star $\star$ denotes the transpose linear map. Then, the quadruple $(V^\star,\mu,\eta,\alpha)$ is a G-Hom-associative algebra. Indeed, $\mu(f_1,f_2)=\mu_{\mathbb{K}}\circ\lambda_2(f_1\otimes f_2)\circ\Delta$ where $\mu_{\mathbb{K}}$ is the multiplication of $\mathbb{K}$ and $f_1,f_2\in V^*$. One has

$$\begin{aligned} \mu\circ(\mu\otimes\alpha)(f_1\otimes f_2\otimes f_3) &= \mu(\mu(f_1\otimes f_2)\otimes\alpha(f_3)) \\ &= \mu_{\mathbb{K}}\circ\lambda_2(\mu(f_1\otimes f_2)\otimes\alpha(f_3))\circ\Delta \\ &= \mu_{\mathbb{K}}\circ\lambda_2(\lambda_2((f_1\otimes f_2)\circ\Delta)\otimes\alpha(f_3))\circ\Delta \\ &= \mu_{\mathbb{K}}\circ(\mu_{\mathbb{K}}\otimes id)\circ\lambda_3(f_1\otimes f_2\otimes f_3)\circ(\Delta\otimes\beta)\circ\Delta. \end{aligned}$$

Similarly

$$\mu\circ(\alpha\otimes\mu)(f_1\otimes f_2\otimes f_3)=\mu_{\mathbb{K}}\circ(id\otimes\mu_{\mathbb{K}})\circ\lambda_3(f_1\otimes f_2\otimes f_3)\circ(\beta\otimes\Delta)\circ\Delta.$$

Using the associativity and the commutativity of $\mu_{\mathbb{K}}$, the α-associator may be written as

$$a_{\alpha,\mu}=\mu_{\mathbb{K}}\circ(id\otimes\mu_{\mathbb{K}})\circ\lambda_3(f_1\otimes f_2\otimes f_3)\circ((\Delta\otimes\beta)\circ\Delta-(\beta\otimes\Delta)\circ\Delta).$$

Then we have the following connection between the α-associator and β-coassociator

$$a_{\alpha,\mu}=\mu_{\mathbb{K}}\circ(id\otimes\mu_{\mathbb{K}})\circ\lambda_3(f_1\otimes f_2\otimes f_3)\circ\mathbf{c}_\beta(\Delta).$$

Therefore if (V,Δ,β) is a G-Hom-coalgebra, then the $(V^\star,\Delta^\star,\beta^\star)$ is a G-Hom-associative algebra.

Proposition 17.4. *Let (V,μ,α) be a finite-dimensional G-Hom-associative algebra where G is a subgroup of $\mathcal{S}_3$. Its dual vector space $V^\star$ is provided with a G-Hom-coalgebra $(V^\star,\mu^\star,\alpha^\star)$, where $\mu^\star,\alpha^\star$ are the transpose maps.*

Proof. Let $\mathcal{A}=(V,\mu,\alpha)$ be a n-dimensional Hom-associative algebra (n finite). Let $\{e_1,\cdots,e_n\}$ be a basis of V and $\{e_1^*,\cdots,e_n^*\}$ be the dual basis. Then $\{e_i^*\otimes e_j^*\}_{i,j}$ is a basis of $\mathcal{A}^\star\otimes\mathcal{A}^\star$. The comultiplication $\Delta=\mu^\star$ on $\mathcal{A}^*$ is defined for $f\in\mathcal{A}^\star$ by

$$\Delta(f)=\sum_{i,j=1}^{n}f(\mu(e_i\otimes e_j))\;e_i^*\otimes e_j^*.$$

Set $\mu(e_i\otimes e_j)=\sum_{k=1}^n C_{ij}^k e_k$ and $\alpha(e_i)=\sum_{k=1}^n\alpha_i^k e_k$. Then $\Delta(e_k^*)=\sum_{i,j=1}^n C_{ij}^k\;e_i^*\otimes e_j^*$ and $\beta(e_i)=\alpha^\star(e_i)=\sum_{k=1}^n\alpha_k^i e_k$.

The condition (17.14) of G-Hom-coassociativity of Δ, applied to any element e_k^* of the basis, is equivalent to

$$\sum_{p,q,s=1}^{n}\sum_{\sigma\in G}(-1)^{\varepsilon(\sigma)}\Big(\sum_{i,j=1}^{n}\alpha_s^jC_{ij}^kC_{pq}^i-\alpha_p^iC_{ij}^kC_{qs}^j\Big)e^*_{\sigma^{-1}(p)}\otimes e^*_{\sigma^{-1}(q)}\otimes e^*_{\sigma-1(s)}=0.$$

Therefore Δ is G-Hom-coassociative if for any $p,q,s,k\in\{1,\cdots,n\}$ one has

$$\sum_{\sigma\in G}(-1)^{\varepsilon(\sigma)}\Big(\sum_{i,j=1}^{n}\alpha_s^jC_{ij}^kC_{pq}^i-\alpha_p^iC_{ij}^kC_{qs}^j\Big)=0.$$

The previous system is exactly the condition (17.15) of G-Hom-associativity of μ, written on $e_{p'}\otimes e_{q'}\otimes e_{s'}$ and setting $p=\sigma(p')$, $q=\sigma(q')$, $s=\sigma(s')$.

Corollary 17.2. *The dual vector space of a Hom-coassociative coalgebra $(V,\Delta,\beta,\varepsilon)$ is a Hom-associative algebra $(V^\star,\Delta^\star,\beta^\star,\varepsilon^\star)$, where $V^\star$ is the dual vector space and the star for the linear maps denotes the transpose map. The dual vector space of finite-dimensional Hom-associative algebra is a Hom-coassociative coalgebra.*

Proof. It is a particular case of the previous Propositions ($G=G_1$).

17.4 Hom-Hopf Algebras

In this section, we introduce a generalization of Hopf algebras and show some relevant properties of the new structure. We also define the module and comodule structure over Hom-associative algebra or Hom-coassociative coalgebra. For classical Hopf algebras theory, we refer to [1, 3, 6, 10, 11, 13]. Let $\mathbb{K}$ be an algebraically closed field of characteristic 0 and V be a vector space over $\mathbb{K}$.

Definition 17.9. A *Hom-bialgebra* is a quintuple $(V,\mu,\alpha,\eta,\Delta,\beta,\varepsilon)$ where

(B1) (V,μ,α,η) is a Hom-associative algebra

(B2) $(V,\Delta,\beta,\varepsilon)$ is a Hom-coassociative coalgebra

(B3) The linear maps Δ and ε are morphisms of algebras (V,μ,α,η).

Remark 17.2. The condition (B3) could be expressed by the following system:

$$\begin{cases} \Delta(e_1) = e_1 \otimes e_1 \qquad \text{where } e_1 = \eta(1) \\ \Delta(\mu(x\otimes y)) = \Delta(x) \bullet \Delta(y) = \sum_{(x)(y)} \mu(x^{(1)} \otimes y^{(1)}) \otimes \mu(x^{(2)} \otimes y^{(2)}) \\ \varepsilon(e_1) = 1 \\ \varepsilon(\mu(x\otimes y)) = \varepsilon(x)\,\varepsilon(y) \end{cases}$$

where the bullet $\bullet$ denotes the multiplication on tensor product and by using the Sweedler's notation $\Delta(x) = \sum_{(x)} x^{(1)} \otimes x^{(2)}$. If there is no ambiguity we denote the multiplication by a dot.

Remark 17.3. One can consider a more restrictive definition where linear maps Δ and ε are morphisms of Hom-associative algebras that is the condition (B3) becomes equivalent to

$$\begin{cases} \Delta(e_1) = e_1 \otimes e_1 \qquad \text{where } e_1 = \eta(1) \\ \Delta(\mu(x\otimes y)) = \Delta(x) \bullet \Delta(y) = \sum_{(x)(y)} \mu(x^{(1)} \otimes y^{(1)}) \otimes \mu(x^{(2)} \otimes y^{(2)}) \\ \varepsilon(e_1) = 1 \\ \varepsilon(\mu(x\otimes y)) = \varepsilon(x)\,\varepsilon(y) \\ \Delta(\alpha(x)) = \sum_{(x)} \alpha(x^{(1)}) \otimes \alpha(x^{(2)}) \\ \varepsilon \circ \alpha(x) = \varepsilon(x) \end{cases}$$

Given a Hom-bialgebra $(V,\mu,\alpha,\eta,\Delta,\beta,\varepsilon)$, we show that the vector space *Hom* (V,V) with the multiplication given by the convolution product carries a structure of Hom-algebra.

Proposition 17.5. *Let $(V,\mu,\alpha,\eta,\Delta,\beta,\varepsilon)$ be a Hom-bialgebra. Then the algebra $Hom(V,V)$ with the multiplication given by the convolution product defined by*

$$f \star g = \mu \circ (f \otimes g) \circ \Delta$$

and the unit being $\eta \circ \varepsilon$ is a Hom-associative algebra with the homomorphism map defined by $\gamma(f) = \alpha \circ f \circ \beta$.

Proof. Let $f,g,h \in Hom(V,V)$. Then

$$\begin{aligned}\gamma(f)*(g*h)) &= \mu\circ(\gamma(f)\otimes(g*h))\Delta\\ &= \mu\circ(\gamma(f)\otimes(\mu\circ(g\otimes h)\circ\Delta))\Delta\\ &= \mu\circ(\alpha\otimes\mu)\circ(f\otimes g\otimes h)\circ(\beta\otimes\Delta))\Delta.\end{aligned}$$

Similarly

$$(f*g)*\gamma(h) = \mu\circ(\mu\otimes\alpha)\circ(f\otimes g\otimes h)\circ(\Delta\otimes\beta))\Delta.$$

Then, the Hom-associativity of μ and the Hom-coassociativity of Δ lead to the Hom-associativity of the convolution product. The unitality is as usual.

Definition 17.10. An endomorphism S of V is said to be an *antipode* if it is the inverse of the identity over V for the Hom-algebra $Hom(V,V)$ with the multiplication given by the convolution product defined by

$$f\star g = \mu\circ(f\otimes g)\Delta$$

and the unit being $\eta\circ\varepsilon$.

The condition being antipode may be expressed by the condition:

$$\mu\circ S\otimes Id\circ\Delta = \mu\circ Id\otimes S\circ\Delta = \eta\circ\varepsilon.$$

Definition 17.11. A *Hom-Hopf algebra* is a Hom-bialgebra with an antipode.

Then, a Hom-Hopf algebra over a $\mathbb{K}$-vector space V is given by

$$\mathscr{H} = (V,\mu,\alpha,\eta,\Delta,\beta,\varepsilon,S)$$

where the following homomorphisms

$$\mu: V\otimes V\rightarrow,\quad \eta:\mathbb{K}\rightarrow V,\quad \alpha: V\rightarrow V$$

$$\Delta: V\rightarrow V\otimes V,\quad \varepsilon: V\rightarrow\mathbb{K},\quad \beta: V\rightarrow V$$

$$S: V\rightarrow\mathbb{K}$$

satisfy the following conditions:

1. (V,μ,α,η) is a unital Hom-associative algebra.
2. $(V,\Delta,\beta,\varepsilon)$ is a counital Hom-coalgebra.
3. Δ and ε are morphisms of algebras, which translate to

$$\begin{cases}\Delta(e_1) = e_1\otimes e_1 \qquad \text{where } e_1 = \eta(1)\\ \Delta(x\cdot y) = \Delta(x)\bullet\Delta(y) = \sum_{(x)(y)} x^{(1)}\cdot y^{(1)}\otimes x^{(2)}\cdot y^{(2)}\\ \varepsilon(e_1) = 1\\ \varepsilon(x\cdot y) = \varepsilon(x)\,\varepsilon(y)\end{cases}$$

4. S is the antipode, so

$$\mu \circ S \otimes Id \circ \Delta = \mu \circ Id \otimes S \circ \Delta = \eta \circ \varepsilon.$$

Remark 17.4. Let V be a finite-dimensional $\mathbb{K}$-vector space.
If $\mathscr{H} = (V, \mu, \alpha, \eta, \Delta, \beta, \varepsilon, S)$ is a Hom-Hopf algebra, then

$$\mathscr{H}^* = (V^*, \Delta^*, \beta^*, \varepsilon^*, \mu^*, \alpha^*, \eta^*, S^*)$$

is also a Hom-Hopf algebra.

17.4.1 Primitive Elements and Generalized Primitive Elements

In the following, we discuss the properties of primitive elements in a Hom-bialgebra.

Let $\mathscr{H} = (V, \mu, \alpha, \eta, \Delta, \beta, \varepsilon)$ be a Hom-bialgebra and $e_1 = \eta(1)$ be the unit.

Definition 17.12. An element $x \in \mathscr{H}$ is called primitive if $\Delta(x) = e_1 \otimes x + x \otimes e_1$.

Let $x \in \mathscr{H}$ be a primitive element. The coassociativity of Δ implies

$$(\beta \otimes \Delta) \circ \Delta(x) = \tau_{13} \circ (\Delta \otimes \beta) \circ \Delta(x)$$

where τ_{13} is a permutation in the symmetric group $\mathscr{S}_3$.

Lemma 17.3. *Let x be a primitive element in $\mathscr{H}$, then $\varepsilon(x) = 0$.*

Proof. By counity property, we have $x = (id \otimes \varepsilon) \circ \Delta(x)$. If $\Delta(x) = e_1 \otimes x + x \otimes e_1$, then $x = \varepsilon(x)e_1 + \varepsilon(e_1)x$, and since $\varepsilon(e_1) = 1$ it implies $\varepsilon(x) = 0$.

Proposition 17.6. *Let $\mathscr{H} = (V, \mu, \alpha, \eta, \Delta, \beta, \varepsilon)$ be a Hom-bialgebra and $e_1 = \eta(1)$ be the unit. If x and y are two primitive elements in $\mathscr{H}$. Then we have $\varepsilon(x) = 0$ and the commutator $[x, y] = \mu(x \otimes y) - \mu(y \otimes x)$ is also a primitive element.*

The set of all primitive elements of $\mathscr{H}$, denoted by $Prim(\mathscr{H})$, has a structure of Hom-Lie algebra.

Proof. By a direct calculation one has

$$\begin{aligned}
\Delta([x,y]) &= \Delta(\mu(x \otimes y) - \mu(y \otimes x)) \\
&= \Delta(x) \bullet \Delta(y) - \Delta(y) \bullet \Delta(x) \\
&= (e_1 \otimes x + x \otimes e_1) \bullet (e_1 \otimes y + y \otimes e_1) - (e_1 \otimes y + y \otimes e_1) \bullet (e_1 \otimes x + x \otimes e_1) \\
&= e_1 \otimes \mu(x \otimes y) + y \otimes x + x \otimes y + \mu(x \otimes y) \otimes e_1 \\
&\quad - e_1 \otimes \mu(y \otimes x) - x \otimes y - y \otimes x - \mu(y \otimes x) \otimes e_1 \\
&= e_1 \otimes (\mu(x \otimes y) - \mu(y \otimes x)) + (\mu(x \otimes y) - \mu(y \otimes x)) \otimes e_1 \\
&= e_1 \otimes [x,y] + [x,y] \otimes e_1
\end{aligned}$$

which means that $Prim(\mathscr{H})$ is closed under the bracket multiplication $[\cdot, \cdot]$.

We have seen in [12] that there is a natural map from the Hom-associative algebras to Hom-Lie algebras. The bracket $[x,y] = \mu(x\otimes y) - \mu(y\otimes x)$ is obviously skewsymmetric and one checks that the Hom-Jacobi condition is satisfied:

$$\begin{aligned}
&[\alpha(x),[y,z]] - [[x,y],\alpha(z)] - [\alpha(y),[x,z]] = \\
&\mu(\alpha(x)\otimes\mu(y\otimes z)) - \mu(\alpha(x)\otimes\mu(z\otimes y)) - \mu(\mu(y\otimes z)\otimes\alpha(x)) \\
&+\mu(\mu(z\otimes y)\otimes\alpha(x)) - \mu(\mu(x\otimes y)\otimes\alpha(z)) + \mu(\mu(y\otimes x)\otimes\alpha(z)) \\
&+\mu(\alpha(z)\otimes\mu(x\otimes y)) - \mu(\alpha(z)\otimes\mu(y\otimes x)) - \mu(\alpha(y)\otimes\mu(x\otimes z)) \\
&+\mu(\alpha(y)\otimes\mu(z\otimes x)) + \mu(\mu(x\otimes z)\otimes\alpha(y)) - \mu(\mu(z\otimes x)\otimes\alpha(y)) = 0
\end{aligned}$$

We introduce now a notion of generalized primitive element.

Definition 17.13. An element $x\in\mathscr{H}$ is called generalized primitive element if it satisfies the conditions

$$(\beta\otimes\Delta)\circ\Delta(x) = \tau_{13}\circ(\Delta\otimes\beta)\circ\Delta(x) \qquad (17.16)$$

$$\Delta^{op}(x) = \Delta(x) \qquad (17.17)$$

where τ_{13} is a permutation in the symmetric group $\mathscr{S}_3$.

Remark 17.5. 1. In particular, a primitive element in $\mathscr{H}$ is a generalized primitive element.
2. The condition (17.16) may be written

$$(\Delta\otimes\beta)\circ\Delta(x) = \tau_{13}\circ(\beta\otimes\Delta)\circ\Delta(x).$$

Proposition 17.7. *Let $\mathscr{H} = (V,\mu,\alpha,\eta,\Delta,\beta,\varepsilon)$ be a Hom-bialgebra and $e_1 = \eta(1)$ be the unit. If x and y are two generalized primitive elements in $\mathscr{H}$. Then, we have $\varepsilon(x) = 0$ and the commutator $[x,y] = \mu(x\otimes y) - \mu(y\otimes x)$ is also a generalized primitive element.*

The set of all generalized primitive elements of $\mathscr{H}$, denoted by GPrim$(\mathscr{H})$, has a structure of Hom-Lie algebra.

Proof. Let x and y be two generalized primitive elements in $\mathscr{H}$. In the following the multiplication μ is denoted by a dot. The following equalities hold:

$$\begin{aligned}
(\Delta\otimes\beta)\circ\Delta(x\cdot y - y\cdot x) &= (\Delta\otimes\beta)\circ\Delta(x\cdot y) - (\Delta\otimes\beta)\circ\Delta(y\cdot x) \\
&= (\Delta\otimes\beta)(\Delta(x)\bullet\Delta(y)) - (\Delta\otimes\beta)(\Delta(y)\bullet\Delta(x)) \\
&= \Delta(x^{(1)}\cdot y^{(1)})\otimes\beta(x^{(2)}\cdot y^{(2)}) - \Delta(y^{(1)}\cdot x^{(1)})\otimes\beta(y^{(2)}\cdot x^{(2)}) \\
&= (x^{(1)(1)}\cdot y^{(1)(1)})\otimes(x^{(1)(2)}\cdot y^{(1)(2)})\otimes\beta(x^{(2)}\cdot y^{(2)}) \\
&\quad -(y^{(1)(1)}\cdot x^{(1)(1)})\otimes(y^{(1)(2)}\cdot x^{(1)(2)})\otimes\beta(y^{(2)}\cdot x^{(2)}).
\end{aligned}$$

Then, using the fact that $\Delta^{op} = \Delta$ for generalized primitive elements one has:

$$\begin{aligned}\tau_{13}\circ(\Delta\otimes\beta)\circ\Delta(x\cdot y-y\cdot x) &= \beta(x^{(2)}\cdot y^{(2)})\otimes(x^{(1)(2)}\cdot y^{(1)(2)})\otimes(x^{(1)(1)}\cdot y^{(1)(1)})\\ &\quad -\beta(y^{(2)}\cdot x^{(2)})\otimes(y^{(1)(2)}\cdot x^{(1)(2)})\otimes(y^{(1)(1)}\cdot x^{(1)(1)})\\ &= (\beta\otimes\Delta)\circ\Delta(x\cdot y-y\cdot x).\end{aligned}$$

The structure of Hom-Lie algebra follows from the same argument as in the primitive elements case.

17.4.2 Antipode's Properties

Let $\mathscr{H}=(V,\mu,\alpha,\eta,\Delta,\beta,\varepsilon,S)$ be a Hom-Hopf algebra.

For any element $x\in V$, using the counity and Sweedler notation, one may write

$$x=\sum_{(x)}x^{(1)}\otimes\varepsilon(x^{(2)})=\sum_{(x)}\varepsilon(x^{(1)})\otimes x^{(2)}. \tag{17.18}$$

Then, for any $f\in End_{\mathbb{K}}(V)$, we have

$$f(x)=\sum_{(x)}f(x^{(1)})\varepsilon(x^{(2)})=\sum_{(x)}\varepsilon(x^{(1)})\otimes f(x^{(2)}). \tag{17.19}$$

Let $f\star g=\mu\circ(f\otimes g)\Delta$ be the convolution product of $f,g\in End_{\mathbb{K}}(V)$. One may write

$$(f\star g)(x)=\sum_{(x)}\mu(f(x^{(1)})\otimes g(x^{(2)})). \tag{17.20}$$

Since the antipode S is the inverse of the identity for the convolution product then S satisfies

$$\varepsilon(x)\eta(1)=\sum_{(x)}\mu(S(x^{(1)})\otimes x^{(2)})=\sum_{(x)}\mu(x^{(1)}\otimes S(x^{(2)})). \tag{17.21}$$

Proposition 17.8. *The antipode S is unique and we have*

- $S(\eta(1))=\eta(1)$.
- $\varepsilon\circ S=\varepsilon$.

Proof. (1) We have $S\star id=id\star S=\eta\circ\varepsilon$. Thus, $(S\star id)\star S=S\star(id\star S)=S$. If S' is another antipode of $\mathscr{H}$ then

$$S'=S'\star id\star S'=S'\star id\star S=S\star id\star S=S.$$

Therefore the antipode when it exists is unique.

(2) Setting $e_1=\eta(1)$ and since $\Delta(e_1)=e_1\otimes e_1$ one has

$$(S\star id)(e_1)=\mu(S(e_1)\otimes e_1)=S(e_1)=\eta(\varepsilon(e_1))=e_1.$$

(3) Applying (17.19) to S, we obtain $S(x)=\sum_{(x)}S(x^{(1)})\varepsilon(x^{(2)})$.

Applying ε to (17.21), we obtain

$$\varepsilon(x) = \varepsilon(\sum_{(x)} \mu(S(x^{(1)}) \otimes x^{(2)})).$$

Since ε is a Hom-algebra morphism, one has

$$\varepsilon(x) = \sum_{(x)} \varepsilon(S(x^{(1)}))\varepsilon(x^{(2)}) = \varepsilon(\sum_{(x)} S(x^{(1)})\varepsilon(x^{(2)})) = \varepsilon(S(x)).$$

Thus $\varepsilon \circ S = \varepsilon$.

17.4.3 Modules and Comodules

We introduce in the following the structure of module and comodule over Hom-associative algebras.

Let $\mathscr{A} = (V, \mu, \alpha)$ be a Hom-associative $\mathbb{K}$-algebra, an $\mathscr{A}$-module (left) is a triple (M, f, γ) where M is $\mathbb{K}$-vector space and f, γ are $\mathbb{K}$-linear maps, $f : M \to M$ and $\gamma : V \otimes M \to M$, such that the following diagram commutes:

$$\begin{array}{ccc} V \otimes V \otimes M & \xrightarrow{\mu \otimes f} & V \otimes M \\ \downarrow{\scriptstyle \alpha \otimes \gamma} & & \downarrow{\scriptstyle \gamma} \\ V \otimes M & \xrightarrow{\gamma} & M \end{array}$$

The dualization leads to comodule definition over a Hom-coassociative coalgebra.

Let $C = (V, \Delta, \beta)$ be a Hom-coassociative coalgebra. A C-comodule (right) is a triple (M, g, ρ) where M is a $\mathbb{K}$-vector space and g, ρ are $\mathbb{K}$-linear maps, $g : M \to M$ and $\rho : M \to M \otimes V$, such that the following diagram commutes:

$$\begin{array}{ccc} M & \xrightarrow{\rho} & M \otimes V \\ \downarrow{\scriptstyle \rho} & & \downarrow{\scriptstyle g \otimes \Delta} \\ M \otimes V & \xrightarrow{\rho \otimes \beta} & M \otimes V \otimes V \end{array}$$

Remark 17.6. A Hom-associative $\mathbb{K}$-algebra $\mathscr{A} = (V, \mu, \alpha)$ is a left $\mathscr{A}$-module with $M = V$, $f = \alpha$ and $\gamma = \mu$. Also, a Hom-coassociative coalgebra $C = (V, \Delta, \beta)$ is a right C-comodule with $M = V$, $g = \beta$ and $\rho = \Delta$. The properties of modules and comodules over Hom-associative algebras or Hom-coassociative algebras will be discussed in a forthcoming paper.

17.4.4 Examples

The classification of two-dimensional Hom-associative algebras, up to isomorphism, yields the following two classes. Let $B = \{e_1, e_2\}$ be a basis where $\eta(1) = e_1$ is the unit.

1. The multiplication μ_1 is defined by $\mu_1(e_1 \otimes e_i) = \mu_1(e_i \otimes e_1) = e_i$ for $i = 1,2$ and $\mu_1(e_2 \otimes e_2) = e_2$, and the homomorphism α_1 is defined, with respect to the basis B, by $\begin{pmatrix} a_1 & 0 \\ a_2 - a_1 & a_2 \end{pmatrix}$.
2. The multiplication μ_2 is defined by $\mu_2(e_1 \otimes e_i) = \mu_2(e_i \otimes e_1) = e_i$ for $i = 1,2$ and $\mu_2(e_2 \otimes e_2) = 0$, and the homomorphism α_2 is defined, with respect to the basis B, by $\begin{pmatrix} a_1 & 0 \\ a_2 & a_1 \end{pmatrix}$.

The Hom-bialgebras corresponding to the Hom-associative algebra defined by μ_1 and α_1 are given in the following table

	Comultiplication	Co-unit	homomorphism
1	$\Delta(e_1) = e_1 \otimes e_1$ $\Delta(e_2) = e_2 \otimes e_2$	$\varepsilon(e_1) = 1$ $\varepsilon(e_2) = 1$	$\begin{pmatrix} b_1 & 0 \\ b_3 & b_2 \end{pmatrix}$
2	$\Delta(e_1) = e_1 \otimes e_1$ $\Delta(e_2) = e_1 \otimes e_2 + e_2 \otimes e_1 - 2e_2 \otimes e_2$	$\varepsilon(e_1) = 1$ $\varepsilon(e_2) = 0$	$\begin{pmatrix} b_1 & \frac{b_1 - b_3}{2} \\ b_2 & b_3 \end{pmatrix}$
3	$\Delta(e_1) = e_1 \otimes e_1$ $\Delta(e_2) = e_1 \otimes e_2 + e_2 \otimes e_1 - e_2 \otimes e_2$	$\varepsilon(e_1) = 1$ $\varepsilon(e_2) = 0$	$\begin{pmatrix} b_1 & b_1 - b_3 \\ b_2 & b_3 \end{pmatrix}$

Only Hom-bialgebra (2) carries a structure of Hom-Hopf algebra with an antipode defined, with respect to a basis B, by the identity matrix.

Remark 17.7. There is no Hom-bialgebra associated to the Hom-associative algebra defined by the multiplication μ_2 and any homomorphism α_2.

Acknowledgements This work was supported by the Swedish Foundation for International Cooperation in Research and Higher Education (STINT), the Crafoord Foundation, the Royal Physiographic Society in Lund, the Royal Swedish Academy of Sciences, the Swedish Research Council and European Erasmus program.

References

1. Drinfel'd V. G.: *Hopf algebras and the quantum Yang–Baxter equation*, Soviet Math. Doklady **32**, 254–258 (1985)
2. Goze M., Remm E.: *Lie-admissible coalgebras*, J. Gen. Lie Theory Appl. **1** , no. 1, 19–28 (2007)
3. Guichardet A.: *Groupes quantiques*, InterEditions / CNRS Editions, Paris (1995)
4. Hartwig J. T., Larsson D., Silvestrov S. D.: *Deformations of Lie algebras using σ-derivations*, J. Algebra **295**, 314–361 (2006)
5. Hellström L., Silvestrov S. D.: *Commuting elements in q-deformed Heisenberg algebras*, World Scientific, Singapore (2000)
6. Kassel C.: *Quantum groups*, Graduate Text in Mathematics, Springer, Berlin (1995)
7. Larsson D., Silvestrov S. D.: *Quasi-hom-Lie algebras, Central Extensions and 2-cocycle-like identities*, J. Algebra **288**, 321–344 (2005)

8. Larsson D., Silvestrov S. D.: *Quasi-Lie algebras*, in "Noncommutative Geometry and Representation Theory in Mathematical Physics", Contemp. Math. **391**, Amer. Math. Soc., Providence, RI, 241–248 (2005)
9. Larsson D., Silvestrov S. D.: *Quasi-deformations of* $sl_2(\mathbb{F})$ *using twisted derivations*, Comm. Algebra **35**, 4303–4318 (2007)
10. Majid S.: *Foundations of quantum group theory*, Cambridge University Press, Cambridge (1995)
11. Makhlouf A.: *Degeneration, rigidity and irreducible components of Hopf algebras*, Algebra Colloquium **12**(2), 241–254 (2005)
12. Makhlouf A., Silvestrov S. D.: *Hom-algebra structures*, J. Gen. Lie Theory, Appl. **2**(2), 51–64 (2008)
13. Montgomery S.: *Hopf algebras and their actions on rings*, AMS Regional Conference Series in Mathematics **82**, (1993)
14. Yau D.: *Enveloping algebra of Hom-Lie algebras*, J. Gen. Lie Theory Appl. **2**(2), 95–108 (2008)

Chapter 18
Bosonisation and Parastatistics

K. Kanakoglou and C. Daskaloyannis

Abstract Definitions of the parastatistics algebras and known results on their Lie (super)algebraic structure are reviewed. The notion of super-Hopf algebra is discussed. The bosonisation technique for switching a Hopf algebra in a braided category ${}_H\mathcal{M}$ (H: a quasitriangular Hopf algebra) into an ordinary Hopf algebra is presented and it is applied in the case of the parabosonic algebra. A bosonisation-like construction is also introduced for the same algebra and the differences are discussed.

18.1 Introduction and Definitions

Throughout this paper we are going to use the following notation conventions:
If x and y are any monomials of the tensor algebra of some k-vector space, we are going to call commutator the following expression:

$$[x,y] = x \otimes y - y \otimes x \equiv xy - yx$$

and anticommutator the following expression:

$$\{x,y\} = x \otimes y + y \otimes x \equiv xy + yx$$

By the field k we shall always mean the field of complex numbers C, and all tensor products will be considered over k unless stated so. Finally we freely use Sweedler's notation for the comultiplication throughout the paper.

K. Kanakoglou
Department of Physics, Aristotle University of Thessaloniki, Thessaloniki 54124, Greece
e-mail: kanakoglou@hotmail.com

C. Daskaloyannis
Department of Mathematics, Aristotle University of Thessaloniki, Thessaloniki 54124, Greece
e-mail: daskalo@math.auth.gr

S. Silvestrov et al. (eds.), *Generalized Lie Theory in Mathematics, Physics and Beyond,*

Parafermionic and parabosonic algebras first appeared in the physics literature by means of generators and relations, in the pioneering works of Green [6] and Greenberg and Messiah [5]. Their purpose was to introduce generalizations of the usual bosonic and fermionic algebras of quantum mechanics, capable of leading to generalized versions of the Bose–Einstein and Fermi–Dirac statistics (see [15]). We start with the definitions of these algebras:

Let us consider the k-vector space V_B freely generated by the elements: b_i^+, b_j^-, $i, j = 1,...,n$. Let $T(V_B)$ denote the tensor algebra of V_B (i.e., the free algebra generated by the elements of the basis). In $T(V_B)$ we consider the two-sided ideal I_B generated by the following elements:

$$\left[\{b_i^{\xi}, b_j^{\eta}\}, b_k^{\varepsilon}\right] - (\varepsilon - \eta)\delta_{jk} b_i^{\xi} - (\varepsilon - \xi)\delta_{ik} b_j^{\eta} \tag{18.1}$$

for all values of $\xi, \eta, \varepsilon = \pm 1$ and $i, j, k = 1, \ldots, n$.
We now have the following:

Definition 18.1. The parabosonic algebra in $2n$ generators $P_B^{(n)}$ (n parabosons) is the quotient algebra of the tensor algebra of V_B with the ideal I_B:

$$P_B^{(n)} = T(V_B)/I_B$$

In a similar way we may describe the parafermionic algebra in $2n$ generators (n parafermions): Let us consider the k-vector space V_F freely generated by the elements: f_i^+, f_j^-, $i, j = 1,...,n$. Let $T(V_F)$ denote the tensor algebra of V_F (i.e., the free algebra generated by the elements of the basis). In $T(V_F)$ we consider the two-sided ideal I_F generated by the following elements:

$$[[f_i^{\xi}, f_j^{\eta}], f_k^{\varepsilon}] - \frac{1}{2}(\varepsilon - \eta)^2 \delta_{jk} f_i^{\xi} + \frac{1}{2}(\varepsilon - \xi)^2 \delta_{ik} f_j^{\eta} \tag{18.2}$$

for all values of $\xi, \eta, \varepsilon = \pm 1$ and $i, j, k = 1, \ldots, n$.
We get the following definition:

Definition 18.2. The parafermionic algebra in $2n$ generators $P_F^{(n)}$ (n parafermions) is the quotient algebra of the tensor algebra of V_F with the ideal I_F:

$$P_F^{(n)} = T(V_F)/I_F$$

18.2 (Super-)Lie and (Super-)Hopf Algebraic Structure of the Parabosonic $P_B^{(n)}$ and Parafermionic $P_F^{(n)}$ Algebras

Due to its simpler nature, parafermionic algebras were the first to be identified as the universal enveloping algebras (UEA) of simple Lie algebras. This was done almost at the same time by Kamefuchi and Takahashi in [9] and by Ryan and Sudarshan in [21]. In fact the following stem from the above mentioned references (see also [16]):

Lemma 18.1. *In the k-vector space $P_F^{(n)}$ we consider the k-subspace generated by the set of elements:*

$$\left\{[f_i^{\xi}, f_j^{\eta}], f_k^{\varepsilon} \;|\xi, \eta, \varepsilon = \pm,\; i, j, k = 1, \ldots, n\right\}$$

The above subspace endowed with a bilinear multiplication $\langle .., .. \rangle$ whose values are determined by the values of the commutator in $P_F^{(n)}$, i.e.

$$\langle f_i^{\xi}, f_j^{\eta} \rangle = [f_i^{\xi}, f_j^{\eta}]$$

and

$$\left\langle [f_i^{\xi}, f_j^{\eta}], f_k^{\varepsilon} \right\rangle = \left[[f_i^{\xi}, f_j^{\eta}], f_k^{\varepsilon} \right] = \frac{1}{2}(\varepsilon - \eta)^2 \delta_{jk} f_i^{\xi} - \frac{1}{2}(\varepsilon - \xi)^2 \delta_{ik} f_j^{\eta}$$

is a simple complex Lie algebra isomorphic to $B_n = so(2n+1)$. The basis in the Cartan subalgebra of B_n can be chosen in such a way that the elements f^+ (respectively: f^-) are negative (respectively: positive) root vectors.

Based on the above observations, the following is finally proved:

Proposition 18.1. *The parafermionic algebra in $2n$ generators is isomorphic to the universal enveloping algebra of the simple complex Lie algebra $B_n = so(2n+1)$ (according to the well known classification of the simple complex Lie algebras), i.e.*

$$P_F^{(n)} \cong U(B_n)$$

An immediate consequence of the above identification is that parafermionic algebras are ordinary Hopf algebras, with the generators $f_i^{\pm}$, $i = 1, ..., n$ being primitive elements. The Hopf algebraic structure of $P_F^{(n)}$ is completely determined by the well known Hopf algebraic structure of the Lie algebras, due to the above isomorphism. For convenience we quote the relations explicitly:

$$\Delta(f_i^{\pm}) = f_i^{\pm} \otimes 1 + 1 \otimes f_i^{\pm} \quad \varepsilon(f_i^{\pm}) = 0 \quad S(f_i^{\pm}) = -f_i^{\pm} \tag{18.3}$$

The algebraic structure of parabosons seemed to be somewhat more complicated. The presence of anticommutators among the trilinear relations defining $P_B^{(n)}$ "breaks" the usual (Lie) antisymmetry and makes impossible the identification of the parabosons with the UEA of any Lie algebra. It was in the early 1980s that was conjectured [15], that due to the mixing of commutators and anticommutators in $P_B^{(n)}$ the proper mathematical "playground" should be some kind of Lie superalgebra (or: Z_2-graded Lie algebra). Starting in the early 1980s, and using the recent (by that time) results in the classification of the finite dimensional simple complex Lie superalgebras which was obtained by Kac (see [7,8]), Palev managed to identify the parabosonic algebra with the UEA of a certain simple complex Lie superalgebra. In [18, 19] (see also [17]), Palev shows the following:

Lemma 18.2. *In the k-vector space $P_B^{(n)}$ we consider the k-subspace generated by the set of elements:*

$$\left\{ \{b_i^\xi, b_j^\eta\}, b_k^\varepsilon \;|\xi, \eta, \varepsilon = \pm,\; i,j,k = 1,\dots,n \right\}$$

This vector space is turned into a superspace (Z_2-graded vector space) by the requirement that b_i^ξ span the odd subspace and $\{b_i^\xi, b_j^\eta\}$ span the even subspace. The above vector space endowed with a bilinear multiplication $\langle .., .. \rangle$ whose values are determined by the values of the anticommutator and the commutator in $P_B^{(n)}$, i.e.

$$\langle b_i^\xi, b_j^\eta \rangle = \{b_i^\xi, b_j^\eta\}$$

and:

$$\left\langle \{b_i^\xi, b_j^\eta\}, b_k^\varepsilon \right\rangle = \left[\{b_i^\xi, b_j^\eta\}, b_k^\varepsilon \right] = (\varepsilon - \eta)\delta_{jk} b_i^\xi + (\varepsilon - \xi)\delta_{ik} b_j^\eta$$

respectively, according to the above mentioned gradation, is a simple, complex super-Lie algebra (or: Z_2-graded Lie algebra) isomorphic to $B(0,n) = osp(1,2n)$. The basis in the Cartan subalgebra of $B(0,n)$ can be chosen in such a way that the elements b^+ (respectively: b^-) are negative (respectively: positive) root vectors.

Note that, according to the above lemma, the even part of $B(0,n)$ is spanned by the elements $\left\{ \{b_i^\xi, b_j^\eta\} \;|\xi, \eta = \pm,\; i,j = 1,\dots,n \right\}$ and is a subalgebra of $B(0,n)$ isomorphic to the Lie algebra $sp(2n)$. Its Lie multiplication can be readily deduced from the above given commutators and reads:

$$\left\langle \{b_i^\xi, b_j^\eta\}, \{b_k^\varepsilon, b_l^\phi\} \right\rangle = \left[\{b_i^\xi, b_j^\eta\}, \{b_k^\varepsilon, b_l^\phi\} \right] =$$

$$(\varepsilon - \eta)\delta_{jk}\{b_i^\xi, b_l^\phi\} + (\varepsilon - \xi)\delta_{ik}\{b_j^\eta, b_l^\phi\} + (\phi - \eta)\delta_{jl}\{b_i^\xi, b_k^\varepsilon\} + (\phi - \xi)\delta_{il}\{b_j^\eta, b_k^\varepsilon\}$$

On the other hand the elements $\left\{ b_k^\varepsilon \;|\varepsilon = \pm,\; k = 1,\dots,n \right\}$ constitute a basis of the odd part of $B(0,n)$.

Note also, that $B(0,n)$ in Kac's notation, is the classical simple complex orthosymplectic Lie superalgebra denoted $osp(1,2n)$ in the notation traditionally used by physicists until then.

Based on the above observations, Palev finally proves (in the above mentioned references):

Proposition 18.2. *The parabosonic algebra in 2n generators is isomorphic to the universal enveloping algebra of the classical simple complex Lie superalgebra $B(0,n)$ (according to the classification of the simple complex Lie superalgebras given by Kac), i.e.*

$$P_B^{(n)} \cong U(B(0,n))$$

The universal enveloping algebra $U(L)$ of a Lie superalgebra L is not a Hopf algebra, at least in the ordinary sense. $U(L)$ is a Z_2-graded associative algebra (or: superalgebra) and it is a super-Hopf algebra in a sense that we briefly describe: First we consider the braided tensor product algebra $U(L)\underline{\otimes}U(L)$, which means the vector space $U(L)\otimes U(L)$ equipped with the associative multiplication:

$$(a\otimes b)\cdot(c\otimes d)=(-1)^{|b||c|}ac\otimes bd$$

for b,c homogeneous elements of $U(L)$, and $|.|$ denotes the degree of an homogeneous element (i.e., $|b|=0$ if b is an even element and $|b|=1$ if b is an odd element). Note that $U(L)\underline{\otimes}U(L)$ is also a superalgebra or: Z_2-graded associative algebra. Then $U(L)$ is equipped with a coproduct

$$\underline{\Delta}:U(L)\rightarrow U(L)\underline{\otimes}U(L)$$

which is an superalgebra homomorphism from $U(L)$ to the braided tensor product algebra $U(L)\underline{\otimes}U(L)$:

$$\underline{\Delta}(ab)=\sum(-1)^{|a_2||b_1|}a_1b_1\otimes a_2b_2=\underline{\Delta}(a)\cdot\underline{\Delta}(b)$$

for any a,b in $U(L)$, with $\Delta(a)=\sum a_1\otimes a_2$, $\Delta(b)=\sum b_1\otimes b_2$, and a_2, b_1 homogeneous. $\underline{\Delta}$ is uniquely determined by its value on the generators of $U(L)$ (i.e., the basis elements of L):

$$\underline{\Delta}(x)=1\otimes x+x\otimes 1$$

Similarly, $U(L)$ is equipped with an antipode $\underline{S}:U(L)\rightarrow U(L)$ which is not an algebra anti-homomorphism (as in ordinary Hopf algebras) but a braided algebra anti-homomorphism (or "twisted" anti-homomorphism) in the following sense:

$$\underline{S}(ab)=(-1)^{|a||b|}\underline{S}(b)\underline{S}(a)$$

for any homogeneous $a,b\in U(L)$.

All the above description is equivalent to saying that $U(L)$ is a Hopf algebra in the braided category of CZ_2-modules ${}_{CZ_2}\mathcal{M}$ or: a braided group where the braiding is induced by the non-trivial quasitriangular structure of the CZ_2 Hopf algebra, i.e., by the non-trivial R-matrix:

$$R_g=\frac{1}{2}(1\otimes 1+1\otimes g+g\otimes 1-g\otimes g)\qquad(18.4)$$

where $1,g$ are the elements of the Z_2 group which is now written multiplicatively. We recall here (see [14]) that if (H,R_H) is a quasitriangular Hopf algebra, then the category of modules ${}_H\mathcal{M}$ is a braided monoidal category, where the braiding is given by a natural family of isomorphisms $\Psi_{V,W}:V\otimes W\cong W\otimes V$, given explicitly by:

$$\Psi_{V,W}(v\otimes w)=\sum(R_H^{(2)}\triangleright w)\otimes(R_H^{(1)}\triangleright v)\qquad(18.5)$$

for any $V,W\in obj({}_H\mathcal{M})$.

Combining (18.4) and (18.5) we immediately get the braiding in the ${}_{CZ_2}\mathcal{M}$ category:

$$\Psi_{V,W}(v \otimes w) = (-1)^{|v||w|} w \otimes v \tag{18.6}$$

This is obviously a symmetric braiding, so we actually have a symmetric monoidal category ${}_{CZ_2}\mathcal{M}$, rather than a truly braided one.

In view of the above description, an immediate consequence of Proposition 18.2, is that the parabosonic algebras $P_B^{(n)}$ are super-Hopf algebras, with the generators $b_i^{\pm}$, $i = 1, ..., n$ being primitive elements. Its super-Hopf algebraic structure is completely determined by the super-Hopf algebraic structure of Lie superalgebras, due to the above mentioned isomorphism. Namely the following relations determine completely the super-Hopf algebraic structure of $P_B^{(n)}$:

$$\underline{\Delta}(b_i^{\pm}) = 1 \otimes b_i^{\pm} + b_i^{\pm} \otimes 1 \quad \underline{\varepsilon}(b_i^{\pm}) = 0 \quad \underline{S}(b_i^{\pm}) = -b_i^{\pm} \tag{18.7}$$

18.3 Bosonisation as a Technique of Reducing Supersymmetry

A general scheme for "transforming" a Hopf algebra B in the braided category ${}_H\mathcal{M}$ (H: some quasitriangular Hopf algebra) into an ordinary one, namely the smash product Hopf algebra: $B \star H$, such that the two algebras have equivalent module categories, has been developed during 1990s. The original reference is [10] (see also [11, 12]). The technique is called bosonisation, the term coming from physics. This technique uses ideas developed in [13, 20]. It is also presented and applied in [1, 3, 4]. We review the main points of the above method:

In general, B being a Hopf algebra in a category, means that its structure maps are morphisms in the category. In particular, if H is some quasitriangular Hopf algebra, B being a Hopf algebra in the braided category ${}_H\mathcal{M}$, means that B is an algebra in ${}_H\mathcal{M}$ (or H-module algebra) and a coalgebra in ${}_H\mathcal{M}$ (or H-module coalgebra) and at the same time Δ_B and ε_B are algebra morphisms in the category ${}_H\mathcal{M}$. (For more details on the above definitions one may consult for example [14]).

Since B is an H-module algebra we can form the cross product algebra $B \rtimes H$ (also called: smash product algebra) which as a k-vector space is $B \otimes H$ (i.e., we write: $b \rtimes h \equiv b \otimes h$ for every $b \in B$, $h \in H$), with multiplication given by:

$$(b \otimes h)(c \otimes g) = \sum b(h_1 \triangleright c) \otimes h_2 g \tag{18.8}$$

$\forall\, b, c \in B$ and $h, g \in H$, and the usual tensor product unit.

On the other hand B is a (left) H-module coalgebra with H: quasitriangular through the R-matrix: $R_H = \sum R_H^{(1)} \otimes R_H^{(2)}$. Quasitriangularity "switches" the (left) action of H on B into a (left) coaction $\rho : B \to H \otimes B$ through:

$$\rho(b) = \sum R_H^{(2)} \otimes (R_H^{(1)} \triangleright b) \tag{18.9}$$

and B endowed with this coaction becomes (see [11, 12]) a (left) H-comodule coalgebra or equivalently a coalgebra in ${}^{H}\mathscr{M}$ (meaning that Δ_B and ε_B are (left) H-comodule morphisms, see [14]).

We recall here (see [11, 12]) that when H is a Hopf algebra and B is a (left) H-comodule coalgebra with the (left) H-coaction given by: $\rho(b) = \sum b^{(1)} \otimes b^{(0)}$, one may form the cross coproduct coalgebra $B \rtimes H$, which as a k-vector space is $B \otimes H$ (i.e., we write: $b \rtimes h \equiv b \otimes h$ for every $b \in B$, $h \in H$), with comultiplication given by:

$$\Delta(b \otimes h) = \sum b_1 \otimes b_2^{(1)} h_1 \otimes b_2^{(0)} \otimes h_2 \tag{18.10}$$

and counit: $\varepsilon(b \otimes h) = \varepsilon_B(b)\varepsilon_H(h)$. (In the above we use in the elements of B upper indices included in parenthesis to denote the components of the coaction according to the Sweedler notation, with the convention that $b^{(i)} \in H$ for $i \neq 0$).

Now we proceed by applying the above described construction of the cross coproduct coalgebra $B \rtimes H$, with the special form of the (left) coaction given by (18.9). Replacing thus (18.9) into (18.10) we get for the special case of the quasitriangular Hopf algebra H the cross coproduct comultiplication:

$$\Delta(b \otimes h) = \sum b_1 \otimes R_H^{(2)} h_1 \otimes (R_H^{(1)} \triangleright b_2) \otimes h_2 \tag{18.11}$$

Finally we can show that the cross product algebra (with multiplication given by (18.8)) and the cross coproduct coalgebra (with comultiplication given by (18.11)) fit together and form a bialgebra (see [11–13,20]). This bialgebra, furnished with an antipode:

$$S(b \otimes h) = (S_H(h_2))u(R^{(1)} \triangleright S_B(b)) \otimes S(R^{(2)} h_1) \tag{18.12}$$

where $u = \sum S_H(R^{(2)})R^{(1)}$, and S_B the (braided) antipode of B, becomes (see [11]) an ordinary Hopf algebra. This is the smash product Hopf algebra denoted $B \star H$. In [10] it is further proved that the category of the braided modules of B (B-modules in ${}_H\mathscr{M}$) is equivalent to the category of the (ordinary) modules of $B \star H$.

18.3.1 An Example of Bosonisation

In the special case that B is some super-Hopf algebra, then: $H = \mathbb{C}\mathbb{Z}_2$, equipped with its non-trivial quasitriangular structure, formerly mentioned. In this case, the technique simplifies and the ordinary Hopf algebra produced is the smash product Hopf algebra $B \star \mathbb{C}\mathbb{Z}_2$. The grading in B is induced by the $\mathbb{C}\mathbb{Z}_2$-action on B:

$$g \triangleright b = (-1)^{|b|} b \tag{18.13}$$

for b homogeneous in B. Utilizing the non-trivial R-matrix R_g and using (18.4) and (18.9) we can readily deduce the form of the induced $\mathbb{C}\mathbb{Z}_2$-coaction on B:

$$\rho(b) = \begin{cases} 1 \otimes b \ , & b : \text{even} \\ g \otimes b \ , & b : \text{odd} \end{cases} \tag{18.14}$$

The above mentioned action and coaction enable us to form the cross product algebra and the cross coproduct coalgebra according to the preceding discussion which finally form the smash product Hopf algebra $B \star \mathbb{C}\mathbb{Z}_2$. The grading of B, is "absorbed" in $B \star \mathbb{C}\mathbb{Z}_2$, and becomes an inner automorphism:

$$gbg = (-1)^{|b|} b$$

where we have identified: $b \star 1 \equiv b$ and $1 \star g \equiv g$ in $B \star \mathbb{C}\mathbb{Z}_2$ and b homogeneous element in B. This inner automorphism is exactly the adjoint action of g on $B \star \mathbb{C}\mathbb{Z}_2$ (as an ordinary Hopf algebra). The following proposition is proved – as an example of the bosonisation technique – in [11]:

Proposition 18.3. *Corresponding to every super-Hopf algebra B there is an ordinary Hopf algebra $B \star \mathbb{C}\mathbb{Z}_2$, its bosonisation, consisting of B extended by adjoining an element g with relations, coproduct, counit and antipode:*

$$\begin{array}{llll} g^2 = 1 & gb = (-1)^{|b|} bg & \Delta(g) = g \otimes g & \Delta(b) = \sum b_1 g^{|b_2|} \otimes b_2 \\ S(g) = g & S(b) = g^{-|b|} \underline{S}(b) & \varepsilon(g) = 1 & \varepsilon(b) = \underline{\varepsilon}(b) \end{array} \tag{18.15}$$

where $\underline{S}$ and $\underline{\varepsilon}$ denote the original maps of the super-Hopf algebra B.
Moreover, the representations of the bosonised Hopf algebra $B \star \mathbb{C}\mathbb{Z}_2$ are precisely the super-representations of the original superalgebra B.

The application of the above proposition in the case of the parabosonic algebra $P_B^{(n)} \cong U(B(0,n))$ is straightforward: we immediately get its bosonised form $P_{B(g)}^{(n)}$ which by definition is:

$$P_{B(g)}^{(n)} \equiv P_B^{(n)} \star \mathbb{C}\mathbb{Z}_2 \cong U(B(0,n)) \star \mathbb{C}\mathbb{Z}_2$$

Utilizing (18.7) which describe the super-Hopf algebraic structure of the parabosonic algebra $P_B^{(n)}$, and replacing them into (18.15) which describe the ordinary Hopf algebra structure of the bosonised superalgebra, we immediately get the explicit form of the (ordinary) Hopf algebra structure of $P_{B(g)}^{(n)} \equiv P_B^{(n)} \star \mathbb{C}\mathbb{Z}_2$ which reads:

$$\begin{array}{ll} \Delta(b_i^{\pm}) = b_i^{\pm} \otimes 1 + g \otimes b_i^{\pm} & \Delta(g) = g \otimes g \\ \varepsilon(b_i^{\pm}) = 0 & \varepsilon(g) = 1 \\ S(b_i^{\pm}) = b_i^{\pm} g = -g b_i^{\pm} & S(g) = g \\ g^2 = 1 & \{g, b_i^{\pm}\} = 0 \end{array} \tag{18.16}$$

where we have again identified $b_i^{\pm} \star 1 \equiv b_i^{\pm}$ and $1 \star g \equiv g$ in $P_B^{(n)} \star \mathbb{C}\mathbb{Z}_2$.

18.3.2 An Alternative Approach

Let us describe now a slightly different construction (see [2]), which achieves the same object: the determination of an ordinary Hopf structure for the parabosonic algebra $P_B^{(n)}$.
Defining:

$$N_{lm} = \frac{1}{2}\{b_l^+, b_m^-\}$$

we notice that these are the generators of the Lie algebra $u(n)$:

$$[N_{kl}, N_{mn}] = \delta_{lm}N_{kn} - \delta_{kn}N_{ml}$$

We introduce now the elements:

$$\mathcal{N} = \sum_{i=1}^{n} N_{ii} = \frac{1}{2}\sum_{i=1}^{n}\{b_i^+, b_i^-\}$$

which are exactly the linear Casimirs of $u(n)$.
We can easily find that they satisfy:

$$[\mathcal{N}, b_i^{\pm}] = \pm b_i^{\pm}$$

Based on the above we inductively prove:

$$[\mathcal{N}^m, b_i^+] = b_i^+((\mathcal{N}+1)^m - \mathcal{N}^m) \tag{18.17}$$

We now introduce the following elements:

$$K^+ = \exp(i\pi\mathcal{N}) \equiv \sum_{m=0}^{\infty}\frac{(i\pi\mathcal{N})^m}{m!}$$

and:

$$K^- = \exp(-i\pi\mathcal{N}) \equiv \sum_{m=0}^{\infty}\frac{(-i\pi\mathcal{N})^m}{m!}$$

Utilizing the above power series expressions and (18.17) we get

$$\{K^+, b_i^{\pm}\} = 0 \; \{K^-, b_i^{\pm}\} = 0 \tag{18.18}$$

A direct application of the Baker–Campbell–Hausdorff formula leads also to:

$$K^+K^- = K^-K^+ = 1 \tag{18.19}$$

We finally have the following proposition:

Proposition 18.4. *Corresponding to the super-Hopf algebra $P_B^{(n)}$ there is an ordinary Hopf algebra $P_{B(K^{\pm})}^{(n)}$, consisting of $P_B^{(n)}$ extended by adjoining two elements*

K^+, K^- with relations, coproduct, counit and antipode:

$$\begin{aligned} \Delta(b_i^{\pm}) &= b_i^{\pm} \otimes 1 + K^{\pm} \otimes b_i^{\pm} & \Delta(K^{\pm}) &= K^{\pm} \otimes K^{\pm} \\ \varepsilon(b_i^{\pm}) &= 0 & \varepsilon(K^{\pm}) &= 1 \\ S(b_i^{\pm}) &= b_i^{\pm} K^{\mp} & S(K^{\pm}) &= K^{\mp} \\ K^+K^- = K^-K^+ &= 1 & \{K^+, b_i^{\pm}\} &= 0 = \{K^-, b_i^{\pm}\} \end{aligned} \tag{18.20}$$

Proof. Consider the k-vector space $k\langle b_i^+, b_j^-, K^{\pm}\rangle$ freely generated by the elements b_i^+, b_j^-, K^+, K^-. Denote $T(b_i^+, b_j^-, K^{\pm})$ its tensor algebra. In the tensor algebra we denote I_{BK} the ideal generated by al the elements of the forms (18.1), (18.18), (18.19). We define:

$$P_{B(K^{\pm})}^{(n)} = T(b_i^+, b_j^-, K^{\pm})/I_{BK}$$

Consider the k-linear map

$$\Delta : k\langle b_i^+, b_j^-, K^{\pm}\rangle \to P_{B(K^{\pm})}^{(n)} \otimes P_{B(K^{\pm})}^{(n)}$$

determined by its values on the basis elements, specified in (18.20). By the universality property of the tensor algebra this map extends to an algebra homomorphism:

$$\Delta : T(b_i^+, b_j^-, K^{\pm}) \to P_{B(K^{\pm})}^{(n)} \otimes P_{B(K^{\pm})}^{(n)}$$

Now we can trivially verify that

$$\Delta(\{K^{\pm}, b_i^{\pm}\}) = \Delta(K^+K^- - 1) = \Delta(K^-K^+ - 1) = 0 \tag{18.21}$$

Considering the usual tensor product algebra $P_{B(K^{\pm})}^{(n)} \otimes P_{B(K^{\pm})}^{(n)}$ with multiplication $(a \otimes b)(c \otimes d) = ac \otimes bd$ for any $a, b, c, d \in P_{B(K^{\pm})}^{(n)}$ we also compute:

$$\Delta([\{b_i^{\xi}, b_j^{\eta}\}, b_k^{\varepsilon}] - (\varepsilon - \eta)\delta_{jk} b_i^{\xi} - (\varepsilon - \xi)\delta_{ik} b_j^{\eta}) = 0 \tag{18.22}$$

Relations (18.21), and (18.22), mean that $I_{BK} \subseteq ker\Delta$ which in turn implies that Δ is uniquely extended as an algebra homomorphism from $P_{B(K^{\pm})}^{(n)}$ to the usual tensor product algebra $P_{B(K^{\pm})}^{(n)} \otimes P_{B(K^{\pm})}^{(n)}$ according to the diagram:

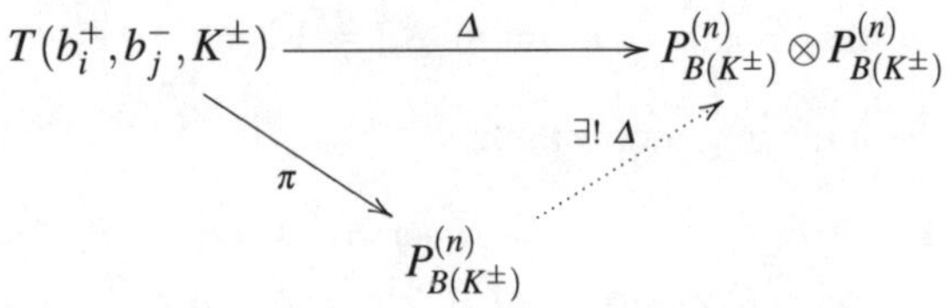

Following the same procedure we construct an algebra homomorphism $\varepsilon : P^{(n)}_{B(K^{\pm})} \to \mathbb{C}$ and an algebra antihomomorphism $S : P^{(n)}_{B(K^{\pm})} \to P^{(n)}_{B(K^{\pm})}$ which are completely determined by their values on the generators of $P^{(n)}_{B(K^{\pm})}$ (i.e., the basis elements of $k\langle b_i^+, b_j^-, K^{\pm}\rangle$). Note that in the case of the antipode we start by defining a linear map S from $k\langle b_i^+, b_j^-, K^{\pm}\rangle$ to the opposite algebra $(P^{(n)}_{B(K^{\pm})})^{op}$, with values determined by (18.20) and following the above described procedure we end up with an algebra anti-homomorphism: $S : P^{(n)}_{B(K^{\pm})} \to P^{(n)}_{B(K^{\pm})}$.
Now it is sufficient to verify the rest of the Hopf algebra axioms (i.e., coassociativity of Δ, counity property for ε, and the compatibility condition which ensures us that S is an antipode) on the generators of $P^{(n)}_{B(K^{\pm})}$. This can be done with straightforward computations (see [2]). □

The above constructed algebra $P^{(n)}_{B(K^{\pm})}$, is an ordinary Hopf algebra in the sense that the comultiplication is extended to the whole of $P^{(n)}_{B(K^{\pm})}$ as an algebra homomorphism

$$\Delta : P^{(n)}_{B(K^{\pm})} \to P^{(n)}_{B(K^{\pm})} \otimes P^{(n)}_{B(K^{\pm})}$$

where $P^{(n)}_{B(K^{\pm})} \otimes P^{(n)}_{B(K^{\pm})}$ is considered as the tensor product algebra with the usual product:

$$(a \otimes b)(c \otimes d) = ac \otimes bd$$

for any $a, b, c, d \in P^{(n)}_{B(K^{\pm})}$ and the antipode extends as usual as an algebra anti-homomorphism.

18.4 Discussion

It is interesting to see the relation between the above constructed Hopf algebras $P^{(n)}_{B(g)}$ and $P^{(n)}_{B(K^{\pm})}$.

From the point of view of the structure, an obvious question arises: While $P^{(n)}_{B(g)}$ is a quasitriangular Hopf algebra through the R-matrix: R_g given in (18.4), there is yet no suitable R-matrix for the Hopf algebra $P^{(n)}_{B(K^{\pm})}$. Thus the question of the quasitriangular structure of $P^{(n)}_{B(K^{\pm})}$ is open.

An other interesting point, concerns the representations of $P^{(n)}_{B(K^{\pm})}$ versus the representations of $P^{(n)}_{B(g)}$. The difference in the comultiplication between the above mentioned Hopf algebras, leads us to the question of whether the tensor product of representations of $P^{(n)}_{B(g)}$ behave differently from the tensor product of representations of $P^{(n)}_{B(K^{\pm})}$.

Finally another open problem which arises from the above mentioned approach, is whether the above construction of $P_{B(K^{\pm})}^{(n)}$ can be extended for the universal enveloping algebra of an arbitrary Lie superalgebra, using power series of suitably chosen Casimirs.

Acknowledgements This paper is part of a project supported by "Pythagoras II", contract number 80897.

References

1. N. Andruskiewitsch, P. Etingof, S. Gelaki, "Triangular Hopf algebras with the Chevalley property", Michigan Math. J., **49**, (2001), p. 277
2. C. Daskaloyannis, K. Kanakoglou, I. Tsohantjis, "Hopf algerbaic structure of the parabosonic and parafermionic algebras and paraparticle generalization of the Jordan–Schwinger map", J. Math. Phys., **41**, 2, (2000), p. 652
3. D. Fischman, "Schur's double centralizer theorem: A Hopf algebra approach", J. Algebra, **157**, (1993), p. 331
4. D. Fischman, S. Montgomery, "A Schur double centralizer theorem for cotriangular Hopf algebras and generalized Lie algebras", J. Algebra, **168**, (1994), p. 594
5. O. W. Greenberg, A.M.L. Messiah, "Selection rules for parafields and the absence of paraparticles in nature", Phys. Rev., **138**, 5B, (1965), p. 1155
6. H. S. Green, "A generalized method of field quantization", Phys. Rev., **90**, 2, (1953), p. 270
7. V. G. Kac, "A sketch of Lie superalgebra theory", Comm. Math. Phys., **53**, (1977), p. 31
8. V. G. Kac, "Lie Superalgebras", Adv. Math., **26**, (1977), p. 8
9. S. Kamefuchi, Y. Takahashi, "A generalisation of field quantization and statistics", Nucl. Phys., **36**, (1962), p. 177
10. S. Majid, "Cross products by braided groups and bosonisation", J. Algebra, **163**, (1994), p. 165
11. S. Majid, "Foundations of Quantum Group Theory", Cambridge University Press, Cambridge, 1995
12. S. Majid, "A quantum groups primer", London Mathematical Society, Lecture Note Series, 292, Cambridge University Press, Cambridge, 2002
13. R. K. Molnar, "Semi-direct products of Hopf Algebras", J. Algebra, **47**, (1977), p. 29
14. S. Montgomery, "Hopf algebras and their actions on rings", CBMS, Regional Conference Series in Mathematics, 82, AMS-NSF, 1993
15. Y. Ohnuki, S. Kamefuchi, "Quantum field theory and parastatistics", University of Tokyo press, Tokyo, 1982
16. T. D. Palev, "Quantization of $U_q(so(2n+1))$ with Deformed Parafermi Operators", Lett. Math. Phys., **31**, (1994), p. 151
17. T. D. Palev, "Quantization of $U_q(osp(1,2n))$ with Deformed Parabose Operators", J. Phys. A: Math. Gen., **26**, (1993), p. L1111
18. T. D. Palev, "A Lie superalgebraic interpretation of the parabose statistics", J. Math. Phys., **21**, 4, (1980), p. 797
19. T. D. Palev, "The quantum superalgebra $U_q(osp(1/2n))$: deformed parabose operators and root of unity representations", J. Phys. A:Math. Gen., **28**, (1995), p. 2605
20. D. E. Radford, "The structure of Hopf algebras with a projection", J. Algebra, **92**, (1985), p. 322
21. C. Ryan, E. C. G. Sudarshan, "Representations of parafermi rings", Nucl. Phys. B, **47**, (1963), p. 207

Chapter 19
Deformations of the Witt, Virasoro, and Current Algebra

Martin Schlichenmaier

Abstract For finite-dimensional Lie algebras there is a tight connection between cohomology theory and deformations of the Lie algebra. This is not the case anymore if the Lie algebra to be deformed is infinite dimensional. Such Lie algebras might be formally rigid but nevertheless allow deformations which are even locally non-trivial. In joint work with Alice Fialowski the author constructed such geometric families for the formally rigid Witt algebra and current Lie algebras. These families are genus one (i.e. elliptic) Lie algebras of Krichever–Novikov type. In this contribution the results are reviewed. The families of algebras are given in explicit form. The constructions are induced by the geometric process of degenerating the elliptic curves to singular cubics.

19.1 Introduction

In this contribution we consider certain infinite dimensional algebras which are of fundamental importance in quite a number of applications. Examples are given by Conformal Field Theory, infinite dimensional integrable systems, and symmetries of partial differential equations.

These algebras can be described geometrically as follows. The Witt algebra is the algebra consisting of those meromorphic vector fields on the Riemann sphere which are holomorphic outside $\{0,\infty\}$. A basis and the associated structure is given by

$$l_n := z^{n+1}\frac{d}{dz},\ n\in\mathbb{Z},\quad \text{with Lie bracket}\quad [l_n,l_m]=(m-n)\,l_{n+m}. \tag{19.1}$$

M. Schlichenmaier
Institute of Mathematics, University of Luxembourg, Campus Limpertsberg,
162 A, Avenue de la Faiencerie, 1511 Luxembourg, Grand-Duchy of Luxembourg
e-mail: Martin.Schlichenmaier@uni.lu

S. Silvestrov et al. (eds.), *Generalized Lie Theory in Mathematics, Physics and Beyond*,

The Virasoro algebra is its universal central extension

$$[l_n, l_m] = (m-n)l_{n+m} + \frac{1}{12}(m^3 - m)\delta_{n,-m}\, t, \qquad [l_n, t] = 0, \tag{19.2}$$

with t an additional basis element which is central.

Furthermore we consider the case of current algebras $\overline{\mathfrak{g}} = \mathfrak{g} \otimes \mathbb{C}[z^{-1}, z]$ and their central extensions $\widehat{\mathfrak{g}}$, the *affine Lie algebras*. Here $\mathfrak{g}$ is a finite-dimensional Lie algebra (which for simplicity we assume to be simple). With the Cartan–Killing form β the central extension $\widehat{\mathfrak{g}}$ is the vector space $\overline{\mathfrak{g}} \oplus t\,\mathbb{C}$ endowed with the Lie bracket

$$[x \otimes z^n, y \otimes z^m] = [x, y] \otimes z^{n+m} - \beta(x, y) \cdot n \cdot \delta_m^{-n} \cdot t, \quad [t, \widehat{\mathfrak{g}}] = 0, \quad x, y \in \mathfrak{g},\ n, m \in \mathbb{Z}. \tag{19.3}$$

It is known that these algebras are rigid in the formal sense. This says that every deformation with a formal parameter will be equivalent to the trivial deformation. In case the Lie algebra is finite-dimensional this yields that every deformation (formal or not) will be locally equivalent to the trivial deformation. Indeed in this case, the deformation problem can be formulated in terms of Lie algebra cohomology. If the relevant cohomology space vanishes the algebra will be rigid in all senses. But the step from finite dimension to infinite dimension changes the picture. Together with Alice Fialowski we constructed in two articles [5, 6] nontrivial geometric deformation families for the Witt algebra (resp. its universal central extension the Virasoro algebra) and for the current algebras (resp. their central extensions the affine algebras), despite the fact that these algebras are rigid [3, 9]. This is a phenomena which in finite dimension cannot occur.

The families we constructed, appear as families of higher-genus multi-point algebras of Krichever–Novikov type, see Sect. 19.3 for their definitions. They are not just abstract families, but families obtained by geometric processes. The results obtained do not have only relevance in deformation theory of algebras, but they also are of importance in areas where vector fields, current, and affine algebras play a role.

A very prominent application of these algebras is two-dimensional conformal field theory (CFT) and its quantization. There the Witt algebra, the Virasoro algebra, the current algebras, the affine algebras, and their representations correspond to CFT on the Riemann sphere (i.e. to genus zero), see [1]. Krichever and Novikov [8] proposed in the case of higher genus Riemann surfaces (with two insertion points) the use of global operator fields which are given with the help of the Lie algebra of vector fields of Krichever–Novikov type, certain related algebras, and their representations (see Sect. 19.3 below).

Their approach was extended by Schlichenmaier to the multi-point situation (i.e. an arbitrary number of insertion points was allowed) [12–15]. The necessary central extensions where constructed. Higher genus multi-point current and affine algebras were introduced [17]. These algebras consist of meromorphic objects on a Riemann surface which are holomorphic outside a finite set A of points. The set A is divided into two disjoint subsets I and O. With respect to some possible interpretation of the Riemann surface as the world-sheet of a string, the points in I are called *in-points*, the points in O are called *out-points*, corresponding to incoming and outgoing free strings. The world-sheet itself corresponds to possible interaction. This splitting introduces an almost-graded structure (see Sect. 19.3) for the algebras and their representations. Such an almost-graded structure is needed to construct representations of relevance in the context of the quantization of CFT, e.g. highest weight representations, fermionic Fock space representations, and Wess–Zumino–Novikov–Witten theory [21,22].

By a maximal degeneration a collection of $\mathbb{P}^1(\mathbb{C})$'s will appear. Indeed, the examples considered in this article are exactly of this type. The deformations appear as families of vector fields and current algebras which are naturally defined over the moduli space of genus one curves (i.e. of elliptic curves, or equivalently of complex one-dimensional tori) with two marked points. These deformations are associated to geometric degenerations of elliptic curves to singular cubic curves. The desingularization (or normalization) of their singularities will yield the projective line as normalization. We will end up with algebras related to the genus zero case. The full geometric picture behind the degeneration was discussed in [16]. In particular, we like to point out, that even if one starts with two marked points, by passing to the boundary of the moduli space one is forced to consider more points (now for a curve of lower genus).

19.2 Deformations of Lie Algebras

A Lie algebra $\mathscr{L}$ can be given as a (complex) vector space (also denoted by $\mathscr{L}$) and a bracket $[.,.]$, which is an anti-symmetric bilinear form

$$\mu_0 : \mathscr{L} \times \mathscr{L} \to \mathscr{L}, \qquad \mu_0(x,y) = [x,y],$$

fulfilling the Jacobi identity

$$\mu(\mu(x,y),z) + \mu(\mu(y,z),x) + \mu(\mu(z,x),y) = 0, \qquad \forall x,y,z \in \mathscr{L}. \tag{19.4}$$

Consider on the same vector space $\mathscr{L}$ is modeled on, a family of Lie algebra structures

$$\mu_t = \mu_0 + t \cdot \phi_1 + t^2 \cdot \phi_2 + \cdots , \tag{19.5}$$

with bilinear maps $\phi_i : \mathscr{L} \times \mathscr{L} \to \mathscr{L}$ such that $\mathscr{L}_t := (\mathscr{L}, \mu_t)$ is a Lie algebra and $\mathscr{L}_0$ is the Lie algebra we started with. The family $\{\mathscr{L}_t\}$ is a *deformation* of $\mathscr{L}_0$ [10,11].

Up to this point we did not specify the "parameter" t. Indeed, different choices are possible:

1. The parameter t might be a variable which allows to plug in numbers $\alpha \in \mathbb{C}$. In this case $\mathscr{L}_\alpha$ is a Lie algebra for every α for which the expression (19.5) is defined. The family can be considered as deformation over the affine line $\mathbb{C}[t]$ or over the convergent power series $\mathbb{C}\{\{t\}\}$. The deformation is called a *geometric* or an *analytic deformation* respectively.

2. We consider t as a formal variable and we allow infinitely many terms in (19.5). It might be the case that μ_t does not exist if we plug in for t any other value different from 0. In this way we obtain deformations over the ring of formal power series $\mathbb{C}[[t]]$. The corresponding deformation is a *formal deformation*.
3. The parameter t is considered as an infinitesimal variable, i.e. we take $t^2 = 0$. We obtain *infinitesimal deformations* defined over the quotient $\mathbb{C}[X]/(X^2) = \mathbb{C}[[X]]/(X^2)$.

More general situations for the parameter space can be considered, see [5].

There is always the trivially deformed family given by $\mu_t = \mu_0$ for all values of t. Two families μ_t and μ_t' deforming the same μ_0 are *equivalent* if there exists a linear automorphism (with the same vagueness about the meaning of t)

$$\psi_t = id + t \cdot \alpha_1 + t^2 \cdot \alpha_2 + \cdots \tag{19.6}$$

with $\alpha_i : \mathscr{L} \to \mathscr{L}$ linear maps such that

$$\mu_t'(x,y) = \psi_t^{-1}(\mu_t(\psi_t(x), \psi_t(y))). \tag{19.7}$$

A Lie algebra $(\mathscr{L}, \mu_0)$ is called *rigid* if every deformation μ_t of μ_0 is locally equivalent to the trivial family. Intuitively, this says that $\mathscr{L}$ cannot be deformed.

The word "locally" in the definition of rigidity means that we only consider the situation for t "near 0". Of course, this depends on the category we consider. As on the formal and the infinitesimal level there exists only one closed point, i.e. the point 0 itself, every deformation over $\mathbb{C}[[t]]$ or $\mathbb{C}[X]/(X^2)$ is already local. This is different on the geometric and analytic level. Here it means that there exists an (etale) open neighbourhood U of 0 such that the family restricted to it is equivalent to the trivial one. In particular, this implies $\mathscr{L}_\alpha \cong \mathscr{L}_0$ for all $\alpha \in U$.

Clearly, a question of fundamental interest is to decide whether a given Lie algebra is rigid. Moreover, the question of rigidity will depend on the category we consider. Depending on the set-up we will have to consider infinitesimal, formal, geometric, and analytic rigidity. If the algebra is not rigid, one would like to know whether there exists a moduli space of (inequivalent) deformations. If so, what is its structure, dimension, etc.?

As explained in the introduction, deformation problems and moduli space problems are related to adapted cohomology theories. To a certain extend (in particular for the finite-dimensional case) this is also true for deformations of Lie algebras. But as far as geometric and analytic deformations are concerned it is wrong for infinite dimensional Lie algebras as our examples show.

For Lie algebra deformations the relevant cohomology space is $\mathbf{H}^2(\mathscr{L}, \mathscr{L})$, the space of Lie algebra two-cohomology classes with values in the adjoint module $\mathscr{L}$.

Recall that these cohomology classes are classes of two-cocycles modulo coboundaries. An antisymmetric bilinear map $\phi : \mathscr{L} \times \mathscr{L} \to \mathscr{L}$ is a Lie algebra *two-cocycle* if $d_2\phi = 0$, or expressed explicitly

$$\phi([x,y],z)+\phi([y,z],x)+\phi([z,x],y)-[x,\phi(y,z)]+[y,\phi(z,x)]-[z,\phi(x,y)]=0. \tag{19.8}$$

The map ϕ will be a *coboundary* if there exists a linear map $\psi : \mathcal{L} \to \mathcal{L}$ with

$$\phi(x,y)=(d_1\psi)(x,y):=\psi([x,y])-[x,\psi(y)]+[y,\psi(x)]. \tag{19.9}$$

If we write down the Jacobi identity for μ_t given by (19.5) then it can be immediately verified that the first non-vanishing ϕ_i has to be a two-cocycle in the above sense. Furthermore, if μ_t and μ_t' are equivalent then the corresponding ϕ_i and ϕ_i' are cohomologous, i.e. their difference is a coboundary.

The following results are well-known:

1. $\mathbf{H}^2(\mathcal{L},\mathcal{L})$ classifies infinitesimal deformations of $\mathcal{L}$ [7].
2. If $\dim \mathbf{H}^2(\mathcal{L},\mathcal{L})<\infty$ then all formal deformations of $\mathcal{L}$ up to equivalence can be realised in this vector space [4].
3. If $\mathbf{H}^2(\mathcal{L},\mathcal{L})=0$ then $\mathcal{L}$ is infinitesimally and formally rigid (this follows directly from (1) and (2)).
4. If $\dim\mathcal{L}<\infty$ then $\mathbf{H}^2(\mathcal{L},\mathcal{L})=0$ implies that $\mathcal{L}$ is also rigid in the geometric and analytic sense [7, 10, 11].

As our examples show, without the condition $\dim\mathcal{L}<\infty$ point (4) is not true anymore.

For the Witt algebra $\mathcal{W}$ one has $\mathbf{H}^2(\mathcal{W},\mathcal{W})=0$ ([3], see also [5]). Hence it is formally rigid. For the classical current algebras $\overline{\mathfrak{g}}=\mathfrak{g}\otimes\mathbb{C}[z^{-1},z]$ with $\mathfrak{g}$ a finite-dimensional simple Lie algebra, Lecomte and Roger [9] showed that $\overline{\mathfrak{g}}$ is formally rigid. Nevertheless, for both types of algebras, including their central extensions, we obtained deformations which are both locally geometrically and analytically non-trivial [5, 6]. Hence they are not rigid in the geometric and analytic sense. These families will be described in the following.

19.3 Krichever–Novikov Algebras

Our geometric families will be families of algebras of Krichever–Novikov type. These algebras play an important role in a global operator approach to higher genus Conformal Field Theory.

They are generalisations of the Virasoro algebra, the current algebras and all their related algebras. Let M be a compact Riemann surface of genus g, or in terms of algebraic geometry, a smooth projective curve over $\mathbb{C}$. Let $N,K\in\mathbb{N}$ with $N\geq 2$ and $1\leq K<N$. Fix

$$I=(P_1,\ldots,P_K), \quad \text{and} \quad O=(Q_1,\ldots,Q_{N-K})$$

disjoint ordered tuples of distinct points ("marked points", "punctures") on the curve. In particular, we assume $P_i\neq Q_j$ for every pair (i,j). The points in I are called the *in-points*, the points in O the *out-points*. Sometimes we consider I and O simply as sets and set $A = I \cup O$ as a set.

Here we will need the following algebras. Let $\mathscr{A}$ be the associative algebra consisting of those meromorphic functions on M which are holomorphic outside the set of points A with point-wise multiplication. Let $\mathscr{L}$ be the Lie algebra consisting of those meromorphic vector fields which are holomorphic outside of A with the usual Lie bracket of vector fields. The algebra $\mathscr{L}$ is called the *vector field algebra of Krichever–Novikov type*. In the two point case they were introduced by Krichever and Novikov [8]. The corresponding generalisation to the multi-point case was done in [12–15]. Obviously, both $\mathscr{A}$ and $\mathscr{L}$ are infinite dimensional algebras.

Furthermore, we will need the *higher-genus multi-point current algebra of Krichever–Novikov type*. We start with a complex finite-dimensional Lie algebra $\mathfrak{g}$ and endow the tensor product $\overline{\mathscr{G}} = \mathfrak{g} \otimes_{\mathbb{C}} \mathscr{A}$ with the Lie bracket

$$[x \otimes f, y \otimes g] = [x,y] \otimes f \cdot g, \qquad x,y \in \mathfrak{g}, \quad f,g \in \mathscr{A}. \tag{19.10}$$

The algebra $\overline{\mathscr{G}}$ is the higher genus current algebra. It is an infinite dimensional Lie algebra and might be considered as the Lie algebra of $\mathfrak{g}$-valued meromorphic functions on the Riemann surface with only poles outside of A.

The classical genus zero and $N = 2$ point case is give by the geometric data

$$M = \mathbb{P}^1(\mathbb{C}) = S^2, \quad I = \{z = 0\}, \quad O = \{z = \infty\}. \tag{19.11}$$

In this case the algebras are the well-known algebras of Conformal Field Theory (CFT). For the function algebra we obtain $\mathscr{A} = \mathbb{C}[z^{-1}, z]$, the algebra of Laurent polynomials. The vector field algebra $\mathscr{L}$ is the Witt algebra $\mathscr{W}$ generated by

$$l_n = z^{n+1}\frac{d}{dz}, \; n \in \mathbb{Z}, \qquad \text{with Lie bracket} \quad [l_n, l_m] = (m-n)\, l_{n+m}. \tag{19.12}$$

The current algebra $\overline{\mathscr{G}}$ is the standard current algebra $\overline{\mathfrak{g}} = \mathfrak{g} \otimes \mathbb{C}[z^{-1}, z]$ with Lie bracket

$$[x \otimes z^n, y \otimes z^m] = [x,y] \otimes z^{n+m}, \qquad x,y \in \mathfrak{g}, \quad n,m \in \mathbb{Z}. \tag{19.13}$$

In the classical situation the algebras are obviously graded by taking as degree $\deg l_n := n$ and $\deg x \otimes z^n := n$. For higher genus there is usually no grading. But it was observed by Krichever and Novikov in the two-point case that a weaker concept, an almost-graded structure, will be enough to develop an interesting theory of representations (Verma modules, etc.). Let $\mathscr{A}$ be an (associative or Lie) algebra admitting a direct decomposition as vector space $\mathscr{A} = \bigoplus_{n \in \mathbb{Z}} \mathscr{A}_n$. The algebra $\mathscr{A}$ is called an *almost-graded* algebra if (1) $\dim \mathscr{A}_n < \infty$ and (2) there are constants R and S such that

$$\mathscr{A}_n \cdot \mathscr{A}_m \subseteq \bigoplus_{h=n+m+R}^{n+m+S} \mathscr{A}_h, \qquad \forall n,m \in \mathbb{Z}\,. \tag{19.14}$$

The elements of $\mathscr{A}_n$ are called *homogeneous elements of degree n*. By exhibiting a special basis, for the multi-point situation such an almost grading was introduced in [12–15]. Essentially, this is done by fixing the order of the basis elements at the points in I in a certain manner and in O in a complementary way to make them unique up to scaling. In the following we will give an explicit description of the basis elements for those genus zero and one situation we need. Hence, we will not recall their general definition but only refer to the above quoted articles.

Proposition 19.1 ([12,15]). *The algebras $\mathscr{L}$, $\mathscr{A}$, and $\overline{\mathscr{G}}$ are almost-graded. The almost-grading depends on the splitting $A = I \cup O$.*

In the construction of infinite dimensional representations of these algebras with certain desired properties (generated by a vacuum, irreducibility, unitarity, etc.) one is typically forced to "regularize" a "naive" action to make it well-defined. Examples of importance in CFT are the fermionic Fock space representations which are constructed by taking semi-infinite forms of a fixed weight.

From the mathematical point of view, with the help of a prescribed procedure one modifies the action to make it well-defined. On the other hand, one has to accept that the modified action in compensation will only be a projective Lie action. Such projective actions are honest Lie actions for suitable centrally extended algebras. In the classical case they are well-known. The unique non-trivial (up to equivalence and rescaling) central extension of the Witt algebra $\mathscr{W}$ is the Virasoro algebra $\mathscr{V}$:

$$[l_n, l_m] = (m-n)l_{n+m} + \tfrac{1}{12}(m^3 - m)\delta_{n,-m}\, t, \qquad [l_n, t] = 0. \tag{19.15}$$

Here t is an additional element of the central extension which commutes with all other elements. For the current algebra $\mathfrak{g} \otimes \mathbb{C}[z^{-1}, z]$ for $\mathfrak{g}$ a simple Lie algebra with Cartan–Killing form β, it is the corresponding affine Lie algebra $\widehat{\mathfrak{g}}$ (or, untwisted affine Kac–Moody algebra):

$$[x \otimes z^n, y \otimes z^m] = [x,y] \otimes z^{n+m} - \beta(x,y) \cdot n \cdot \delta_m^{-n} \cdot t, \quad [t, \widehat{\mathfrak{g}}] = 0, \quad x,y \in \mathfrak{g},\ n,m \in \mathbb{Z}. \tag{19.16}$$

The additional terms in front of the elements t are 2-cocycles of the Lie algebras with values in the trivial module $\mathbb{C}$. Indeed for a Lie algebra V central extensions are classified (up to equivalence) by the second Lie algebra cohomology $\mathbf{H}^2(V, \mathbb{C})$ of V with values in the trivial module $\mathbb{C}$. Similar to the above, the bilinear form $\psi : V \times V \to \mathbb{C}$ is called Lie algebra 2-cocycle iff ψ is antisymmetric and fulfills the cocycle condition

$$0 = d_2\psi(x,y,z) := \psi([x,y],z) + \psi([y,z],x) + \psi([z,x],y). \tag{19.17}$$

It will be a coboundary if there exists a linear form $\kappa : V \to \mathbb{C}$ such that

$$\psi(x,y) = (d_1\kappa)(x,y) := \kappa([x,y]). \tag{19.18}$$

To extend the classical cocycles to the Krichever–Novikov type algebras they first have to be given in geometric terms. Geometric versions of the 2-cocycles are given as follows (see [18] and [19] for details). For the vector field algebra $\mathscr{L}$ we take

$$\gamma_{S,R}(e,f) := \frac{1}{24\pi\mathrm{i}} \int_{C_S} \left(\frac{1}{2}(e'''f - ef''') - R\cdot(e'f - ef') \right) dz\,. \tag{19.19}$$

Here the integration path C_S is a loop separating the in-points from the out-points and R is a holomorphic projective connection (see [5, Definition 4.2]) to make the integrand well-defined. For the current algebra $\overline{\mathscr{G}}$ we take

$$\gamma_S(x\otimes f, y\otimes g) = \beta(x,y)\frac{1}{2\pi\mathrm{i}}\int_{C_S} f dg. \tag{19.20}$$

The reader should be warned. For the classical algebras, i.e. the Witt and the current algebras for the simple Lie algebras $\mathfrak{g}$, there exists up to rescaling and equivalence only one non-trivial central extension. This will be the Virasoro algebra for the Witt algebra and the affine Kac–Moody algebra for the current algebra respectively. This is not true anymore for higher genus or/and the multi-point situation. But it was shown in [18] and [19] that (again up to equivalence and rescaling) there exists only one non-trivial central extension which allows to extend the almost-grading by giving the element t a degree in such a way that it will also be almost-graded. This unique extension will be given by the geometric cocycles (19.19), (19.20).

19.4 The Geometric Families

19.4.1 Complex Torus

Let $\tau \in \mathbb{C}$ with $\Im\tau > 0$ and L be the lattice

$$L = \langle 1, \tau\rangle_{\mathbb{Z}} := \{m + n\cdot\tau \mid m,n \in \mathbb{Z}\} \subset \mathbb{C}. \tag{19.21}$$

The complex one-dimensional torus is the quotient $T = \mathbb{C}/L$. It carries a natural structure of a complex manifold coming from the structure of $\mathbb{C}$. It will be a compact Riemann surface of genus 1.

The field of meromorphic functions on T is generated by the doubly-periodic Weierstraß $\wp$ function and its derivative $\wp'$ fulfilling the differential equation

$$(\wp')^2 = 4(\wp - e_1)(\wp - e_2)(\wp - e_3) = 4\wp^3 - g_2\wp - g_3, \tag{19.22}$$

with

$$\Delta := {g_2}^3 - 27{g_3}^2 = 16(e_1 - e_2)^2(e_1 - e_3)^2(e_2 - e_3)^2 \neq 0. \tag{19.23}$$

Furthermore,

$$g_2 = -4(e_1e_2 + e_1e_3 + e_2e_3), \quad g_3 = 4(e_1e_2e_3). \tag{19.24}$$

The numbers e_i are pairwise distinct, can be given as

$$\wp(\frac{1}{2}) = e_1, \qquad \wp(\frac{\tau}{2}) = e_2, \qquad \wp(\frac{\tau+1}{2}) = e_3, \tag{19.25}$$

and fulfill

$$e_1 + e_2 + e_3 = 0. \tag{19.26}$$

The function $\wp$ is an even meromorphic function with poles of order two at the points of the lattice and holomorphic elsewhere. The function $\wp'$ is an odd meromorphic function with poles of order three at the points of the lattice and holomorphic elsewhere. $\wp'$ has zeros of order one at the points $1/2, \tau/2$ and $(1+\tau)/2$ and all their translates under the lattice.

We have to pass here to the algebraic-geometric picture. The map

$$T \to \mathbb{P}^2(\mathbb{C}), \quad z \bmod L \mapsto \begin{cases} (\wp(z) : \wp'(z) : 1), & z \notin L \\ (0:1:0), & z \in L \end{cases} \tag{19.27}$$

realizes T as a complex-algebraic smooth curve in the projective plane. As its genus is one it is an elliptic curve. The affine coordinates are $X = \wp(z,\tau)$ and $Y = \wp'(z,\tau)$. From (19.22) it follows that the affine part of the curve can be given by the smooth cubic curve defined by

$$Y^2 = 4(X-e_1)(X-e_2)(X-e_3) = 4X^3 - g_2X - g_3 =: f(X). \tag{19.28}$$

The point at infinity on the curve is the point $\infty = (0:1:0)$.

We consider the algebras of Krichever–Novikov type corresponding to the elliptic curve and possible poles at $\bar{z} = \bar{0}$ and $\bar{z} = \overline{1/2}$ (respectively in the algebraic-geometric picture, at the points ∞ and $(e_1, 0)$).[1]

19.4.2 Vector Field Algebra

First we consider the vector field algebra $\mathscr{L}$. A basis of the vector field algebra is given by

$$V_{2k+1} := (X-e_1)^k Y \frac{d}{dX}, \quad V_{2k} := \frac{1}{2} f(X)(X-e_1)^{k-2} \frac{d}{dX}, \qquad k \in \mathbb{Z}. \tag{19.29}$$

If we vary the points e_1 and e_2 (and accordingly $e_3 = -(e_1+e_2)$) we obtain families of curves and associated families of vector field algebras. At least this is the case as long as the curves are non-singular. To describe the families in detail consider the following straight lines

$$D_s := \{(e_1, e_2) \in \mathbb{C}^2 \mid e_2 = s \cdot e_1\},\ s \in \mathbb{C}, \qquad D_\infty := \{(0, e_2) \in \mathbb{C}^2\}, \tag{19.30}$$

and the open subset

[1] Here $\bar{z}$ does not denote conjugation, but taking the residue class modulo the lattice.

$$B = \mathbb{C}^2 \setminus (D_1 \cup D_{-1/2} \cup D_{-2}) \subset \mathbb{C}^2. \tag{19.31}$$

The curves are non-singular exactly over the points of B. Over the exceptional D_s at least two of the e_i are the same. For the vector field algebra we obtain

$$[V_n, V_m] = \begin{cases} (m-n)V_{n+m}, & n,m \text{ odd}, \\ (m-n)\big(V_{n+m} + 3e_1 V_{n+m-2} \\ \qquad +(e_1-e_2)(e_1-e_3)V_{n+m-4}\big), & n,m \text{ even}, \\ (m-n)V_{n+m} + (m-n-1)3e_1 V_{n+m-2} \\ \qquad +(m-n-2)(e_1-e_2)(e_1-e_3)V_{n+m-4}, & n \text{ odd, } m \text{ even}. \end{cases} \tag{19.32}$$

In fact these relations define Lie algebras for every pair $(e_1, e_2) \in \mathbb{C}^2$. We denote by $\mathcal{L}^{(e_1,e_2)}$ the Lie algebra corresponding to (e_1, e_2). Obviously, $\mathcal{L}^{(0,0)} \cong \mathcal{W}$.

Proposition 19.2 ([5], Proposition 5.1). *For $(e_1, e_2) \neq (0,0)$ the algebras $\mathcal{L}^{(e_1,e_2)}$ are not isomorphic to the Witt algebra $\mathcal{W}$, but $\mathcal{L}^{(0,0)} \cong \mathcal{W}$.*

If we restrict our two-dimensional family to a line D_s ($s \neq \infty$) then we obtain a one-dimension family

$$[V_n, V_m] = \begin{cases} (m-n)V_{n+m}, & n,m \text{ odd}, \\ (m-n)\big(V_{n+m} + 3e_1 V_{n+m-2} \\ \qquad +e_1^2(1-s)(2+s)V_{n+m-4}\big), & n,m \text{ even}, \\ (m-n)V_{n+m} + (m-n-1)3e_1 V_{n+m-2} \\ \qquad +(m-n-2)e_1^2(1-s)(2+s)V_{n+m-4}, & n \text{ odd, } m \text{ even}. \end{cases} \tag{19.33}$$

Here s has a fixed value and e_1 is the deformation parameter. (A similar family exists for $s = \infty$.) It can be shown that as long as $e_1 \neq 0$ the algebras over two points in D_s are pairwise isomorphic but not isomorphic to the algebra over 0, which is the Witt algebra. Using the result $\mathbf{H}^2(\mathcal{W}, \mathcal{W}) = \{0\}$ of Fialowski [3] we get

Theorem 19.1. *Despite its infinitesimal and formal rigidity the Witt algebra $\mathcal{W}$ admits deformations $\mathcal{L}_t$ over the affine line with $\mathcal{L}_0 \cong \mathcal{W}$ which restricted to every (Zariski or analytic) neighbourhood of $t = 0$ are non-trivial.*

For the one-dimensional families (19.33) we have that $\mathcal{L}_t \cong \mathcal{L}_{t'}$ for $t, t' \neq 0$. The isomorphism is given by rescaling the basis elements. This is possible as long as $e_1 \neq 0$. In fact, using $V_n^* = (\sqrt{e_1})^{-n} V_n$ (for $s \neq \infty$) we obtain for e_1 always the algebra with $e_1 = 1$ in the structure equations (19.33).

Using the cocycle (19.19) in the families (19.32), (19.33) a central term can be easily incorporated. With respect to the flat coordinate $z - a$ we can take the projective connection $R \equiv 0$. The integral along a separating cocycle C_S is obtained by taking the residue at $z = 0$. In this way we obtain geometric families of deformations for the Virasoro algebra. They are locally non-trivial despite the fact that the Virasoro algebra is formally rigid.

19.4.3 The Current Algebra

Let $\mathfrak{g}$ be a simple finite-dimensional Lie algebra (similar results are true for general Lie algebras) and $\mathscr{A}$ the algebra of meromorphic functions corresponding to the geometric situation discussed above. A basis for $\mathscr{A}$ is given by

$$A_{2k} = (X-e_1)^k, \qquad A_{2k+1} = \frac{1}{2} Y \cdot (X-e_1)^{k-1}, \qquad k \in \mathbb{Z}. \tag{19.34}$$

We calculate for the elements of $\overline{\mathscr{G}}$

$$[x \otimes A_n, y \otimes A_m] = \begin{cases} [x,y] \otimes A_{n+m}, & n \text{ or } m \text{ even}, \\ [x,y] \otimes A_{n+m} + 3e_1 [x,y] \otimes A_{n+m-2} & \\ +(e_1-e_2)(2e_1+e_2)[x,y] \otimes A_{n+m-4}, & n \text{ and } m \text{ odd}. \end{cases} \tag{19.35}$$

If we let e_1 and e_2 (and hence also e_3) go to zero we obtain the classical current algebra as degeneration. Again it can be shown that the family, even if restricted on D_s, is locally non-trivial, see [6]. Recall that by results of Lecomte and Roger [9] the current algebra is formally rigid if $\mathfrak{g}$ is simple. But our families show that it is neither geometrically nor analytically rigid.

Also in this case we can construct families of centrally extended algebras by considering the cocycle (19.20). In this way we obtain non-trivial deformation families for the formally rigid classical affine algebras of Kac–Moody type. The cocycle (19.20) is

$$\gamma(x \otimes A_n, y \otimes A_m) = p(e_1,e_2) \cdot \beta(x,y) \cdot \frac{1}{2\pi \mathrm{i}} \int_{C_S} A_n dA_m. \tag{19.36}$$

Here $p(e_1,e_2)$ is an arbitrary polynomial in the variables e_1 and e_2. and β the Cartan–Killing form. The integral can be calculated [6, Theorem 4.6] as

$$\frac{1}{2\pi \mathrm{i}} \int_{C_S} A_n dA_m = \begin{cases} -n\delta_m^{-n}, & n, m \text{ even}, \\ 0, & n, m \text{ different parity}, \\ -n\delta_m^{-n} + 3e_1(-n+1)\delta_m^{-n+2} + & \\ +(e_1-e_2)(2e_1+e_2)(-n+2)\delta_m^{-n+4}, & n, m \text{ odd}. \end{cases} \tag{19.37}$$

19.5 The Geometric Background

If we take $e_1 = e_2 = e_3$ in the definition of the cubic curve (19.28) we obtain the cuspidal cubic E_C with affine part given by the polynomial $Y^2 = 4X^3$. It has a singularity at $(0,0)$ and the desingularization is given by the projective line $\mathbb{P}^1(\mathbb{C})$. This says there exists a surjective (algebraic) map $\pi_C : \mathbb{P}^1(\mathbb{C}) \to E_C$ which outside the singular point is $1:1$. Over the cusp lies exactly one point. The vector fields, resp.

the functions, resp. the $\mathfrak{g}$-valued functions can be degenerated to E_C and pull-backed to vector fields, resp. functions, resp. $\mathfrak{g}$-valued functions on $\mathbb{P}^1(\mathbb{C})$. The point $(e_1,0)$ where a pole is allowed moves to the cusp. The other point stays at infinity. In particular by pulling back the degenerated vector field algebra we obtain the algebra of vector fields with two possible poles, which is the Witt algebra. And by pulling back the degenerated current algebra we obtain the classical current algebra.

The exceptional lines D_s for $s = 1, -1/2, -2$ are related to interesting geometric situations. Above $D_s \setminus \{(0,0)\}$ with these values of s, two of the e_i are the same, the third one remains distinct. The curve will be a nodal cubic E_N defined by $Y^2 = 4(X-e)^2(X-e)$. The singularity will be a node with the coordinates $(e,0)$. Again the desingularization will be the projective line $\pi_N : \mathbb{P}^1(\mathbb{C}) \to E_N$. But now above the node there will be two points in $\mathbb{P}^1(\mathbb{C})$. For the pull-backs we have the following two situations:

(1) If $s = 1$ or $s = -2$ then $e = e_1$ and the node is a possible point for a pole. We obtain objects on $\mathbb{P}^1(\mathbb{C})$ which beside the pole at ∞ might have poles at two other places. Hence, we obtain a three-point Krichever–Novikov algebras of genus 0.

(2) If $s = -1/2$ then at the node there is no pole. The number of possible poles for the pull-back remains two. We obtain certain subalgebras of the classical two point case. For example, it is required that the functions have the same value at the two points lying above the node. Additionally, for the vector field case we have to pay attention to the fact that the vector fields obtained by pull-back acquire zeros at the points lying above the node.

As these algebras give important examples of infinite dimensional Lie algebras they will be discussed in the next section.

19.6 Examples for the Degenerated Situations

We consider the Krichever–Novikov type algebras for the Riemann sphere $S^2 = \mathbb{P}^1(\mathbb{C})$ (i.e. for genus 0) and three marked points. Given any triple of three points there exists always an analytic automorphism of S^2 mapping this triple to $\{a, -a, \infty\}$, with $a \neq 0$. In fact $a = 1$ would suffice. Without restriction we can take

$$I := \{a, -a\}, \quad O := \{\infty\}.$$

19.6.1 Three-Point Genus 0 Current Algebra

Due to the symmetry of the situation it is more convenient to take a symmetrized basis of $\mathscr{A}$:

$$A_{2k} := (z-a)^k(z+a)^k, \qquad A_{2k+1} := z(z-a)^k(z+a)^k, \qquad k \in \mathbb{Z}. \tag{19.38}$$

It is shown in [16] that it is a basis. By more or less direct calculations one can show the structure equation for the current algebra $\overline{\mathscr{G}}$

$$[x\otimes A_n, y\otimes A_m] = \begin{cases} [x,y]\otimes A_{n+m}, & n \text{ or } m \text{ even}, \\ [x,y]\otimes A_{n+m} + a^2[x,y]\otimes A_{n+m-2}, & n \text{ and } m \text{ odd}, \end{cases} \tag{19.39}$$

Again $a=1$ could be set. The reason to keep a is that if we vary a over the affine line we obtain for $a=0$ the classical current algebra. In particular, this family gives again a deformation. In [6] it was shown that this deformation is locally not equivalent to the trivial family. In fact, if we consider the family (19.35) over the lines D_1 or D_{-2} then those families are isomorphic to (19.39).

For the central extension $\widehat{\mathscr{G}}$ obtained via the defining cocycle we get (see [6], A.13 and A.14)

$$\gamma(x\otimes A_n, y\otimes A_m) = \beta(x,y)\cdot \frac{1}{2\pi \mathrm{i}}\int_{C_S} A_n dA_m, \tag{19.40}$$

with

$$\frac{1}{2\pi \mathrm{i}}\int_{C_S} A_n dA_m = \begin{cases} -n\delta_m^{-n}, & n, m \text{ even}, \\ 0, & n, m \text{ different parity}, \\ -n\delta_m^{-n} + a^2(-n+1)\delta_m^{-n+2}, & n, m \text{ odd}. \end{cases} \tag{19.41}$$

This is again in accordance with the families over D_1 and D_{-2}.

19.6.2 *Three-Point* $\mathfrak{sl}(2,\mathbb{C})$*-Current Algebra for Genus 0*

Of course, given a simple Lie algebra $\mathfrak{g}$ with generators and structure equations the relations above can be written in these terms. An important example is $\mathfrak{sl}(2,\mathbb{C})$ with the standard generators

$$h := \begin{pmatrix} 1 & 0 \\ 0 & -1 \end{pmatrix}, \quad e := \begin{pmatrix} 0 & 1 \\ 0 & 0 \end{pmatrix}, \quad f := \begin{pmatrix} 0 & 0 \\ 1 & 0 \end{pmatrix}.$$

We set $e_n := e\otimes A_n$, $n\in\mathbb{Z}$ and in the same way f_n and h_n. Recall that $\beta(x,y) = \operatorname{tr}(x\cdot y)$. We calculate

$$[e_n, f_m] = \begin{cases} h_{n+m}, & n \text{ or } m \text{ even}, \\ h_{n+m} + a^2 h_{n+m-2}, & n \text{ and } m \text{ odd}, \end{cases} \tag{19.42}$$

$$[h_n, e_m] = \begin{cases} 2e_{n+m}, & n \text{ or } m \text{ even}, \\ 2e_{n+m} + 2a^2 e_{n+m-2}, & n \text{ and } m \text{ odd}, \end{cases} \tag{19.43}$$

$$[h_n, f_m] = \begin{cases} -2f_{n+m}, & n \text{ or } m \text{ even}, \\ -2f_{n+m} - 2a^2 f_{n+m-2}, & n \text{ and } m \text{ odd}. \end{cases} \tag{19.44}$$

For the central extension we get

$$[e_n, f_m] = \begin{cases} h_{n+m} - n\delta_m^{-n}, & n \text{ or } m \text{ even}, \\ h_{n+m} + a^2 h_{n+m-2} - n\delta_m^{-n} - a^2(n-1)\delta_m^{-n+2}, & n \text{ and } m \text{ odd}. \end{cases} \tag{19.45}$$

For the other commutators we do not have contributions from the center.

19.6.3 Three-Points Genus 0 Vector Field Algebra

As shown in [16] a basis of the corresponding Krichever–Novikov algebra is given by

$$V_{2k} := z(z-\alpha)^k (z+\alpha)^k \frac{d}{dz}, \qquad V_{2k+1} := (z-\alpha)^{k+1}(z+\alpha)^{k+1}\frac{d}{dz}, \qquad k \in \mathbb{Z}. \tag{19.46}$$

One calculates

$$[V_n, V_m] = \begin{cases} (m-n)V_{n+m}, & n,m \text{ odd}, \\ (m-n)\big(V_{n+m} + \alpha^2 V_{n+m-2}\big), & n,m \text{ even}, \\ (m-n)V_{n+m} + (m-n-1)\alpha^2 V_{n+m-2}, & n \text{ odd}, \ m \text{ even}. \end{cases} \tag{19.47}$$

We get exactly the structure for the algebras (19.33) over D_1 and D_{-2}.

19.6.4 The Degenerated Current Algebra

If $e_1 \neq e_2 = e_3$, then the point of a possible pole will remain non-singular. This appears if we approach a point of $D^*_{-1/2}$. For the pullbacks of the functions it is now necessary that they have the same value at the points α and $-\alpha$. Hence, all elements of the algebra generated by the pullbacks will have the some property. We will describe this algebra in the following.

The set of elements

$$A_n := \begin{cases} z^n, & n \text{ even}, \\ z^n - \alpha^2 z^{n-2} = z^{n-2}(z^2 - \alpha^2), & n \text{ odd}, \end{cases} \tag{19.48}$$

for $n \in \mathbb{Z}$ is a basis of the subalgebra of meromorphic functions on $\mathbb{P}^1$ which are holomorphic outside 0 and ∞ and have the same value at α and $-\alpha$. The corresponding current algebra is

$$[x \otimes A_n, y \otimes A_m] = \begin{cases} [x,y]\,A_{n+m}, & \text{for } n \text{ or } m \text{ even}, \\ [x,y]\,A_{n+m} - 2\alpha^2[x,y]\,A_{n+m-2} & \\ \qquad + \alpha^4[x,y]\,A_{n+m-4}, & \text{for } n \text{ and } m \text{ both odd}. \end{cases} \tag{19.49}$$

A two-cocycle can be easily given. The obtained algebra is isomorphic to the current algebras lying above $D^*_{-1/2}$.

19.6.5 The Degenerate Vector Field Algebra

Finally, we consider the subalgebra of the Witt algebra defined by the basis elements

$$\begin{aligned} V_{2k} &= z^{2k-3}(z^2-\alpha^2)^2\frac{d}{dz} = l_{2k} - 2\alpha^2 l_{2k-2} + \alpha^4 l_{2k-4}, \\ V_{2k+1} &= z^{2k}(z^2-\alpha^2)\frac{d}{dz} = l_{2k+1} - \alpha^2 l_{2k-1}. \end{aligned} \tag{19.50}$$

One calculates

$$[V_n, V_m] = \begin{cases} (m-n)V_{n+m}, & n,m \text{ odd}, \\ (m-n)\left(V_{n+m} - 2\alpha V_{n+m-2} + \alpha^2 V_{n+m-4}\right), & n,m \text{ even}, \\ (m-n)V_{n+m} + (m-n-1)(-2\alpha)V_{n+m-2} & \\ \qquad +(m-n-2)\alpha^2 V_{n+m-4}, & n \text{ odd},\ m \text{ even}. \end{cases} \tag{19.51}$$

This subalgebra can be described as the subalgebra of meromorphic vector fields vanishing at α and $-\alpha$, with possible poles at 0 and ∞ and such that in the representation of $V(z) = f(z)(z^2-\alpha^2)\frac{d}{dz}$ the function f fulfills $f(\alpha) = f(-\alpha)$.

In accordance with the geometric picture, the obtained algebra is isomorphic to the vector field algebras lying above $D^*_{-1/2}$.

References

1. Belavin, A.A., Polyakov, A.M., Zamolodchikov, A.B.: *Infinite conformal symmetry in two-dimensional quantum field theory*. Nucl. Phys. B **241**, 333–380 (1984).
2. Fialowski, A.: *An example of formal deformations of Lie algebras*. NATO Conference on Deformation Theory of Algebras and Applications, Proceedings, Kluwer 1988, 375–401.
3. Fialowski, A.: *Deformations of some Infinite Dimensional Lie Algebras*. J. Math. Phys. **31**, 1340–1343 (1990).
4. Fialowski, A., Fuchs, D.: *Construction of Miniversal Deformations of Lie Algebras*. J. Funct. Anal. **161**, 76–110 (1999).
5. Fialowski, A., Schlichenmaier, M.: *Global Deformations of the Witt algebra of Krichever–Novikov Type*. Comm. Contemp. Math. **5**, 921–945 (2003).
6. Fialowski, A., Schlichenmaier, M.: *Global Geometric Deformations of Current Algebras as Krichever–Novikov Type Algebras*. Comm. Math. Phys. **260**, 579–612 (2005).
7. Gerstenhaber, M.: *On the Deformation of Rings and Algebras I,II,III* Ann. Math. **79**, 59–10 (1964); **84**, 1–19 (1966); **88**, 1–34 (1968).
8. Krichever I.M., Novikov S.P.: *Algebras of Virasoro type, Riemann surfaces and structures of the theory of solitons*. Funktional Anal. i. Prilozhen. **21**, 46–63 (1987); *Virasoro type algebras, Riemann surfaces and strings in Minkowski space*. Funktional Anal. i. Prilozhen. **21**, 47–61 (1987); *Algebras of Virasoro type, energy-momentum tensors and decompositions of operators on Riemann surfaces*, Funktional Anal. i. Prilozhen. **23**, 46–63 (1989).

9. Lecomte, P., Roger, C.: *Rigidity of Current Lie Algebras of Complex Simple Type*. J. London Math. Soc. (2) **37**, 232–240 (1988).
10. Nijenhuis, A., Richardson, R.: *Cohomology and Deformations of Algebraic Structures*. Bull. Amer. Math. Soc. **70**, 406–411 (1964).
11. Nijenhuis, A., Richardson, R.: *Cohomology and Deformations in Graded Lie Algebras*. Bull. Amer. Math. Soc. **72**, 1–29 (1966).
12. Schlichenmaier, M.: *Verallgemeinerte Krichever - Novikov Algebren und deren Darstellungen*, University of Mannheim, June 1990.
13. Schlichenmaier, M.: *Krichever–Novikov algebras for more than two points*. Lett. Math. Phys. **19**, 151–165 (1990).
14. Schlichenmaier, M.: *Krichever–Novikov algebras for more than two points: explicit generators*. Lett. Math. Phys. **19**, 327–336 (1990).
15. Schlichenmaier, M.: *Central Extensions and Semi-Infinite Wedge Representations of Krichever-Novikov Algebras for More than Two Points*. Lett. Math. Phys. **20**, 33–46 (1991).
16. Schlichenmaier, M.: *Degenerations of Generalized Krichever–Novikov Algebras on Tori*. Jour. Math. Phys. **34**, 3809–3824 (1993).
17. Schlichenmaier, M.: *Differential operator algebras on compact Riemann surfaces*. Generalized Symmetries in Physics (Clausthal 1993, Germany) (H.-D. Doebner, V.K. Dobrev, A.G. Ushveridze, eds.), World Scientific, Singapore, 1994, pp. 425–434.
18. Schlichenmaier, M.: *Local Cocycles and Central Extensions for Multi-Point Algebras of Krichever–Novikov Type*. J. reine angew. Math. **559**, 53–94 (2003).
19. Schlichenmaier, M.: *Higher Genus Affine Lie algebras of Krichever–Novikov Type*, Moscow Math. Jour. **3**, 1395–142 (2003).
20. Schlichenmaier, M., Sheinman, O.K.: *The Sugawara construction and Casimir operators for Krichever-Novikov algebras*. J. Math. Sci., New York **92**, no. 2, 3807–3834 (1998), q-alg/9512016.
21. Schlichenmaier, M., Sheinman, O.: *Wess-Zumino-Witten-Novikov Theory, Knizhnik-Zamolodchikov Equations, and Krichever–Novikov Algebras*. Russian Math. Surv. **54**, 213–250 (1999).
22. Schlichenmaier, M., Sheinman, O.: *Knizhnik–Zamolodchikov Equations for Positive Genus and Krichever–Novikov Algebras*. Russian Math. Surv. **59**, 737–770 (2004).

Chapter 20
Conformal Algebras in the Context of Linear Algebraic Groups

Pavel Kolesnikov

Abstract We consider a category of algebras related to a pair (G,V), where G is a linear algebraic group that acts on an affine algebraic variety V. As a particular case, we obtain the category of pseudo-algebras over a commutative affine Hopf algebra. We apply the introduced language to describe irreducible subalgebras of the conformal endomorphism algebra over a linear algebraic group.

20.1 Introduction

Conformal algebras and similar structures usually appear in the context of operator product expansion (OPE) [7]. In algebra, these structures are related to vertex operator algebras (see, e.g., [4]) and their numerous applications.

From the algebraic point of view, a conformal algebra is a linear space C with countably many operations of multiplication, satisfying certain axioms (see [6–8]). The most important one is the axiom of locality which states that for every $a,b \in C$ only a finite number of their products are nonzero.

In [1], a generalization of the notion of a conformal algebra was proposed in terms of pseudo-tensor categories [2] associated with Hopf algebras. The natural source of Hopf algebras is provided by linear algebraic groups. Hence, it is natural to expect that conformal algebras and some of their generalizations have a natural interpretation in terms of linear algebraic groups.

The purpose of this note is to present the corresponding construction explicitly: given an affine algebraic variety V equipped with a continuous (left) action of a linear algebraic group G, we build a category of (G,V)-conformal algebras. If $V = G$

P. Kolesnikov
Sobolev Institute of Mathematics, 630090 Novosibirsk, Russia
e-mail: pavelsk@math.nsc.ru

S. Silvestrov et al. (eds.), *Generalized Lie Theory in Mathematics, Physics and Beyond,*

and the action is given by the group product on G then the category of (G,G)-conformal (or (G)-conformal) algebras is equivalent to the category of H^{op}-pseudo-algebras, where H is the coordinate Hopf algebra of G. In particular, if G is the affine line $\mathbb{A}_1$ then (G)-conformal algebra is the same as conformal algebra in the sense of the definition in [8], stated in terms of λ-product.

Throughout the paper, $\Bbbk$ is an algebraically closed field, G is a linear algebraic group. By a G-set we mean a Zariski closed subset V of an affine space endowed with a (left) continuous action of the group G. Denote by $H = \Bbbk[G]$ the coordinate Hopf algebra of G, and let Δ, ε, S stand for the coproduct, counit, antipode, respectively. If V is a G-set then the coordinate algebra $A = \Bbbk[V]$ is an H-comodule algebra with coaction $\Delta_A : A \to H \otimes A$.

We will use the following notation (omitting the summation symbol):

$$\Delta(h) = h_{(1)} \otimes h_{(2)}, \quad (\Delta \otimes \mathrm{id})\Delta(h) = h_{(1)} \otimes h_{(2)} \otimes h_{(3)},$$
$$(S \otimes \mathrm{id})\Delta(h) = h_{(-1)} \otimes h_{(2)}, \quad (\mathrm{id} \otimes S)\Delta(h) = h_{(1)} \otimes h_{(-2)},$$
$$\Delta_A(a) = a_{(1)} \otimes a_{(2)}, \quad (\Delta \otimes \mathrm{id})\Delta_A = a_{(1)} \otimes a_{(2)} \otimes a_{(3)}, \quad \text{etc.}$$

By H^{op} we denote the same algebra H with coproduct $\Delta^{\mathrm{op}}(h) = h_{(2)} \otimes h_{(1)}$ (i.e., $H^{\mathrm{op}} = \Bbbk[G^{\mathrm{op}}]$).

In Sect. 20.2 we state a general definition of a conformal algebra associated with a pair (G,V) of a linear algebraic group G and a G-set V, i.e., a (G,V)-conformal algebra. The idea is fairly simple. An ordinary algebra $\mathscr{A}$ over a field can be completely described by the operator of (left) multiplication $\mu : \mathscr{A} \to \operatorname{End} \mathscr{A}$, $\mu(a) : b \mapsto ab$. For (G,V)-conformal algebras we just need to understand what is End in this case. The construction obtained is similar to what is called a "conformal endomorphism" in [8].

If $V = G$ (action is given by the left multiplication) then the category obtained turns to be the same as the category of H^{op}-pseudo-algebras by [1].

Section 20.3 is devoted to the definition of associative conformal algebras (that has sense for $V = G$). This turns to be the same as associative pseudo-algebra.

An important example of an associative (G)-conformal algebra is provided by an action of G on a G-set V. In a few words, this algebra $\operatorname{Cend}_n^{G,V}$ consists of all "algebraic" rules of transformation of the space of vector-valued regular functions $V \to \Bbbk^n$ by means of G. Here the word "algebraic" stands for two properties: local regularity and translation invariance. In particular, the left-shift operator L_g, $g \in G$, is algebraic in this sense.

The main result of this paper is stated in Sect. 20.4. In [9], the complete description of irreducible (G)-conformal subalgebras of $\operatorname{Cend}_n^{G,G}$ was obtained for $G = \mathbb{A}_1$ (that proved a conjecture from [3]). The crucial statement in [9] is that an irreducible conformal subalgebra together with scalar regular functions generate a left ideal in $\operatorname{Cend}_n^{G,G}$. In the present paper, we generalize this statement to the case of an arbitrary linear algebraic group.

20.2 Categories of Conformal Algebras

Let M, N be two linear spaces over $\Bbbk$, and let S stands for $\mathrm{Hom}(M,N)$. Suppose V is a G-set.

Definition 20.1. A map $a : V \to S$ is said to be *locally regular* if for any $u \in M$ the map $z \mapsto a(z)u$ is a regular N-valued map on V.

Suppose, in addition, that S is a left A-module. A map $a : V \to S$ is *translation-invariant* (or T-invariant, for short) if

$$fa(z) = f(z)a(z), \quad f \in A,\ z \in V. \tag{20.1}$$

The main example we are going to consider is the following one. Let M be a left H-module, and let N be a left A-module. Then $S = \mathrm{Hom}(M,N)$ can be considered as a left A-module with respect to

$$(f\varphi)(u) = f_{(2)}\varphi(f_{(-1)}u), \quad f \in A,\ u \in M,\ \varphi \in S. \tag{20.2}$$

Definition 20.2. A *conformal homomorphism* from M to N is a locally regular and T-invariant map $a : V \to \mathrm{Hom}(M,N)$.

The set $\mathrm{Chom}(M,N)$ of all conformal homomorphisms is a linear space which is also an A-module with respect to the action defined by

$$(fa)(z) = f(z)a(z), \quad f \in A,\ a \in \mathrm{Chom}(M,N),\ z \in V.$$

If a linear space M is A- and H-module then $\mathrm{Chom}(M,M)$ is denoted by $\mathrm{Cend}\,M$ (conformal endomorphisms of M).

Definition 20.3. A (G,V)*-conformal algebra* C is an A- and H-module endowed with an A-linear map $\mu : C \to \mathrm{Cend}\,C$.

A morphism of (G,V)-conformal algebras C_1 and C_2 is an A-linear and H-linear map $\varphi : C_1 \to C_2$ such that the following diagram is commutative:

$$\begin{array}{ccccc}
C_1 & \xrightarrow{\mu_1} & \mathrm{Cend}\,C_1 & \xrightarrow{\varphi_0} & \mathrm{Chom}(C_1,C_2) \\
\downarrow{\scriptstyle\varphi} & & & & \downarrow{\scriptstyle\mathrm{id}} \\
C_2 & \xrightarrow{\mu_2} & \mathrm{Cend}\,C_2 & \xrightarrow{\varphi^0} & \mathrm{Chom}(C_1,C_2)
\end{array}$$

Here $\varphi_0(a)(z) = \varphi a(z)$, $\varphi^0(b)(z) = b(z)\varphi$, $a \in \mathrm{Cend}\,C_1$, $b \in \mathrm{Cend}\,C_2$, $z \in V$.

Example 20.1. Let $G = V = \{e\}$. Then $\mathrm{Chom}(M,N) = \mathrm{Hom}(M,N)$, and it is clear that a (G,V)-conformal algebra is just the same as ordinary algebra over a field $\Bbbk$.

Example 20.2. Let $G = V = \mathbb{A}_1 \simeq (\Bbbk,+)$, $\mathrm{char}\,\Bbbk = 0$. Then $H = A = \Bbbk[T]$ with the canonical Hopf algebra structure, and the notion of a (G,V)-conformal algebra coincides with the notion of a conformal algebra in [7].

The purpose of this section is to show that for any linear algebraic group G the category of (G,G)-conformal algebras (assuming $\Delta = \Delta_A$, i.e., the group acts on itself by multiplication) coincides with the category of H^{op}-pseudo-algebras [1].

It follows from the definition that a (G,V)-conformal algebra C can be endowed with a family of bilinear operations $(\cdot_z \cdot) : C \times C \to C$, $z \in V$, such that:

(C1) For every $a,b \in C$ the map $z \mapsto (a_z b)$ is regular.
(C2) $(fa_z b) = f(z)(a_z b) = f_{(2)}(a_z f_{(-1)} b)$, $f \in A$.

Indeed, $(a_z b) = \mu(a)(z)b$, $z \in V$, satisfy (C1), (C2). Converse is also clear: given a family of operations $(\cdot_z \cdot)$ satisfying (C1), (C2), one may define $\mu : C \to \operatorname{Cend} C$ by $\mu(a) : z \mapsto (a_z \cdot) \in \operatorname{End} C$.

Morphisms of (G,V)-conformal algebras preserve operations $(\cdot_z \cdot)$ for all $z \in V$. Conversely, an A- and H-linear map that preserves these operations is a morphism of (G,V)-conformal algebras.

Let us fix an ascending chain of subspaces

$$0 = A_{-1} \subset A_0 \subset A_1 \subset \dots \tag{20.3}$$

such that $\dim A_n < \infty$, $\bigcup\limits_{n\geq 0} A_n = A$. Then the descending sequence

$$A^* = A_{-1}^{\perp} \supset A_0^{\perp} \supset A_1^{\perp} \supset \dots$$

form a basis of neighborhoods of zero in A^*, such that $\operatorname{codim}_{A^*} A_n^{\perp} < \infty$, $\bigcap_{n\geq 0} A_n^{\perp} = \{0\}$. Thus, we get a topology on A^* which does not depend on the choice of $\{A_n\}_{n\geq 0}$.

Let us consider A^* as an A-module via $\langle g\xi, f\rangle = \langle \xi, gf\rangle$, $f,g \in A$, $\xi \in A^*$. This is indeed a module since A is commutative.

If M, N are some linear spaces then $\operatorname{Hom}(M,N)$ can also be considered as a topological linear space with respect to so called finite topology (see, e.g., [5]). Namely, the system of neighborhoods of zero in $\operatorname{Hom}(M,N)$ is given by $U_{u_1,\dots,u_n} = \{\varphi \mid \varphi(u_i) = 0, i = 1,\dots,n\}$, $u_1,\dots,u_n \in M$, $n \geq 0$.

Proposition 20.1. *Let M be an A-module, N be an H-module. Then there is a linear isomorphism between* $\operatorname{Chom}(M,N)$ *and the space of A-linear continuous maps α : $A^* \to \operatorname{Hom}(M,N)$*

Proof. Suppose $a \in \operatorname{Chom}(M,N)$. For an element $u \in M$ denote by $\hat{a}_u$ the function $z \mapsto a(z)u$. Since a is locally regular, $\hat{a}_u \in A \otimes N$. Therefore, for any $\xi \in A^*$ we may compute $(a_{(\xi)} u) = (\langle \xi, \cdot\rangle \otimes \mathrm{id})\hat{a}_u$.

The map $\xi \mapsto a_{(\xi)} = (a_{(\xi)} \cdot)$ is continuous. Indeed, consider a sequence $\xi_n \in A^*$, $n \geq 0$, such that $\xi_n \underset{n\to\infty}{\longrightarrow} 0$. Then for any $u_1,\dots,u_m \in M$ there exist $f_1,\dots,f_k \in A$ such that $\hat{a}_{u_i}(z) \in \operatorname{Span}\{f_1(z),\dots,f_k(z)\} \otimes N$. Therefore, $\langle \xi_n, f_j\rangle = 0$, $j = 1,\dots,k$, for sufficiently large n, and so $a_{(\xi_n)} \to 0$ in $\operatorname{Hom}(M,N)$.

Conversely, any continuous linear map $\alpha : A^* \to \operatorname{Hom}(M,N)$ defines a locally regular map $a : V \to \operatorname{Hom}(M,N)$ in a natural way (for every $z \in V$ the corresponding evaluation map is an element of A^*). Indeed, suppose $\{g_i\}_{i\in I}$ is a basis of A agreed with the filtration (20.3). Denote by ξ_i such an element of A^* that $\langle \xi_i, g_j\rangle = \delta_{ij}$. Then for any $u \in M$ there exist only a finite number of $i \in I$ such that $v_i = \alpha(\xi_i)u \neq 0$. Therefore, $(a(z)u) = \sum_{i\in I} g_i(z) \otimes v_i$ is a regular function, hence, a is locally regular. We obtain a one-to-one correspondence between locally regular maps $a : V \to \mathrm{Hom}(M,N)$ and continuous maps $\alpha : A^* \to \mathrm{Hom}(M,N)$.

If a is T-invariant then

$$\begin{aligned}\alpha(f\xi)u &= (\langle f\xi,\cdot\rangle \otimes \mathrm{id})\hat{a}_u = (\langle \xi,\cdot\rangle \otimes \mathrm{id})(f\otimes 1)\hat{a}_u \\ &= (\langle \xi,\cdot\rangle \otimes \mathrm{id})(1\otimes f_{(2)})\hat{a}_{f_{(-1)}u} = f_{(2)}\alpha(\xi)f_{(-1)}u,\end{aligned}$$

for $f \in A$, $u \in M$, $\xi \in A^*$. The converse is obvious: if α is A-linear then a is T-invariant.

Definition 20.4. An H- and A-module C endowed with linear operations $(\cdot_{(\xi)}\cdot) : C\otimes C \to C$, $\xi \in A^*$, is called an *(H,A)-conformal algebra* if the following properties hold:

(P1) $\operatorname{codim}\{\xi \in A^* \mid (a_{(\xi)}b) = 0\} < \infty$ for every $a, b \in C$.
(P2) $f_{(2)}(a_{(\xi)}f_{(-1)}b) = (a_{(f\xi)}b) = (fa_{(\xi)}b)$, $f \in A$.

A morphism of (H,A)-conformal algebras C_1 and C_2 is an A- and H-linear map $\varphi : C_1 \to C_2$ such that $\varphi(a_{(\xi)}b) = \varphi(a)_{(\xi)}\varphi(b)$, $a, b \in C_1$, $\xi \in A^*$.

Proposition 20.2. *The category of (G,V)-conformal algebras C coincides with the category of (H,A)-conformal algebras.*

Proof. It remains to note that (P1) is equivalent to continuity of the map $\xi \mapsto (a_{(\xi)}\cdot)$, $a \in C$, (P2) is equivalent to T-invariance of the corresponding map $z \mapsto (a_z\cdot)$, $z \in V$, and A-linearity of the map $\mu : a \mapsto (a_z\cdot)$.

It follows from Proposition 20.1 that an A- and H-linear map φ is a morphism of (G,V)-conformal algebras iff φ is a morphism of (H,A)-conformal algebras.

Consider the case when $V = G$ ($A = H$) with $\Delta_A = \Delta$. The notion of a (G,G)-conformal algebra (or just (G)-conformal, for short) generalizes the notion of a conformal algebra in the frames of pseudo-algebras [1, 2].

Definition 20.5 ([1]). Let H be a Hopf algebra. An *H-pseudo-algebra* is a left H-module C endowed with an $(H\otimes H)$-linear map

$$* : C\otimes C \to (H\otimes H)\otimes_H C,$$

called *pseudo-product*. Here $H\otimes H$ is considered as the outer product of regular right H-modules.

A morphism of H-pseudo-algebras C_1 and C_2 is an H-linear map $\varphi : C_1 \to C_2$ such that

$$\varphi(a) * \varphi(b) = (\mathrm{id}\otimes \mathrm{id}\otimes_H \varphi)(a*b), \quad a, b \in C_1.$$

The operation $*$ on a pseudo-algebra C can be extended to a map

$$*: (H^{\otimes n} \otimes_H C) \otimes (H^{\otimes m} \otimes_H C) \to H^{\otimes (n+m)} \otimes_H C$$

by the rule

$$(F_1 \otimes_H a_1) * (F_2 \otimes_H a_2) = (F_1 \otimes F_2 \otimes_H 1)(\Delta^{n-1} \otimes \Delta^{m-1} \otimes_H \mathrm{id})(a_1 * a_2),$$

$F_1 \in H^{\otimes n}, F_2 \in H^{\otimes m}, a_1, a_2 \in C, \Delta^k(h) = h_{(1)} \otimes \ldots \otimes h_{(k+1)}$ for $h \in H, k \geq 1$. Pseudo-algebra is said to be associative if for every $a,b,c \in C$ the following relation holds:

$$a * (b * c) = (a * b) * c. \tag{20.4}$$

Proposition 20.3. *The category of (G)-conformal algebras coincides with the category of H^{op}-pseudo-algebras.*

Proof. Suppose C is a (G)-conformal algebra. By Proposition 20.2 this is an (H,H)-conformal algebra. Relation (P2) implies

$$f_{(1)}(a_{(f_{(-2)}\xi)} b) = (a_{(\xi)} f b), \quad f \in H, \; \xi \in H^*. \tag{20.5}$$

Given $a, b \in C$, consider $(a_z b) = \sum_i f_i \otimes c_i$, $f_i \in H$, $c_i \in C$, z is a variable that ranges over G. Define

$$(a * b) = \sum_i (f_i \otimes 1) \otimes_{H^{\mathrm{op}}} c_i \in (H^{\mathrm{op}} \otimes H^{\mathrm{op}}) \otimes_{H^{\mathrm{op}}} C. \tag{20.6}$$

It is now straightforward to check that $*$ is an H^{op}-bilinear map, i.e.,

$$(fa * gb) = (f \otimes g \otimes_{H^{\mathrm{op}}} 1)(a * b), \quad f, g \in H.$$

Indeed, $(fa_z gb) = f(z) g_{(-2)}(z) g_{(1)}(a_z b)$, so $(fa * gb) = (f g_{(-2)} \otimes 1) \otimes_{H^{\mathrm{op}}} g_{(1)} c_i$. It remains to note that $(g_{(-2)} \otimes 1) \otimes_{H^{\mathrm{op}}} g_{(1)} = (1 \otimes g) \otimes_{H^{\mathrm{op}}} 1$.

The same relation (20.6) makes clear how to obtain a (G)-conformal algebra structure on an H^{op}-pseudo-algebra.

In particular, if $G = \mathbb{A}_1$ then $H \simeq H^{\mathrm{op}} = \Bbbk[T]$, and a (G)-conformal algebra is the same as a $\Bbbk[T]$-pseudo-algebra and the same as a conformal algebra in the sense of [7].

20.3 Associative (G)-Conformal Algebras

Throughout the rest of the paper, we consider (G)-conformal algebras.

Recall that an ordinary algebra $\mathscr{A}$ is said to be associative if the map $\mu : \mathscr{A} \to \operatorname{End} \mathscr{A}$, $\mu(a) : b \mapsto ab$, is a homomorphism of algebras. In the same way one may define associative (G)-conformal algebras. The definition obtained is equivalent to associativity of a pseudo-algebra.

Suppose M is a left H-module, and define a family of binary operations $(\cdot_z\cdot)$, $z \in G$, on the space $\operatorname{Cend} M$ as follows:

$$(a_z b)(g) = a(z)b(gz^{-1}), \quad a,b \in \operatorname{Cend} M,\ g \in G. \tag{20.7}$$

For every $z \in G$, the map $g \mapsto (a_z b)(g)$ is locally regular, and it is easy to see that it is also T-invariant.

Remark 20.1. In general, $\operatorname{Cend} M$ is not a (G)-conformal algebra since the map $z \mapsto (a_z b)$ is not regular. But if M is a finitely generated H-module then $\operatorname{Cend} M$ is a (G)-conformal algebra, as follows from [1, Proposition 10.5] and Proposition 20.3.

Definition 20.6. A (G)-conformal algebra C is said to be associative if $\mu : C \to \operatorname{Cend} C$ preserves the operations $(\cdot_z\cdot)$, $z \in G$. This is equivalent to the following identity on C:

$$((a_z b)_g c) = (a_z (b_{gz^{-1}} c)), \quad g,z \in G, \tag{20.8}$$

or

$$(a_z (b_g c)) = ((a_z b)_{gz} c)).$$

A (G)-conformal algebra C is associative iff C is an associative H^{op}-pseudo-algebra, i.e., the relation (20.4) holds for the pseudo-product (20.6).

A representation of an associative (G)-conformal algebra C on an H-module M is an H-linear map $\rho : C \to \operatorname{Cend} M$ that preserves the operations $(\cdot_z\cdot)$, $z \in G$. In this case, M is said to be a (conformal) C-module.

An interesting series of examples of associative (G)-conformal algebras appears from consideration of G-sets.

Let V be a G-set, and let A be the algebra of all regular functions on V. Suppose M is a finitely generated A-module, and denote by $\operatorname{Cend}^{G,V} M$ the space of all locally regular maps $a : G \to \operatorname{End} M$ such that

$$a(g)(fu) = L_g f(a(g)u), \quad u \in M,\ f \in A,\ g \in G.$$

For example, the left-shift action of G on A given by $g \times f \mapsto L_g f$, $g \in G$, $f \in A$, where $(L_g f)(v) = f(g^{-1}v)$, $v \in V$, belongs to $\operatorname{Cend}^{G,V} M$.

Define H-module structure on $C = \operatorname{Cend}^{G,V} M$ by the rule

$$(fa)(g) = f(g)a(g), \quad f \in H,\ a \in C,\ g \in G.$$

Also, define operations $(\cdot_g\cdot)$, $g \in G$, as follows:

$$(a_g b)(z) = a(g)b(g^{-1}z), \quad a,b \in C,\ g,z \in G. \tag{20.9}$$

It is straightforward to check that both fa and $(a_g b)$, belong to C. Moreover, we have

$$(ha_g b) = h(g)(a_g b) = h_{(1)}(a_g h_{(-2)} b), \quad h \in H,\ a,b \in C,\ g \in G.$$

Therefore, $C = \mathrm{Cend}^{G,V} M$ is an associative (G^{op})-conformal algebra (regularity of $(a_z b)$ can be shown similarly to [1, Proposition 10.5]; recall that M is a finitely generated A-module). By Proposition 20.3, $\mathrm{Cend}^{G,V} M$ is an H-pseudo-algebra.

If M is a free n-generated A-module then $\mathrm{Cend}^{G,V} M$ is denoted by $\mathrm{Cend}_n^{G,V}$. The structure of $\mathrm{Cend}_n^{G,V}$ can be completely described as follows (cf. [1, 3]). Consider $C = H \otimes A \otimes \mathbb{M}_n(\Bbbk)$ endowed with operations $(\cdot_z\cdot)$, $z \in G$, given by

$$(h_1 \otimes f_1 \otimes a_1)_z(h_2 \otimes f_2 \otimes a_2) = (h_1 L_z h_2 \otimes f_1 L_z f_2 \otimes a_1 a_2),$$

where $h_i \in H$, $f_i \in A$, $a_i \in \mathbb{M}_n(\Bbbk)$. The (G^{op})-conformal algebra obtained is isomorphic to $\mathrm{Cend}_n^{G,V}$. In terms of pseudo-product, the operations on C are given by the relation

$$(h_1 \otimes f_1 \otimes a_1) * (h_2 \otimes f_2 \otimes a_2) = (h_1 f_{2(-1)} \otimes h_2) \otimes_H (1 \otimes f_1 f_{2(2)} \otimes a_1 a_2). \quad (20.10)$$

Proposition 20.4 (cf. [3]). *Right ideals of* $\mathrm{Cend}_n^{G,V}$ *are of the form* $I = H \otimes I_0$, I_0 *is a right ideal of* $\mathbb{M}_n(A)$. *Left ideals of* $\mathrm{Cend}_n^{G,V}$ *are of the form* $I = \mathscr{F}(H \otimes I_0)$, *where* I_0 *is a left ideal of* $\mathbb{M}_n(A)$, $\mathscr{F} : \mathrm{Cend}_n \to \mathrm{Cend}_n$ *is defined by*

$$\mathscr{F}(h \otimes f \otimes a) = f_{(-1)} h \otimes f_{(2)} \otimes a.$$

(G^{op})*-Conformal algebra* $\mathrm{Cend}_n^{G,V}$ *is simple iff* V *is an irreducible* G*-set.*

Proof. The H-linear map $\mathscr{F}$ is invertible:

$$\mathscr{F}^{-1}(h \otimes f \otimes a) = f_{(1)} h \otimes f_{(2)} \otimes a.$$

Let us consider only left ideals of $\mathrm{Cend}_n^{G,V}$, since right ideals are similar to describe. Suppose I is a left ideal of $\mathrm{Cend}_n^{G,V}$. Consider $a \in I$, $a = \mathscr{F}(b)$,

$$b = \sum_{i=1}^{m} \sum_j h_i \otimes f_j \otimes b_{ij}, \quad a = \sum_{i=1}^{m} \sum_j h_i f_{j(-1)} \otimes f_{j(2)} \otimes b_{ij},$$

where h_i, $i = 1, \dots, m$, are linearly independent. Then

$$\begin{aligned} I \ni x = (1 \otimes 1 \otimes E)_g a \quad & \sum_{i,j} (1 \otimes 1 \otimes E)_g (h_i f_{j(-1)} \otimes f_{j(2)} \otimes b_{ij}) \\ &= \sum_{i,j} (L_g h_i)(L_g f_{j(-1)}) \otimes L_g f_{j(2)} \otimes b_{ij} = \sum_{i,j} (L_g h_i) f_{j(-1)} \otimes f_{j(2)} \otimes b_{ij} \end{aligned}$$

since $L_g f_{(-1)} \otimes L_g f_{(2)} = f_{(-1)} \otimes f_{(2)}$. Therefore,

$$\mathscr{F}^{-1}(x) = \sum_{i,j} (L_g h_i) \otimes f_j \otimes b_{ij} \in \mathscr{F}^{-1}(I).$$

Since $g \in G$ is an arbitrary element, there exist $g_1, \dots, g_m \in G$ such that $h = |h_i(g_k)|_{i,k=1,\dots,m} \neq 0$. Hence, $\sum_j h \otimes f_j \otimes b_{ij} \in \mathscr{F}^{-1}(I)$ for all $i = 1, \dots, m$. Consider the ideal $\mathfrak{h}$ of H that consists of all such $h \in H$. This ideal is obviously L_g-invariant for all $g \in G$ (one may multiply with $(1 \otimes 1 \otimes E)_g$). Since G acts on itself irreducibly, we have $\mathfrak{h} = H$ and, therefore, $\sum_j f_{j(-1)} \otimes f_{j(2)} \otimes b_{ij} \in I$ for every i.

Finally, consider $I_0 = \left\{ x = \sum_j f_j \otimes a_j \in A \otimes \mathbb{M}_n(\Bbbk) \mid \mathscr{F}(1 \otimes x) \in I \right\}$. The following computation shows that I_0 is a left ideal of $A \otimes \mathbb{M}_n(\Bbbk)$:

$$f_{1(-1)}((1 \otimes f_{1(2)} \otimes a)_e (f_{2(-1)} \otimes f_{2(2)} \otimes b)) = f_{1(-1)} f_{2(-1)} \otimes f_{1(2)} f_{2(2)} \otimes ab.$$

Hence, $I = \mathscr{F}(H \otimes I_0)$.

Now the last statement is obvious: a two-sided ideal of $\mathrm{Cend}_n^{G,V}$ corresponds to an ideal of $\mathbb{M}_n(A)$ which is L_g-invariant for all $g \in G$.

20.4 Conformal Endomorphism Algebra over a Linear Algebraic Group

In this section, we consider (G^{op})-conformal algebra $\mathrm{Cend}_n = \mathrm{Cend}_n^{G,G}$. This algebra consists of those transformation rules of the space M_n of vector-valued regular functions $G \to \Bbbk^n$ that are similar to the left-shift action, as described in the previous section.

Denote by $W_n(G)$ the $\Bbbk$-linear span of all operators from $\mathrm{End}\, M_n$ of the form $a(g)$, $a \in \mathrm{Cend}_n$, $g \in G$. This is a subalgebra of $\mathrm{End}\, M_n$ since (20.9) implies

$$a(g_1)b(g_2) = (a_{g_1} b)(g_1 g_2), \quad g_1, g_2 \in G. \tag{20.11}$$

We will simply denote $W_n(G)$ by W_n. The following statement allows to consider a "normal form" of an element of W_n.

Lemma 20.1. *If $a(g) = 0$ then $\left(a_g \mathrm{Cend}_n\right) = 0$; the converse is also true. If $\sum_i a_i(g_i) = 0$, $g_i \in G$ are pairwise different, then $a_i(g_i) = 0$ for all i.*

Proof. The first statement immediately follows from (20.11). Let us prove the second one. If $\sum_{i=1}^m a_i(g_i) = 0$ then $0 = \sum_{i=1}^m a_i(g_i) f = \sum_i (L_{g_i} f) a(g_i)$ for all $f \in H$. Assume there exists $u \in M_n$ such that $a(g_i)u \neq 0$ for some i. Denote by U the set of all $z \in G$ such that $(a(g_i)u)(z) \neq 0$. This is a Zariski open set, so we may choose $f_1, \dots, f_m \in H$ in such a way that $\det |L_{g_i} f_j|(z) \neq 0$ for some $z \in U$. But $\sum_i (L_{g_i} f_j)(z)(a(g_i)u)(z) = 0$ and hence $(a(g_i)u)(z) = 0$ for all $i = 1, \dots, m$, $z \in U$.

Lemma 20.2. (1) *Algebra W_n can be considered as a topological left H-module with respect to $h \cdot a(g) = h(g)a(g)$, $a \in \mathrm{Cend}_n$, $h \in H$, $g \in G$, assuming discrete topology on H and finite topology on W_n.*

(2) *(G^{op})-Conformal algebra Cend_n can be considered as a topological left W_n-module with respect to $a(g) \cdot b = (a_g b)$, $a, b \in \mathrm{Cend}_n$, $g \in G$, assuming finite topology on W_n and discrete topology on Cend_n.*

Proof. It follows from Lemma 20.1 that H-module structure on W_n and W_n-module structure on Cend_n are well-defined. Associativity of the action is obvious.

(1) Suppose $W_n \ni \alpha_k \to 0$ as $k \to \infty$, each α_k can be presented in a form given by Lemma 20.1. We need to check that $(h \cdot \alpha_k) \to 0$ for a fixed $h \in H$. Indeed, direct computation shows that

$$h_{(1)}(\alpha_k h_{(-2)} v) = (h \cdot \alpha_k)(v), \quad v \in M_n.$$

Thus, $(h \cdot \alpha_k)$ converges to zero in W_n.

(2) Suppose $x \in \mathrm{Cend}_n$, and check that $\alpha_k \cdot x = 0 \in \mathrm{Cend}_n$ for $k \gg 0$. Indeed, if $x = 1 \otimes f \otimes a$, $f \in H$, $a \in \mathbb{M}_n(\Bbbk)$, then the claim follows from the definition of finite topology. For a generic $x \in \mathrm{Cend}_n$, note that

$$\alpha_k \cdot hx = h_{(2)}((h_{(-1)} \cdot \alpha_k) \cdot x), \quad h \in H.$$

It remains to use the statement (1).

Theorem 20.1. *Algebra W_n is a dense subalgebra of $\mathrm{End}\, M_n$ with respect to the finite topology, i.e., for any $\Bbbk$-linearly independent $u_1, \ldots, u_m \in M_n$ and for any $v_1, \ldots, v_m \in M_n$, $m \geq 1$, there exists $\alpha \in W_n$ such that $\alpha u_i = v_i$, $i = 1, \ldots, m$.*

Proof. It is easy to show that W_n acts on M_n irreducibly, i.e., for any $0 \neq u \in M_n$ and for any $v \in M_n$ there exists $\alpha \in W_n$ such that $\alpha u = v$.

Also, note that W_n is closed under multiplication by any $f \in H$ from the right and from the left:

$$fa(g) = ((1 \otimes f \otimes E)a)(g), \quad a(g)f = L_g fa(g), \quad f \in H,\ g \in G.$$

(This multiplication is just the composition of linear maps in $\mathrm{End}_{\Bbbk} M_n$.)

The following statement is important.

Proposition 20.5. *Let S be a subalgebra of W_n that acts irreducibly on M_n. If $HS, SH \subseteq S$ then $\mathrm{End}_S M_n \subseteq \mathrm{End}_H M_n$.*

Proof. Denote

$$D = \mathrm{End}_S M_n = \{\varphi \in \mathrm{End}\, M_n \mid \varphi\alpha = \alpha\varphi \text{ for all } \alpha \in S\}.$$

This is a division algebra by the Schur's Lemma. Consider an arbitrary $0 \neq \alpha \in S$, $f \in H$, $\varphi \in D$. Since $f\alpha, \alpha f \in S$, we have $(\varphi f)\alpha = \varphi(f\alpha) = (f\alpha)\varphi = (f\varphi)\alpha$. Therefore,

$$(\varphi f - f\varphi)\alpha = 0. \tag{20.12}$$

Since $[[\varphi, f], \beta] = [[\varphi, \beta], f] + [\varphi, [f, \beta]] = 0$ for any $\beta \in S$, the map $[\varphi, f]$ belongs to D. Hence, (20.12) implies $[\varphi, f] = 0$, so $\varphi \in \operatorname{End}_H M_n$.

Lemma 20.3. *If D is a division subalgebra of* $\operatorname{End}_H M_n$ *that contains* id_{M_n} *then* $D = \Bbbk$.

Proof. The endomorphism algebra $\operatorname{End}_H M_n$ is isomorphic to the algebra $\mathbb{M}_n(H)$. Suppose $D \subset \mathbb{M}_n(H)$ is a division algebra, and let x be an element of $D \setminus \Bbbk$.

Since D contains λE, $\lambda \in \Bbbk$, E is the identity matrix, the matrix $x + \lambda E$ is invertible for any $\lambda \in \Bbbk$. Denote $f(z, \lambda) = \det(x + \lambda E)(z)$. It is clear that $f(z, \lambda) = \lambda^n + \cdots + \det x$, $(\det x)(z) \neq 0$ at any point $z \in G$. Hence, for any $z \in G$ there exists $\lambda \in \Bbbk$ such that $f(z, \lambda) = 0$ that is impossible since $x + \lambda E$ is invertible.

It remains to note that by the Jacobson's Density Theorem (see, e.g., [5]) W_n is dense in $\operatorname{End}_D M_n$, where $D = \operatorname{End}_{W_n} M_n$. By Proposition 20.5 $D \subset \operatorname{End}_H M_n$, so Lemma 20.3 implies $D = \Bbbk$. This proves Theorem 20.1.

Corollary 20.1. *If $S \subseteq W_n$ is a subalgebra that acts irreducibly on M_n and $HS, SH \subset S$, then S is a dense (over $\Bbbk$) subalgebra of* $\operatorname{End} M_n$.

If C is a (G^{op})-conformal subalgebra of Cend_n then $W_n(C) = \operatorname{Span}_{\Bbbk}\{a(g) \mid a \in C, g \in G\}$ is a subalgebra of W_n.

Definition 20.7. A subalgebra $C \subseteq \mathrm{Cend}_n$ is called *irreducible* if there are no $W_n(C)$-invariant H-submodules of M_n except for $\{0\}$ and entire M_n.

The main result of [9] is a description of irreducible subalgebras of Cend_n for $G = \mathbb{A}_1$. The following statement generalizes the crucial observation of [9] to the case of an arbitrary linear algebraic group.

Theorem 20.2. *Let C be an irreducible subalgebra of* Cend_n. *Then $C_1 = (1 \otimes H \otimes E)C$ is a left ideal of* Cend_n, *$C_1 = \mathscr{F}(H \otimes I_0)$, where I_0 is an essential left ideal of* $\mathbb{M}_n(H)$.

Proof. Let $S = W_n(C)$. Note that $S_1 = HS$ is also a subalgebra, that can be expressed as $S_1 = W_n(C_1)$, $C_1 = (1 \otimes H \otimes E)C$. It is clear that C_1 is a conformal subalgebra of Cend_n. Then for any $0 \neq u \in M_n$ and for any $v \in M_n$ there exists $\alpha \in S_1$ such that $\alpha u = v$.

By Corollary 20.1, S_1 is dense in $\operatorname{End} M_n$. Now recall that C_1 is a left S_1-module by Lemma 20.2(2).

Suppose $\{\alpha_k\}_{k \geq 0}$ is a sequence in S_1 converging to $\alpha \in W_n$. Then for any $a \in \mathrm{Cend}_n$ there exists N such that $\alpha_k \cdot a = \alpha \cdot a$ for all $k \geq N$. Hence, C_1 is a left ideal of Cend_n. By Proposition 20.4 $C_1 = \mathscr{F}(H \otimes I_0)$.

Since $\mathbb{M}_n(H)$ is a semiprime ring which is left and right Goldie, a non-essential ideal has a non-zero right annihilator. Hence, I_0 must be essential. Conversely, if I_0 is essential then there exists a regular element $a \in I_0$, thus, $\mathbb{M}_n(H)a$ gives rise to an irreducible conformal subalgebra of Cend_n.

Acknowledgements This work was partially supported by RFBR 05–01–00230. The author gratefully acknowledges the support of the Pierre Deligne fund based on his 2004 Balzan prize in mathematics.

References

1. Bakalov, B., D'Andrea A., Kac, V.G.: Theory of finite pseudoalgebras. Adv. Math. **162**, no 1, 1–140 (2001).
2. Beilinson, A., Drinfeld, V.: Chiral algebras, American Mathematical Society Colloquium Publications **51**. AMS, Providence, RI (2004).
3. Boyallian, C., Kac, V.G., Liberati, J.I.: On the classification of subalgebras of Cend_N and gc_N. J. Algebra **260**, no 1, 32–63 (2003).
4. Frenkel, I., Lepowsky, J., Meurman, A.: Vertex operator algebras and the Monster, Pure and Applied Mathematics **134**. Academic, Boston, (1988).
5. Jacobson, N.: Structure of rings, American Mathematical Society Colloquium Publications **37**. AMS, Providence, RI (1964).
6. Kac, V.G.: The idea of locality. In: Physical applications and mathematical aspects of geometry, groups and algebras, pp. 16–32. World Scientific, Singapore (1997).
7. Kac, V.G.: Vertex algebras for beginners (second edition), University Lecture Series **10**. AMS, Providence, RI (1998).
8. Kac, V.G.: Formal distribution algebras and conformal algebras. In: XIIth International Congress of Mathematical Physics (ICMP 1997) (Brisbane), pp. 80–97. International, Cambrige, MA (1999).
9. Kolesnikov, P.: Associative conformal algebras with finite faithful representation. Adv. Math. **202**, no 2, 602–637 (2006).

Chapter 21
Lie Color and Hom-Lie Algebras of Witt Type and Their Central Extensions

Gunnar Sigurdsson and Sergei Silvestrov

Abstract In this article, two classes of Γ-graded Witt-type algebras, Lie color and hom-Lie algebras of Witt type, are considered. These algebras can be seen as generalizations of Lie algebras of Witt type. One-dimensional central extensions of Lie color and hom-Lie algebras of Witt type are investigated.

21.1 Introduction

Since 1980s, when research on quantum deformations (or q-deformations) of Lie algebras began a period of rapid expansion in connection with the introduction of quantum groups motivated by applications to the Yang–Baxter equation and quantum inverse scattering methods, several other versions of (q-) deformed Lie algebras have appeared, especially in physical contexts such as string theory. The main objects for these deformations were infinite-dimensional algebras, primarily the Heisenberg algebras (oscillator algebras) and the Virasoro algebra, see [1–4] and the references therein.

An important question about these algebras is whether they obey some deformed (twisted) versions of skew-symmetry or Jacobi identity. At the same time, in a well-known direct generalization of Lie algebras and Lie superalgebras to general commutative grading groups, the class of Lie color algebras, generalized skew-symmetry and Jacobi type identities, graded by a commutative group and twisted by a scalar bicharacter, hold (see [8]).

G. Sigurdsson
Department of Theoretical Physics, School of Engineering Sciences, Royal Institute of Technology, AlbaNova University Center, 106 91 Stockholm, Sweden
e-mail: gunnsi@kth.se

S. Silvestrov
Centre for Mathematical Sciences, Lund University, Box 118, 221 00 Lund, Sweden
e-mail: ssilvest@maths.lth.se

Throughout we let $\mathbb{F}$ be a field of characteristic zero, $\mathbb{F}^* = \mathbb{F}\setminus\{0\}$ and $(\Gamma,+)$ an abelian group. Recall that a *Lie color algebra* or *Γ-graded ε-Lie algebra* (see [10, 11]) is a Γ-graded linear space L with a bilinear multiplication $\langle\cdot,\cdot\rangle$ satisfying:

- $\langle x,y\rangle = -\varepsilon(\gamma_x,\gamma_y)\langle y,x\rangle$
- $\varepsilon(\gamma_z,\gamma_x)\langle x,\langle y,z\rangle\rangle + \varepsilon(\gamma_x,\gamma_y)\langle y,\langle z,x\rangle\rangle + \varepsilon(\gamma_y,\gamma_z)\langle z,\langle x,y\rangle\rangle = 0$

for $x \in L_{\gamma_x}, y \in L_{\gamma_y}$ and $z \in L_{\gamma_z}$ where $L = \bigoplus_{\gamma\in\Gamma} L_\gamma$. Here γ_x, γ_y and γ_z are the graded degrees of x, y and z respectively . The map $\varepsilon : \Gamma\times\Gamma \to \mathbb{F}^*$, called a commutation factor, is a bicharacter on Γ with a symmetry property $\varepsilon(\gamma_x,\gamma_y)\varepsilon(\gamma_y,\gamma_x) = 1$. Notice that Lie color algebras include both Γ-graded Lie algebras (with $\varepsilon(\gamma_x,\gamma_y) = 1$ for all $\gamma_x,\gamma_y \in \Gamma$) and Lie superalgebras (with $\Gamma = \mathbb{Z}_2$ and $\varepsilon(\gamma_x,\gamma_y) = (-1)^{\gamma_x\gamma_y}$). Lie color algebras are examples of Γ-graded quasi-Lie algebras (see [7]).

A remarkable and not yet fully exploited and understood feature of Lie superalgebras and Lie color algebras is that they often appear simultaneously with usual Lie algebras in various deformation families of algebras for initial, final or other important special values of the deformation parameters.

Let $s : \Gamma \to \mathbb{F}$ be a function defined over Γ and consider the Γ-graded vector space $L = \bigoplus_{g\in\Gamma} \mathbb{F}\mathbf{e}_g$ over $\mathbb{F}$ with a basis $\{\mathbf{e}_g \,|\, g \in \Gamma\}$. Assume that $\varepsilon : \Gamma\times\Gamma \to \mathbb{F}^*$ is a commutation factor on Γ. The linear space L endowed with the bracket product

$$[\mathbf{e}_g,\mathbf{e}_h] = (s(h) - \varepsilon(g,h)s(g))\,\mathbf{e}_{g+h} \tag{21.1}$$

can be shown to be a Lie color algebra $L(\Gamma,\varepsilon,s)$ under certain conditions on the grading group Γ and the mappings ε and s. $L(\Gamma,\varepsilon,s)$ is then said to be a *Lie color algebra of Witt type*. For further details, we refer to [13]. The special case given by $\varepsilon(g,h) = 1$ for all $g,h \in \Gamma$ and with s satisfying the condition

$$(s(g) - s(h))(s(g+h) - s(g) - s(h) + s(0)) = 0 \;\text{ for all }\; g,h \in \Gamma,$$

is a Γ-graded Lie algebra and called a *Lie algebra of Witt type* (see [12] for instance).

The main purpose of this article is to study Lie color algebra central extensions of Witt-type Lie color algebras by a one-dimensional central part $\mathbb{F}\mathbf{c}$ and central extensions of Γ-graded hom-Lie algebras of Witt type.

21.2 Central Extensions of Witt-Type Lie Color Algebras

Throughout this section, we assume that the additively written gradation group $(\Gamma,+)$ is 2-torsion free. Moreover, the commutation factor ε is supposed to be *non-degenerate*, which means that the set $\{g \in \Gamma \mid \varepsilon(g,h) = 1 \text{ for all } h \in \Gamma\} = \{0\}$. Notice that the mapping $\varepsilon(\cdot,\cdot) : \Gamma \to \mathbb{F}^*$, $g \mapsto \varepsilon(g,g)$ is a homomorphism of the group Γ into the multiplicative group $\{-1,1\}$. A group element $g \in \Gamma$ is called *even* or *odd* if $\varepsilon(g,g)$ is equal to 1 or -1, respectively. For a given commutation factor ε on Γ, it is common to define the subsets

$$\Gamma_+ = \{g \in \Gamma \mid \varepsilon(g,g) = 1\}, \quad \Gamma_- = \{g \in \Gamma \mid \varepsilon(g,g) = -1\}.$$

Clearly there are two possibilities, either one has $\Gamma_+ = \Gamma$, or Γ_+ is a subgroup of index 2 in Γ. In the following, $\circlearrowleft_{g,h,k}$ will be used to concisely denote summation over the cyclic permutation on g,h,k.

On the graded linear space $L = \bigoplus_{g\in\Gamma} \mathbb{F}\mathbf{e}_g$ with one-dimensional homogeneous components $\mathbb{F}\mathbf{e}_g$, we introduce a bracket product given by (21.1). Putting $S(g,h) = (s(h) - \varepsilon(g,h)s(g))$, we have

$$[\mathbf{e}_g, \mathbf{e}_h] = (s(h) - \varepsilon(g,h)s(g))\,\mathbf{e}_{g+h} = S(g,h)\,\mathbf{e}_{g+h}.$$

Under the given restrictions on Γ and ε, it can be proved (see [13]) that L is a Lie color algebra if and only if the function $s : \Gamma \to \mathbb{F}$ satisfies the following conditions

$$s(g) = 0 \text{ or } s(g) = s(0), \tag{21.2}$$
$$(s(h) - \varepsilon(g,h)s(g))(s(g+h) - s(g) - s(h) + s(0)) = 0, \tag{21.3}$$

for all $g,h \in \Gamma$. We then obtain a Lie color algebra of Witt type denoted as $L(\Gamma,\varepsilon,s)$. Note that if $s(0) \neq 0$ (the non-abelian case) then by (21.2) the image $\mathrm{Im}(s) = \{0, s(0)\}$ of the map $s : \Gamma \to \mathbb{F}$ consists of two elements only. Hence s can be expressed as $s(g) = s(0)\chi_{\mathrm{supp(s)}}(g)$ for any $g \in \Gamma$, the support of s being defined as $\mathrm{supp}(s) = \{g \in \Gamma \mid s(g) = s(0)\}$ and where χ_A denotes the characteristic function of a subset $A \subseteq \Gamma$.

The following simple lemma is easy to prove by methods of basic group theory.

Lemma 21.1. *Let G be a subset of an abelian group $(\Gamma,+)$. Then G is a subgroup of Γ if and only if:*

(1) The identity element $0 \in G$
(2) $g \in G,\ h \notin G \Rightarrow g+h \notin G$

By the conditions (21.2) and (21.3) together with Lemma 21.1, it immediately follows that either $\mathrm{supp}(s)$ is empty (in case $s(0) = 0$) or $\mathrm{supp}(s)$ is a subgroup of the gradation group Γ. For more details on these matters we refer to [13].

We want to investigate Lie color algebra extensions $\hat{L}(\Gamma,\varepsilon,s)$ of $L(\Gamma,\varepsilon,s)$ by a one-dimensional center $\mathbb{F}\mathbf{c}$. The center can be considered as the one-dimensional trivial Lie color algebra with singleton basis $\{\mathbf{c}\}$ (the charge). We say that a Lie color algebra $\hat{L}(\Gamma,\varepsilon,s)$ is a one-dimensional central extension of $L(\Gamma,\varepsilon,s)$ if $\hat{L}(\Gamma,\varepsilon,s)$ can be written as a direct sum $L(\Gamma,\varepsilon,s) \oplus \mathbb{F}\mathbf{c}$ of $L(\Gamma,\varepsilon,s)$ and $\mathbb{F}\mathbf{c}$ as vector spaces, where the Lie color bracket $\langle\cdot,\cdot\rangle$ in $\hat{L}(\Gamma,\varepsilon,s)$ is given by

$$\langle \mathbf{e}_g, \mathbf{e}_h \rangle = [\mathbf{e}_g, \mathbf{e}_h] + \varphi(\mathbf{e}_g, \mathbf{e}_h)\,\mathbf{c}, \tag{21.4}$$
$$\langle \mathbf{e}_g, \mathbf{c} \rangle = 0 \tag{21.5}$$

for all $g,h \in \Gamma$. Here $[\cdot,\cdot]$ is the Lie color bracket (21.1) and $\varphi : L \times L \to \mathbb{F}$ is a bilinear form (the 2-cocycle) on $L(\Gamma,\varepsilon,s)$ such that the bracket (21.4) satisfies the axioms of a Lie color bracket. The charge $\mathbf{c}$ is assumed to be homogeneous of degree zero. For further reference see [5,6,9].

Proposition 21.1. *Let $L(\Gamma,\varepsilon,s)$ be a Lie color algebra of Witt type over a field $\mathbb{F}$ of characteristic zero. Assume that the gradation group Γ is 2-torsion free, and the commutation factor ε is non-degenerate. Then $L(\Gamma,\varepsilon,s)$ admits a one-dimensional central extension $\hat{L}(\Gamma,\varepsilon,s)=L(\Gamma,\varepsilon,s)\oplus\mathbb{F}\mathbf{c}$ defined by (21.4), (21.5) if and only if φ is given by*

$$\varphi(\mathbf{e}_g,\mathbf{e}_h)=\delta_{g,-h}\kappa(h), \qquad g,h\in\Gamma,$$

where κ is an $\mathbb{F}$-valued function defined on $\Gamma=\Gamma_+\cup\Gamma_-$ such that:

(1) $\kappa(-g)=-\kappa(g)$ *if* $g\in\Gamma_+$, $\kappa(-g)=\kappa(g)$ *if* $g\in\Gamma_-$
(2) $\kappa(g)=\varepsilon(g,h)\kappa(h)$ *for all* $g,h\notin\operatorname{supp}(s)$ *such that* $g+h\in\operatorname{supp}(s)$
(3) $\circlearrowleft_{g,h,k}\varepsilon(k,g)(1-\varepsilon(h,k))\kappa(g)=0$ *for all* $g,h,k\in supp(s)$ *with* $g+h+k=0$.

Proof. Given a basis $\{\mathbf{e}_g\}_{g\in\Gamma}$ of the linear space L, there is a well-defined map $b:\Gamma\times\Gamma\to\mathbb{F}$, if we set $b(g,h):=\varphi(\mathbf{e}_g,\mathbf{e}_h)$ for all $g,h\in\Gamma$. Then we have

$$\langle\mathbf{e}_g,\mathbf{e}_h\rangle=[\mathbf{e}_g,\mathbf{e}_h]+b(g,h)\,\mathbf{c}=S(g,h)\,\mathbf{e}_{g+h}+b(g,h)\,\mathbf{c}. \tag{21.6}$$

Applying (21.6), the double bracket $\langle\mathbf{e}_g,\langle\mathbf{e}_h,\mathbf{e}_k\rangle\rangle$ can be expressed as

$$\begin{aligned}\langle\mathbf{e}_g,\langle\mathbf{e}_h,\mathbf{e}_k\rangle\rangle&=\langle\mathbf{e}_g,S(h,k)\mathbf{e}_{h+k}+b(h,k)\,\mathbf{c}\rangle\\&=S(h,k)S(g,h+k)\mathbf{e}_{g+h+k}+S(h,k)b(g,h+k)\,\mathbf{c}\end{aligned} \tag{21.7}$$

for all $g,h,k\in\Gamma$. Since $L(\Gamma,\varepsilon,s)$ is a Lie color algebra, the ε-Jacobi identity is satisfied by the bracket $[\cdot,\cdot]$ and hence by (21.7)

$$\circlearrowleft_{g,h,k}\varepsilon(k,g)S(h,k)S(g,h+k)\mathbf{e}_{g+h+k}=0.$$

Furthermore, by the ε-Jacobi identity for $\langle\cdot,\cdot\rangle$, $\circlearrowleft_{g,h,k}\varepsilon(k,g)\langle\mathbf{e}_g,\langle\mathbf{e}_h,\mathbf{e}_k\rangle\rangle=0$, this yields

$$\circlearrowleft_{g,h,k}\varepsilon(k,g)S(h,k)b(g,h+k)\mathbf{c}=0,$$

or equivalently

$$\circlearrowleft_{g,h,k}\varepsilon(k,g)(s(k)-\varepsilon(h,k)s(h))b(g,h+k)=0. \tag{21.8}$$

Write

$$b(g,h):=\delta_{g,-h}\kappa(h). \tag{21.9}$$

This means that

$$\langle\mathbf{e}_g,\mathbf{e}_{-g}\rangle=S(g,-g)\,\mathbf{e}_0+\kappa(-g)\,\mathbf{c}$$

for all $g\in\Gamma$ so the grading condition is satisfied.

Note that the ε-skew-symmetry condition implies

$$\varepsilon(g,h)b(h,g)=\varepsilon(g,h)\delta_{h,-g}\kappa(g)=-\delta_{g,-h}\kappa(h).$$

For $g+h\neq0$ this is trivially satisfied. Taking $h=-g$ yields $\kappa(-g)=-\varepsilon(g,-g)\kappa(g)$. Recall that $\varepsilon(g,g)\varepsilon(g,-g)=\varepsilon(g,0)=1$. In the case where $\Gamma_+=\Gamma$ then also

$\varepsilon(g,-g)=1$ and the mapping κ is an odd function on Γ, $\kappa(-g)=-\kappa(g)$ for all $g\in\Gamma$. On the other hand, if $\Gamma_-\neq\varnothing$ it follows that $\kappa(-g)=\kappa(g)$ for all $g\in\Gamma_-$, while we still have $\kappa(-g)=-\kappa(g)$ on Γ_+. Since $\varepsilon(0,0)=1$, it is always true that $\kappa(0)=1$, if we assume that the characteristic of the field $\mathbb{F}$ is different from 2.

Going back to (21.8), this can be rewritten as

$$\circlearrowleft_{g,h,k}\varepsilon(k,g)(s(k)-\varepsilon(h,k)s(h))\delta_{g,-(h+k)}\kappa(h+k)=0. \tag{21.10}$$

The identity is trivially satisfied if $g+h+k\neq 0$. For all $g,h,k\in\Gamma$ such that $g+h+k=0$ we obtain

$$\circlearrowleft_{g,h,k}\varepsilon(k,-h-k)(s(k)-\varepsilon(h,k)s(h))\kappa(h+k)=0.$$

Recalling that ε is a bihomomorphism on $\Gamma\times\Gamma$, we have

$$\circlearrowleft_{g,h,k}\varepsilon(k,-h)\varepsilon(k,-k)(s(k)-\varepsilon(h,k)s(h))\kappa(h+k)=0. \tag{21.11}$$

Let $s(0)\neq 0$ so that $\text{supp}(s)$ is non-empty. For all $g,h,k\in\Gamma$ such that $g,h\notin\text{supp}(s)$, $k\in\text{supp}(s)$ and $g+h+k=0$ it follows that (21.11) reduces to

$$\varepsilon(k,-k)\varepsilon(k,-h)\kappa(h+k)-\varepsilon(g,-g)\varepsilon(g,-k)\varepsilon(k,g)\kappa(k+g)=0. \tag{21.12}$$

This equality holds if and only if

$$\kappa(-g)-\varepsilon(g,h)\kappa(-h)=0,$$

for all $g,h\notin\text{supp}(s)$ such that $g+h\in\text{supp}(s)$. We notice that $\varepsilon(g,h)=\varepsilon(-g,-h)$ for all $g,h\in\Gamma$. Moreover, $g,h\notin\text{supp}(s)$ with $g+h\in\text{supp}(s)$ if and only if $-g,-h\notin\text{supp}(s)$ and $-(g+h)\in\text{supp}(s)$. The conclusion is that (21.12) holds iff

$$\kappa(g)-\varepsilon(g,h)\kappa(h)=0,$$

for all $g,h\notin\text{supp}(s)$ such that $g+h\in\text{supp}(s)$, i.e. condition (2). Note that in this case the factor $(s(h)-\varepsilon(g,h)s(g))$ in (21.3) vanishes.

In the next step, we consider the case where more than one of the elements g,h,k are contained in $\text{supp}(s)$. Recall that we assume they satisfy the constraint $g+h+k=0$. Since $\text{supp}(s)$ is a subgroup of Γ, this means in particular that if two of the elements, say h and k, are inside $\text{supp}(s)$ then also the third, $g=-h-k$ must belong to $\text{supp}(s)$. So it remains to consider the case where $g,h,k\in\text{supp}(s)$, i.e. $s(g)=s(h)=s(k)=s(0)\neq 0$. Then obviously also $s(g+h)=s(0)$ so (21.3) is truly satisfied in this case. The identity (21.10) now yields

$$\circlearrowleft_{g,h,k}\varepsilon(k,g)(1-\varepsilon(h,k))\kappa(-g)=0.$$

Now by the symmetry property, $\varepsilon(g,h)=\varepsilon(-g,-h)$ for all $g,h\in\Gamma$, this shows that $\circlearrowleft_{g,h,k}\varepsilon(k,g)(1-\varepsilon(h,k))\kappa(g)=0$ for all $g,h,k\in\Gamma$ such that $g+h+k=0$.

Remark 21.1. By condition (2) of Proposition 21.1 we have $\kappa(g) = \varepsilon(g,h)\kappa(h)$ for all $g,h \notin \mathrm{supp}(s)$ such that $g+h \in \mathrm{supp}(s)$. Now if $g \notin \mathrm{supp}(s)$, then also $-g \notin \mathrm{supp}(s)$ and $g+(-g) = 0 \in \mathrm{supp}(s)$, since $\mathrm{supp}(s)$ is a subgroup of Γ. This means that $\kappa(g) = \varepsilon(g,-g)\kappa(-g)$. As $\varepsilon(g,g)\varepsilon(g,-g) = 1$, we have $\kappa(-g) = \varepsilon(g,g)\kappa(g)$. If g is even, then $\varepsilon(g,g) = 1$ and $\kappa(g)$ is odd by condition (1) implying that $\kappa(g) = 0$ for all $g \notin \mathrm{supp}(s)$, i.e. $\mathrm{supp}(\kappa) \subseteq \mathrm{supp}(s)$.

21.3 Central Extensions of Γ-Graded Hom-Lie Algebras of Witt Type

Definition 21.1. A *Γ-graded hom-Lie algebra* is a triple $(L, \langle\cdot,\cdot\rangle_L, \alpha)$ where

- $L = \bigoplus_{\gamma\in\Gamma} L_\gamma$ is a Γ-graded linear space over $\mathbb{F}$
- $\langle\cdot,\cdot\rangle_L : L\times L \to L$ is a bilinear map called a product or bracket in L
- $\alpha : L \to L$ is a linear map mapping $\cup_{\gamma\in\Gamma} L_\gamma$ to $\cup_{\gamma\in\Gamma} L_\gamma$

such that the following conditions hold:

- (Γ-Grading axiom) $\langle L_{\gamma_1}, L_{\gamma_2}\rangle_L \subseteq L_{\gamma_1+\gamma_2}$, for all $\gamma_1,\gamma_2 \in \Gamma$
- (Skew-symmetry) $\langle x,y\rangle_L = -\langle y,x\rangle_L$, for all $(x,y) \in \cup_{\gamma\in\Gamma} L_\gamma \times \cup_{\gamma\in\Gamma} L_\gamma$
- (α-Twisted Jacobi identity) $\circlearrowleft_{x,y,z} \langle\alpha(x), \langle y,z\rangle_L\rangle_L = 0$, whenever $x \in L_{\gamma_x}, y \in L_{\gamma_y}$ and $z \in L_{\gamma_z}$ where $\gamma_x,\gamma_y,\gamma_z \in \Gamma$.

We form the Γ-graded $\mathbb{F}$-vector space $W = \bigoplus_{\gamma\in\Gamma} \mathbb{F}\mathbf{e}_\gamma$, where $\mathbb{F}$ is our ground field and $\{\mathbf{e}_\gamma \mid \gamma\in\Gamma\}$ is a basis of W. Let $s : \Gamma \to \mathbb{F}$ be a function on Γ and equip W with the bracket product

$$[\mathbf{e}_g, \mathbf{e}_h] = (s(h) - s(g))\,\mathbf{e}_{g+h} = S(g,h)\,\mathbf{e}_{g+h}. \tag{21.13}$$

Without loss of generality, we can set $s(0) = 0$. When $(W, [\cdot,\cdot], \alpha)$ is a hom-Lie algebra, then we call it a hom-Lie algebra of Witt type. Such a Γ-graded hom-Lie algebra will be denoted as $W(\Gamma,\alpha,s)$. Obviously, the skew-symmetry condition $[\mathbf{e}_g, \mathbf{e}_h] = -[\mathbf{e}_h, \mathbf{e}_g]$ is automatically satisfied since by construction $S(g,h) = -S(h,g)$ for all $g,h \in \Gamma$. The α-twisted Jacobi identity yields

$$\circlearrowleft_{g,h,k} [\alpha(\mathbf{e}_g), [\mathbf{e}_h, \mathbf{e}_k]] = \circlearrowleft_{g,h,k} S(h,k)[\alpha(\mathbf{e}_g), \mathbf{e}_{h+k}] = 0 \tag{21.14}$$

for all $g,h,k \in \Gamma$.

We know that each homogeneous subspace $W_\gamma = \mathbb{F}\mathbf{e}_\gamma$ is one-dimensional. Given that $\alpha : W \to W$ is a linear map, mapping $\cup_{\gamma\in\Gamma} W_\gamma$ to $\cup_{\gamma\in\Gamma} W_\gamma$, it follows that there exists a function $p : \Gamma \to \Gamma$ such that $\alpha(\mathbf{e}_\gamma) = \alpha_\gamma^{p(\gamma)} \mathbf{e}_{p(\gamma)}$, where $\alpha_\gamma^{p(\gamma)} \in \mathbb{F}$, for every $\gamma \in \Gamma$. By (21.14) we thus have

$$\circlearrowleft_{g,h,k} \alpha_g^{p(g)} S(h,k) S(p(g), h+k) \mathbf{e}_{p(g)+h+k} = 0 \tag{21.15}$$

for all $g,h,k \in \Gamma$.

A special case of interest is when the linear map α is homogeneous of degree t, where t is a fixed element of the group Γ. Then we have $p(\gamma)=\gamma+t$ for every $\gamma\in\Gamma$ and the identity (21.15) can be written

$$\circlearrowleft_{g,h,k}\, \alpha_g^{g+t} S(h,k)S(g+t,h+k)\mathbf{e}_{g+t+h+k}=0.$$

This holds if and only if

$$\circlearrowleft_{g,h,k}\, \alpha_g^{g+t}(s(k)-s(g))(s(h+k)-s(g+t))=0 \tag{21.16}$$

for all $g,h,k\in\Gamma$.

We look for hom-Lie algebra extensions $\hat{W}(\Gamma,\alpha,s)$ of $W(\Gamma,\alpha,s)$ by a one-dimensional center $\mathbb{F}\mathbf{c}$. A hom-Lie algebra $\hat{W}(\Gamma,\alpha,s)$ is said to be a one-dimensional central extension of $W(\Gamma,\alpha,s)$ if $\hat{W}(\Gamma,\alpha,s)$ can be written as a direct sum of $W(\Gamma,\alpha,s)$ and $\mathbb{F}\mathbf{c}$ as vector spaces, $W(\Gamma,\alpha,s)\oplus\mathbb{F}\mathbf{c}$, where the hom-Lie bracket $\langle\cdot,\cdot\rangle$ in $\hat{W}(\Gamma,\alpha,s)$ is given by

$$\langle\mathbf{e}_g,\mathbf{e}_h\rangle=[\mathbf{e}_g,\mathbf{e}_h]+\psi(\mathbf{e}_g,\mathbf{e}_h)\,\mathbf{c}, \tag{21.17}$$
$$\langle\mathbf{e}_g,\mathbf{c}\rangle=0 \tag{21.18}$$

for all $g,h\in\Gamma$. By $[\cdot,\cdot]$ we mean the hom-Lie bracket (21.13) and $\psi: L\times L\to\mathbb{F}$ denotes a bilinear form (the 2-cocycle) on $W(\Gamma,\alpha,s)$ such that the bracket (21.17) satisfies the axioms of a hom-Lie bracket. The charge $\mathbf{c}$ is assumed to be homogeneous of degree zero.

Let $\{\mathbf{e}_g\}_{g\in\Gamma}$ be a basis of the linear space W and define a map $d:\Gamma\times\Gamma\to\mathbb{F}$, by setting $d(g,h):=\psi(\mathbf{e}_g,\mathbf{e}_h)$ for all $g,h\in\Gamma$. Then we have by (21.17) and (21.13)

$$\langle\mathbf{e}_g,\mathbf{e}_h\rangle=[\mathbf{e}_g,\mathbf{e}_h]+d(g,h)\,\mathbf{c}=S(g,h)\,\mathbf{e}_{g+h}+d(g,h)\,\mathbf{c}. \tag{21.19}$$

Applying (21.19), the double bracket $\langle\mathbf{e}_{p(g)},\langle\mathbf{e}_h,\mathbf{e}_k\rangle\rangle$ can be expressed as

$$\begin{aligned}\langle\mathbf{e}_{p(g)},\langle\mathbf{e}_h,\mathbf{e}_k\rangle\rangle&=\langle\mathbf{e}_{p(g)},S(h,k)\mathbf{e}_{h+k}+d(h,k)\,\mathbf{c}\rangle\\&=S(h,k)S(p(g),h+k)\mathbf{e}_{p(g)+h+k}+S(h,k)d(p(g),h+k)\,\mathbf{c}\end{aligned} \tag{21.20}$$

for all $g,h,k\in\Gamma$.

Knowing that $W(\Gamma,\alpha,s)$ is a Γ-graded hom-Lie algebra, the α-twisted Jacobi identity is satisfied by the bracket $[\cdot,\cdot]$ and hence, for all $g,h,k\in\Gamma$,

$$\circlearrowleft_{g,h,k}\, \alpha_g^{p(g)} S(h,k)S(p(g),h+k)\mathbf{e}_{p(g)+h+k}=0.$$

By the α-twisted Jacobi identity for $\langle\cdot,\cdot\rangle$, $\circlearrowleft_{g,h,k}\, \alpha_g^{p(g)}\langle\mathbf{e}_{p(g)},\langle\mathbf{e}_h,\mathbf{e}_k\rangle\rangle=0$, we thus obtain

$$\circlearrowleft_{g,h,k}\, \alpha_g^{p(g)} S(h,k)d(p(g),h+k)\mathbf{c}=0,$$

or expressed differently

$$\circlearrowleft_{g,h,k}\, \alpha_g^{p(g)}(s(k)-s(h))d(p(g),h+k)=0. \tag{21.21}$$

We write
$$d(g,h) := \delta_{g,-h}\omega(h).$$

This means that
$$\langle \mathbf{e}_g, \mathbf{e}_{-g} \rangle = S(g,-g)\,\mathbf{e}_0 + \omega(-g)\,\mathbf{c}$$

for all $g \in \Gamma$, so clearly the grading condition is fulfilled. We notice that the skew-symmetry condition $\langle \mathbf{e}_g, \mathbf{e}_h \rangle = -\langle \mathbf{e}_h, \mathbf{e}_g \rangle$ implies $d(h,g) = -d(g,h)$, i.e.
$$\delta_{h,-g}\omega(g) = -\delta_{g,-h}\omega(h).$$

For $g+h \neq 0$ this is trivially satisfied. Taking $h = -g$ this yields $\omega(g) = -\omega(-g)$ so $\omega : \Gamma \to \mathbb{F}$ has to be an odd function on Γ. Going back to (21.21), it can be rewritten as
$$\circlearrowleft_{g,h,k} \alpha_g^{p(g)}(s(k)-s(h))\delta_{p(g),-(h+k)}\omega(h+k) = 0.$$

The identity is trivially satisfied if $p(g)+h+k \neq 0$. For all $g,h,k \in \Gamma$ such that $p(g)+h+k = 0$ we thus obtain
$$\circlearrowleft_{g,h,k} \alpha_g^{p(g)}(s(-p(g)-h)-s(h))\omega(p(g)) = 0.$$

We summarize our results as the following proposition.

Proposition 21.2. *Let $W(\Gamma,\alpha,s)$ be a Γ-graded hom-Lie algebra of Witt type over a field $\mathbb{F}$ of zero characteristic. Then $W(\Gamma,\alpha,s)$ admits a one-dimensional central extension $\hat{W}(\Gamma,\alpha,s) = W(\Gamma,\alpha,s) \oplus \mathbb{F}\mathbf{c}$ defined by (21.17), (21.18) if and only if ψ is given by*
$$\psi(\mathbf{e}_g, \mathbf{e}_h) = \delta_{g,-h}\omega(h), \qquad g,h \in \Gamma,$$

where ω is an $\mathbb{F}$-valued function on Γ satisfying

(1) $\omega(-g) = -\omega(g)$ for all $g \in \Gamma$

(2) $\circlearrowleft_{g,h,k} \alpha_g^{p(g)}(s(k)-s(h))\omega(p(g)) = 0$ for all $g,h,k \in \Gamma$ with $p(g)+h+k = 0$,

where $p : \Gamma \to \Gamma$ is a map such that $\alpha(\mathbf{e}_\gamma) = \alpha_\gamma^{p(\gamma)}\mathbf{e}_{p(\gamma)}$ for all $\gamma \in \Gamma$.

Acknowledgements This work was supported by The Crafoord Foundation, The Royal Swedish Academy of Sciences, The Royal Physiographic Society in Lund and The Swedish Foundation for International Cooperation in Research and Higher Education (STINT). The first author is grateful to The Centre for Mathematical Sciences, Lund University for hospitality and support during his visits in Lund.

References

1. Chaichian M., Kulish P.P., Lukierski J.: *q-Deformed Jacobi identity, q-oscillators and q-deformed infinite-dimensional algebras*, Phys. Lett. B, **237** (1990), 401.
2. Curtright T.L., Zachos C.K.: *Deforming maps for quantum algebras*, Phys. Lett. B, **243** (1990), 237.

3. Hartwig J.T., Larsson D., Silvestrov S.D.: *Deformations of Lie algebras using σ-derivations*, J. Algebra **295** (2006), 314.
4. Hellström L., Silvestrov S.D.: *Commuting Elements in q-Deformed Heisenberg Algebras*, World Scientific, Singapore, 2000.
5. Kac V.G., Raina A.K.: *Highest weight representations of infinite-dimensional Lie algebras*, World Scientific, Singapore, 1987, 145 pp.
6. Larsson D., Silvestrov S.D.: *Quasi-hom-Lie algebras, Central Extensions and 2-cocycle-like identities*, J. Algebra **288** (2005), 321.
7. Larsson D., Silvestrov S.D.: *Graded quasi-Lie algebras*, Czech. J. Phys. **55** (2005), 1473.
8. Larsson D., Silvestrov S.D.: *Quasi-Deformations of $\mathfrak{sl}_2(\mathbb{F})$ using twisted derivations*, Comm. Algebra **35**(12) (2007), 4303.
9. Li W., Wilson R.L.: *Central Extensions of Some Algebras*, Proc. Am. Math. Soc. **126** (1998), 2569.
10. Rittenberg V., Wyler D.: *Generalized superalgebras*, Nucl. Phys. B **139** (1978), 189.
11. Scheunert M.: *Generalized Lie algebras*, J. Math. Phys. **20** (1979), 712.
12. Yu R.W.T.: *Algèbre de Lie de type Witt*, Comm. Algebra **25**(5) (1997), 1471.
13. Zhou J.: *Lie color algebras of Witt type*, J. Nanjing Univ. **21** (2004), 219.

Chapter 22
A Note on Quasi-Lie and Hom-Lie Structures of σ-Derivations of $\mathbb{C}[z_1^{\pm 1}, \ldots, z_n^{\pm 1}]$

Lionel Richard and Sergei Silvestrov

Abstract In a previous paper we studied the properties of the bracket defined by Hartwig, Larsson and the second author in (J. Algebra 295, 2006) on σ-derivations of Laurent polynomials in one variable. Here we consider the case of several variables, and emphasize on the question of when this bracket defines a hom-Lie structure rather than a quasi-Lie one.

22.1 Introduction

In [1–4] a new class of algebras called quasi-Lie algebras and its subclasses, quasi-hom-Lie algebras and hom-Lie algebras, have been introduced. An important characteristic feature of those algebras is that they obey some deformed or twisted versions of skew-symmetry and Jacobi identity with respect to some possibly deformed or twisted bilinear bracket multiplication. Quasi-Lie algebras include color Lie algebras, and in particular Lie algebras and Lie superalgebras, as well as various interesting quantum deformations of Lie algebras, in particular of the Heisenberg Lie algebra, oscillator algebras, sl_2 and other finite-dimensional Lie algebras as well as of infinite-dimensional Lie algebras of Witt and Virasoro type applied in physics within the string theory, vertex operator models, quantum scattering, lattice models and other contexts (for more details and precise references see the Introduction of [5] and references therein). Many of these quantum deformations of Lie algebras can be shown to play role of underlying algebraic objects for calculi of

L. Richard
School of Mathematics of the University of Edinburgh and Maxwell Institute for Mathematical Sciences, JCMB – King's Buildings, EH9 3JZ Edinburgh, UK
e-mail: lionel.richard@ed.ac.uk

S. Silvestrov
Centre for Mathematical Sciences, Lund University, Box 118, 221 00 Lund, Sweden
e-mail: Sergei.Silvestrov@math.lth.se

S. Silvestrov et al. (eds.), *Generalized Lie Theory in Mathematics, Physics and Beyond,*

twisted, discretized or deformed derivations and difference type operators and thus in corresponding general non-commutative differential calculi.

When derivations are replaced by twisted derivations, vector fields become replaced by twisted vector fields. It was proved in [1] that under some general assumptions, those are closed under a natural twisted skew-symmetric bracket multiplication satisfying a twisted 6-term Jacobi identity generalizing the usual Lie algebras 3-term Jacobi identity that is recovered when no twisting is present. This result is shown to be instrumental for construction of various examples and classes of quasi-Lie algebras. Both known and new one-parameter and multi-parameter deformations of Witt and Virasoro algebras and other Lie and color Lie algebras has been constructed within this framework in [1–4].

In [5], we gained further insight in the particular class of quasi Lie algebra deformations of the Witt algebra, introduced in [1] via the general twisted bracket construction, and associated with twisted discretization of derivations generalizing the Jackson q-derivatives to the case of twistings by general endomorphisms of Laurent polynomials in one variable. Here we consider the case of several variables, and emphasize on the question of when this bracket defines a hom-Lie structure rather than a more general quasi-Lie one.

22.2 Framework

Recall the following definitions.

Definition 22.1. Let $\mathscr{A}$ be an algebra, and σ an endomorphism of $\mathscr{A}$. A σ-derivation is a linear map D satisfying $D(ab) = \sigma(a)D(b) + D(a)b$ for all $a, b \in \mathscr{A}$. We denote the set of all σ-derivations by $\mathscr{D}_\sigma(\mathscr{A})$.

Example 22.1. It is easy to check that for any $p \in \mathscr{A}$, the k-linear map Δ_p defined by $\Delta_p(a) = pa - \sigma(a)p$ for all $a \in \mathscr{A}$ is a σ-derivation of $\mathscr{A}$. Note that if $\mathscr{A}$ is commutative, then we have $\Delta_p = p(\mathrm{id} - \sigma)$.

Definition 22.2. The map Δ_p defined above is called the inner σ-derivation associated to p. The set of all inner derivations of $\mathscr{A}$ will be denoted $\mathscr{I}nn_\sigma(\mathscr{A})$.

Recall that for a commutative unique factorisation domain (UFD) we have the following.

Theorem 22.1 ([1], Theorems 4 and 6, Example 12). *Let σ be an endomorphism of a UFD $\mathscr{A}$, and $\sigma \neq \mathrm{id}$. Then $\mathscr{D}_\sigma(\mathscr{A})$ is free of rank one as an $\mathscr{A}$-module with generator*

$$\Delta = \frac{\mathrm{id} - \sigma}{g}, \text{with } g = \gcd((\mathrm{id} - \sigma)(\mathscr{A})).$$

Moreover, the σ-derivation Δ satisfies $\Delta \circ \sigma = \delta\sigma \circ \Delta$, with $\delta = \sigma(g)/g \in \mathscr{A}$, and the map

$$[\cdot,\cdot]_\sigma : \mathscr{A}\Delta \times \mathscr{A}\Delta \to \mathscr{A}\Delta$$

defined by setting

$$[a\Delta, b\Delta]_\sigma = (\sigma(a)\Delta)\circ(b\Delta) - (\sigma(b)\Delta)\circ(a\Delta), \quad \textit{for } a,b \in \mathscr{A}, \tag{22.1}$$

is a well-defined $\mathbb{C}$-algebra product on the $\mathbb{C}$-linear space $\mathscr{A}\Delta$, satisfying the following identities for $a,b \in \mathscr{A}$:

$$[a\Delta, b\Delta]_\sigma = \big(\sigma(a)\Delta(b) - \sigma(b)\Delta(a)\big)\Delta, \tag{22.2}$$

$$[a\Delta, b\Delta]_\sigma = -[b\Delta, a\Delta]_\sigma. \tag{22.3}$$

In addition, for $a,b,c \in \mathscr{A}$:

$$\begin{aligned}[\sigma(a)\Delta,[b\Delta,c\Delta]_\sigma]_\sigma + \delta[a\Delta,[b\Delta,c\Delta]_\sigma]_\sigma + \\ +[\sigma(b)\Delta,[c\Delta,a\Delta]_\sigma]_\sigma + \delta[b\Delta,[c\Delta,a\Delta]_\sigma]_\sigma + \\ +[\sigma(c)\Delta,[a\Delta,b\Delta]_\sigma]_\sigma + \delta[c\Delta,[a\Delta,b\Delta]_\sigma]_\sigma = 0.\end{aligned} \tag{22.4}$$

According to [3, Definition 2.1], this defines a *quasi-Lie* structure on $\mathscr{D}_\sigma(\mathscr{A})$. The question we are interested in is: when does this become a hom-Lie structure? That is, when does the 6-term Jacobi-like identity (22.4) become a 3-term identity of the form

$$\circlearrowleft_{a,b,c} [(\mathrm{id}+\zeta)(a\Delta),[b\Delta,c\Delta]_\sigma]_\sigma, \tag{22.5}$$

with ζ an endomorphism of $(\mathscr{D}_\sigma(\mathscr{A}),[.,.]_\sigma)$? We shall mention here the following difficulty. The expression of the bracket depends on the choice of the generator Δ, and should be denoted by $[.,.]_{\sigma,\Delta}$. So the question should rather be: when does there exist a generator Δ such that the 6-terms Jacobi-like identity is of the form (22.5)? We shall keep in mind the following, coming from [1, Remark 6 and Proposition 6].

Remark 22.1. Let $\mathscr{A}$ be a k-algebra which is also a UFD, and g, g' two gcd's of $(1-\sigma)(\mathscr{A})$. Then each one of the two σ-derivations $\Delta = (1-\sigma)/g$ and $\Delta' = (1-\sigma)/g'$ is a generator of $\mathscr{D}_\sigma(\mathscr{A})$. Let $u \in \mathscr{A}$ be such that $g = ug'$. Then u is a unit, and for all $a,b \in \mathscr{A}$ one has $[a,b]_{\sigma,\Delta} = \alpha[a,b]_{\sigma,\Delta'}$, with $\alpha = \sigma(u)/u$ also invertible in $\mathscr{A}$. Moreover, if $\alpha \in \mathbb{C}^*$ then the two (nonassociative) algebra structures defined on $\mathscr{D}_\sigma(\mathscr{A})$ by Δ and Δ' are naturally isomorphic.

In order to compare Δ and Δ', we need to know how the bracket behaves with respect to linearity. It is clear that it is k-linear, and *not* $\mathscr{A}$-linear. Actually one has the following.

Proposition 22.1. *Under the hypothesis of Theorem 22.1, the following assertions are equivalent for an element $\alpha \in \mathscr{A}$:*

1. $\forall a,b \in \mathscr{A}$: $[\alpha a\Delta, b\Delta]_\sigma = \alpha[a\Delta, b\Delta]_\sigma$
2. $\sigma(\alpha) = \alpha$

In other words, the algebra of σ-invariants $\mathscr{A}^\sigma$ is the greatest subalgebra B of $\mathscr{A}$ such that $[.,.]_\sigma$ is B-bilinear.

Proof. The left-hand side of the first equality can be rewritten as follows:

$$\begin{aligned}[\alpha a\Delta, b\Delta]_\sigma &= (\sigma(\alpha a)\Delta(b) - \sigma(b)\Delta(\alpha a))\Delta \\ &= (1/g)(\sigma(\alpha)\sigma(a)(b-\sigma(b)) - \sigma(b)(\alpha a - \sigma(\alpha)\sigma(a)))\Delta \\ &= (1/g)(\sigma(\alpha)\sigma(a)b - \alpha a\sigma(b))\Delta,\end{aligned}$$

and in the same way one shows that the left-hand side is equal to

$$\alpha[a\Delta, b\Delta]_\sigma = (1/g)\alpha(\sigma(a)b - \sigma(b)a)\Delta.$$

The Proposition follows immediately from this. □

One particular case where the bracket appears to be hom-Lie is the following, [1, Example 17]. Assume that the element δ in Theorem 22.1 is invertible and σ-invariant. Then $\overline{\sigma}$ defined by $\overline{\sigma}(a\Delta) = \sigma(a)\Delta$ is a morphism of $(\mathscr{D}_\sigma(\mathscr{A}), [.,.]_\sigma)$, and this algebra is actually a hom-Lie algebra, with homomorphism $\zeta = \delta^{-1}\overline{\sigma}$.

22.3 Sufficient Condition

In view of Theorem 22.1 we need to obtain a greatest common divisor g of $(1-\sigma)(\mathscr{A})$. In the setting of Laurent polynomials in next section, one can use the following easy lemma.

Lemma 22.1. *Let g be an element of $\mathscr{A}$ such that g is a common divisor of $(1-\sigma)(\mathscr{A})$ and $\sigma(g) = \lambda g$ with $\lambda \in \mathbb{C}\setminus\{0,1\}$. Then g is a greatest common divisor of $(1-\sigma)(\mathscr{A})$.*

Let us recall here the following link between the element g of $\mathscr{A}$ and inner σ-derivations.

Proposition 22.2 ([5], Proposition 3.1). *The σ-derivation $a\Delta$ is inner if and only if g divides a. In other words, $\mathscr{I}nn_\sigma(\mathscr{A}) = g\mathscr{A}\Delta$. In particular, Δ itself is inner, and $\mathscr{D}_\sigma(\mathscr{A}) = \mathscr{I}nn_\sigma(\mathscr{A})$, if and only if g is a unit.*

In the latter case, one can choose $\Delta = 1-\sigma$, that is $g = 1$, which leads easily to the fact that if all σ-derivations of $\mathscr{A}$ are inner, then $\mathscr{D}_\sigma(\mathscr{A})$ is a hom-Lie algebra (since $\delta = \sigma(g)/g$).

This condition is certainly too strong. One actually has the following.

Proposition 22.3. *Let $\mathscr{A}$ be a UFD, and σ a non-identity endomorphism of $\mathscr{A}$. With the notations of Theorem 22.1, assume there exists a $\mathbb{C}$-linear map γ such that*

$$\forall a,b \in \mathscr{A} \quad \delta[a\Delta, b\Delta]_\sigma = [\gamma(a)\Delta, b\Delta]_\sigma \tag{22.6}$$

$$\forall a,b \in \mathscr{A} \quad [\gamma(a)\Delta, \gamma(b)\Delta]_\sigma = \gamma(\sigma(a)\Delta(b) - \sigma(b)\Delta(a))\Delta. \tag{22.7}$$

Then $\mathscr{D}_\sigma(\mathscr{A})$ is a hom-Lie algebra.

Proof. Follows directly from (22.4) and (22.2). □

Note that (22.7) is just saying that $a\Delta \mapsto \gamma(a)\Delta$ is an endomorphism of $(\mathscr{D}_\sigma(\mathscr{A}),[.,.]_\sigma)$. Then thanks to (22.6) and (22.3) the left-hand side of (22.7) can be read as $\delta^2[a\Delta,b\Delta]_\sigma$. Assume that $\mathrm{Ann}\Delta = 0$, so that we can omit Δ in the formulas. This leads then to $\delta^2(\sigma(a)\Delta(b) - \Delta(a)\sigma(b)) = \gamma(\sigma(a)\Delta(b) - \sigma(b)\Delta(a))$, so

$$(\gamma - \delta^2)(\sigma(a)\Delta(b) - \sigma(b)\Delta(a)) = 0. \tag{22.8}$$

Note that from the definitions it is clear that γ must be linear. Using $\Delta = (1-\sigma)/g$ we get $\sigma(a)\Delta(g) - \sigma(g)\Delta(a) = \sigma(a) - \delta a$, so replacing b by g in (22.8) gives $(\gamma - \delta^2)(\sigma(a) - \delta a) = 0$. This should be true for all $a \in \mathscr{A}$, and finally:

Proposition 22.4. *Let γ be a $\mathbb{C}$-linear endomorphism of $\mathscr{A}$ satisfying conditions (22.6) and (22.7). Then:*

$$(\gamma - \delta^2)(\sigma - \delta) = 0.$$

Let us see now where (22.6) leads us. First, applying the definition of $[.,.]_\sigma$, we get

$$\delta(\sigma(a)\Delta(b) - \sigma(b)\Delta(a))\Delta = (\sigma(\gamma(a))\Delta(b) - \sigma(b)\Delta(\gamma(a)))\Delta.$$

Then using the fact that $\Delta = (1-\sigma)/g$ and $\mathrm{Ann}(\Delta) = 0$, we get

$$\delta(\sigma(a)b - \sigma(b)a)/g = (\sigma(\gamma(a))b - \sigma(b)\gamma(a))/g,$$

which could be written in the following "functional equation" way

$$(\sigma(\gamma(a)) - \delta\sigma(a))b = \sigma(b)(\gamma(a) - \delta a).$$

For $b = g$ one gets $\sigma(\gamma(a)) - \delta\sigma(a) = \delta(\gamma(a) - \delta a)$, so $(\sigma - \delta)(\gamma(a)) = \delta(\sigma - \delta)(a)$. One sees here that it is enough to assume that δ is such that $\sigma(\delta) = \delta$, and to put $\gamma(a) = \delta a$. This leads to the following:

Question 22.1. Assume that $\delta \in \mathscr{A}$ such that $\Delta \circ \sigma = \delta\sigma \circ \Delta$ satisfies $\sigma(\delta) = \delta$. Is there an element $g \in \mathscr{A}$ such that $\sigma(g) = \delta g$ and g is a gcd of $(1-\sigma)(\mathscr{A})$?

22.4 Laurent Polynomials

Consider the algebra $\mathscr{A} = \mathbb{C}[z_1^{\pm 1},\dots,z_n^{\pm 1}]$. Since $\mathscr{A}$ is a unique factorisation domain, Theorem 22.1 applies.

In one variable $n = 1$, it appears that, up to a unit (i.e. a monomial), the polynomial g is equal to $z - \sigma(z)$ (see [1, Sect. 3.2]). For n variables we have the following.

Lemma 22.2. *Define $g_i = z_i - \sigma(z_i)$ for all $1 \le i \le n$. Then $\gcd(g_1,\dots,g_n)$ is a greatest common divisor of $(1-\sigma)(\mathscr{A})$.*

Proof. Note that by linearity it is enough to show that this gcd divides any monomial. Let g be a gcd of $(1-\sigma)(\mathscr{A})$. For consistency we start the proof for $n = 1$.

Then by definition g divides $\tilde{g} = z - \sigma(z)$. For any $s > 0$, it is well-known that $z - \sigma(z)$ is a factor of $z^s - \sigma(z)^s$. For $s < 0$, just use the fact that $\sigma(z)$ is a unit, and $z^s - \sigma(z)^s = -z^s\sigma(z)^s(z^{-s} - \sigma(z)^{-s})$. Now assume the Lemma is true for $n-1$ variables. Again by definition g divides all of the g_i s, so g divides their gcd. Reciprocally, let h be a common divisor of the g_i s. In particular it is a common divisor of the g_i s for $1 \leq i \leq n$, so by assumption it is a common divisor of $f - \sigma(f)$ for all $f \in \mathbb{C}[z_1^{\pm 1}, \ldots, z_{n-1}^{\pm 1}]$. Now for any monomial $z^a = z_1^{a_1} \ldots z_n^{a_n}$ write $z^a - \sigma(z^a) = (z_1^{a_1} \ldots z_{n-1}^{a_{n-1}} - \sigma(z_1^{a_1} \ldots z_{n-1}^{a_{n-1}}))z_n^{a_n} + \sigma(z_1^{a_1} \ldots z_{n-1}^{a_{n-1}})(z_n^{a_n} - \sigma(z_n^{a_n}))$. It is clear that h divides both terms of the right-hand side, and this ends the proof. □

Remark 22.2. It is clear that the preceding proof works for any finitely generated algebra which is a UFD.

An endomorphism σ of $\mathscr{A}$ is defined by an n-uple $\underline{q} = (q_1, \ldots, q_n) \in (\mathbb{C}^*)^n$ and a matrix $S = (S_{i,j}) \in M_n(\mathbb{Z})$ such that $\sigma(z_i) = q_i z_1^{S_{1,i}} \ldots z_n^{S_{n,i}}$ for all $1 \leq i \leq n$. Note that in order to simplify later notations, the matrix S here is the transpose of the one defined in [1].

Proposition 22.5. *With the preceding notations, assume that 1 is an eigenvalue of S, such that there exists an eigenvector $G = (G_1, \ldots, G_n) \in \mathbb{Z}^n \setminus 0$. Assume moreover that $\prod_{i=1}^n q_i^{G_i} \neq 1$ (in particular G must not be the zero vector). Then $\mathscr{D}_\sigma(\mathscr{A}) = \mathscr{I}nn_\sigma(\mathscr{A})$, and the bracket defined using $g = z^G$ on this space induces a hom-Lie algebra structure.*

Proof. It results from the hypothesis that $(1 - \sigma)(z^G) = (1 - Q)z^G$ with $Q \in \mathbb{C} \setminus \{0, 1\}$. From Lemma 22.1 and Proposition 22.2 we deduce that $\mathscr{D}_\sigma(\mathscr{A}) = \mathscr{I}nn_\sigma(\mathscr{A})$. Now the rest of the statement is [1, Theorem 37]. □

Acknowledgements The first author is supported by EPSRC Grant EP/D034167/1. Both authors are grateful to the Crafoord foundation, the Swedish Foundation for International Cooperation in Research and Higher Education (STINT), the Royal Physiographic Society in Lund, the Swedish Royal Academy of Sciences and the Swedish Research Council for support of this work.

References

1. Hartwig, J.T., Larsson, D., Silvestrov, S.D.: Deformations of Lie algebras using σ-derivations. J. Algebra **295** (2006), 314–361.
2. Larsson, D., Silvestrov, S.D.: Quasi-hom-Lie algebras, Central Extensions and 2-cocycle-like identities. J. Algebra **288** (2005), 321–344
3. Larsson, D., Silvestrov, S.D.: Quasi-Lie algebras. In: Noncommutative Geometry and Representation Theory in Mathematical Physics, pp. 241–248. Contemp. Math. **391** (2005). Am. Math. Soc., Providence, RI.
4. Larsson D., Silvestrov S. D.: Quasi-deformations of $sl_2(\mathbb{F})$ using twisted derivations, Comm. Algebra, **35** (2007), 4303–4318.
5. Richard, L., Silvestrov, S.D.: Quasi-Lie structure of σ-derivations of $\mathbb{C}[t^{\pm 1}]$. J. Algebra **319** (2008), 1285–1304.

Part V
Commutative Subalgebras in Noncommutative Algebras

Chapter 23
Algebraic Dependence of Commuting Elements in Algebras

Sergei Silvestrov, Christian Svensson, and Marcel de Jeu

Abstract The aim of this paper to draw attention to several aspects of the algebraic dependence in algebras. The article starts with discussions of the algebraic dependence problem in commutative algebras. Then the Burchnall–Chaundy construction for proving algebraic dependence and obtaining the corresponding algebraic curves for commuting differential operators in the Heisenberg algebra is reviewed. Next some old and new results on algebraic dependence of commuting q-difference operators and elements in q-deformed Heisenberg algebras are reviewed. The main ideas and essence of two proofs of this are reviewed and compared. One is the algorithmic dimension growth existence proof. The other is the recent proof extending the Burchnall–Chaundy approach from differential operators and the Heisenberg algebra to the q-deformed Heisenberg algebra, showing that the Burchnall–Chaundy eliminant construction indeed provides annihilating curves for commuting elements in the q-deformed Heisenberg algebras for q not a root of unity.

23.1 Introduction

In 1994, one of the authors of the present paper, S. Silvestrov, based on consideration of the previous literature and a series of trial computations, made the following three part conjecture:

S. Silvestrov
Centre for Mathematical Sciences, Lund University, Box 118, 221 00 Lund, Sweden
e-mail: Sergei.Silvestrov@math.lth.se

C. Svensson
Mathematical Institute, Leiden University, P.O. Box 9512, 2300 RA Leiden, The Netherlands
and
Centre for Mathematical Sciences, Lund University, Box 118, SE-22100 Lund, Sweden
e-mail: chriss@math.leidenuniv.nl

M. de Jeu
Mathematical Institute, Leiden University, P.O. Box 9512, 2300 RA Leiden, The Netherlands
e-mail: mdejeu@math.leidenuniv.nl

S. Silvestrov et al. (eds.), *Generalized Lie Theory in Mathematics, Physics and Beyond,*

- The first part of the conjecture stated that the Burchnall–Chaundy type result on algebraic dependence of commuting elements can be proved in greater generality, that is for much more general classes of non-commutative algebras and rings than the Heisenberg algebra and related algebras of differential operators treated by Burchnall and Chaundy and in subsequent literature.
- The second part stated that the Burchnall–Chaundy eliminant construction of annihilating algebraic curves formulated in determinant (resultant) form works well after some appropriate modifications for the most or possibly for all classes of algebras where the Burchnall–Chaundy type result on algebraic dependence of commuting elements can be proved.
- Finally, the third part of the conjecture stated that the construction and the proof of the vanishing of the corresponding determinant algebraic curves on the commuting elements can be performed for all classes of algebras or rings where this fact is true, using only the internal structure and calculations for the elements in the corresponding algebras or rings and the algebraic combinatorial expansion formulas and methods for the corresponding determinants, that is, without any need of passing to operator representations and use of analytic methods as used in the Burchnall–Chaundy type proofs.

This third part of the conjecture remains widely open with no general such proofs available for any classes of algebras and rings, even in the case of the usual Heisenberg algebra and differential operators, and with only a series of examples calculated for the Heisenberg algebra, q-Heisenberg algebra and some more general algebras, all supporting the conjecture. In the first and the second part of the conjecture progress has been made. In [6], the key Burchnall–Chaundy type theorem on algebraic dependence of commuting elements in q-deformed Heisenberg algebras (and thus as a corollary for q-difference operators as operators representing q-deformed Heisenberg algebras) was obtained. The result and the methods have been extended to more general algebras and rings generalizing q-deformed Heisenberg algebras (generalized Weyl structures and graded rings) in [7]. The proof in [6] is totally different from the Burchnall–Chaundy type proof. It is an existence argument based only on the intrinsic properties of the elements and internal structure of q-deformed Heisenberg algebras, thus supporting the first part of the conjecture. It can be used successfully for an algorithmic implementation for computing the corresponding algebraic curves for given commuting elements. However, it does not give any specific information on the structure or properties of such algebraic curves or any general formulae. It is thus important to have a way of describing such algebraic curves by some explicit formulae, as for example those obtained using the Burchnall–Chaundy eliminant construction for the $q = 1$ case, i.e., for the classical Heisenberg algebra. In [11], a step in that direction was made by offering a number of examples all supporting the claim that the eliminant determinant method should work in the general case. However, no general proof for this was provided. The complete proof following the footsteps of the Burchnall–Chaundy approach in the case of q not a root of unity has been recently obtained [8], by showing that the determinant eliminant construction, properly adjusted for the q-deformed Heisenberg algebras, gives annihilating curves for commuting elements in the q-deformed

Heisenberg algebra when q is not a root of unity, thus confirming the second part of the conjecture for these algebras. In our proof we adapt the Burchnall–Chaundy eliminant determinant method of the case $q = 1$ of differential operators to the q-deformed case, after passing to a specific faithful representation of the q-deformed Heisenberg algebra on Laurent series and then performing a detailed analysis of the kernels of arbitrary operators in the image of this representation. While exploring the determinant eliminant construction of the annihilating curves, we also obtain some further information on such curves and some other results on dimensions and bases in the eigenspaces of the q-difference operators in the image of the chosen representation of the q-deformed Heisenberg algebra. In the case of q being a root of unity the algebraic dependence of commuting elements holds only over the center of the q-deformed Heisenberg algebra [6], and it is unknown yet how to modify the eliminant determinant construction to yield annihilating curves for this case.

The present article starts with discussions of the algebraic dependence problem in algebras. Then the Burchnall–Chaundy construction for proving algebraic dependence and obtaining corresponding algebraic curves for commuting differential operators and commuting elements in the Heisenberg algebra is reviewed. Next some old and new results on algebraic dependence of commuting q-difference operators and elements in the q-deformed Heisenberg algebra are discussed. In the final two subsections we review two proofs for algebraic dependence of commuting elements in the q-deformed Heisenberg algebra. The first one is the recent proof from [8] extending the Burchnall–Chaundy approach from differential operators and the Heisenberg algebra to q-difference operators and the q-deformed Heisenberg algebra, showing that the Burchnall–Chaundy eliminant construction indeed provides annihilating curves for q-difference operators and for commuting elements in q-deformed Heisenberg algebras for q not a root of unity. The second one is the algorithmic dimension growth existence proof from [6].

23.2 Description of the Problem: Commuting Elements in an Algebra Are Given, Then Find Curves They Lie on

Any two elements α and β in a field k lie on an algebraic curve of the second degree $F(x,y) = (x-\alpha)(y-\beta) = xy - \alpha y - \beta x + \alpha\beta = 0$. The important feature of this curve is that its coefficients are also elements in the field k. The same holds if the field k is replaced by any commutative k-algebra R, except that then the coefficients in the *annihilating polynomial* F are elements in R, and hence it becomes of interest from the side of building interplay with the algebraic geometry over the field k to determine whether one may find an annihilating polynomial with coefficients from k for any two elements in the commutative algebra R. It is well-known that this is not always possible even in the case of the ordinary commutative polynomial algebras over a field unless some special conditions are imposed on the considered polynomials. The situation is similar of course for rational functions or many other commutative algebras of functions. The ideals of polynomials annihilating subsets in an algebra are also well-known to be fundamental for algebraic geometry, for Gröbner basis analysis in computational algebra and consequently for various applications in Physics and Engineering.

Another appearance of this type of problems worth mentioning in our context comes from Galois theory and number theory in connection to algebraic and transcendental field extensions. If L is a field extension $K \subset L$ of a field K, then using Zorn's lemma one can show that there always exists a subset of L which is maximal algebraically independent over K. Any such subset is called transcendence basis of the field L over the subfield K (or of the extension $K \subset L$). All transcendence bases have the same cardinality called the transcendence degree of a field extension. If $B = \{b_1, \ldots, b_n\}$ is a finite transcendence basis of $K \subset L$, then L is an algebraic extension of the subfield $K(B)$ in L generated by B over the subfield K. This means in particular that for any $l \in L$ the set $\{B, l\}$ is algebraically dependent over K, that is, there exists a polynomial F in $n+1$ indeterminates over K such that $F(b_1, \ldots, b_n, a) = 0$. Investigation of algebraic dependence, transcendence basis and transcendence degree is a highly non-trivial important direction in number theory and theory of fields with striking results and many longstanding open problems. For example, if algebraic numbers $a_1, ..., a_n$ are linearly independent over $\mathbb{Q}$, then $e^{a_1}, ..., e^{a_n}$ are algebraically independent over $\mathbb{Q}$ (Lindemann–Weierstrass theorem, 1880s). Whether e and π are algebraically dependent over $\mathbb{Q}$ or not is unknown, and only relatively recently (1996) a long-standing conjecture on algebraic independence of π and e^{π} was confirmed [13]. Comprehensive overviews and references in this direction can be found in [14].

In view of the above, naturally important problems are to describe, analyze and classify:

(C1) Commutative algebras over K in which any pair of elements is algebraically dependent over K.

(C2) Pairs (A, B) of a commutative algebra A and a subalgebra B over a field K such that any pair of elements in A is algebraically dependent over B.

Problem (C1) is of course a special case of (C2).

In the polynomial algebra $K[x_1, \ldots, x_n]$ generated by n independent commuting indeterminates, for instance the set $\{x_1, \ldots, x_n\}$ as well as any of its non-empty subsets are algebraically independent over K. Thus $K[x_1, \ldots, x_n]$ does not belong to the class of algebras in the problem (C1). The same holds of course for any algebra containing $K[x_1, \ldots, x_n]$. In general algebraic dependence of polynomials happens only under some restrictive conditions on the rank (or vanishing) of their Jacobian and other fine aspects.

This indicates that some principal changes have to be introduced in order to be able to get examples of algebras satisfying (C1) or (C2). It turns out that the range of possibilities expands dramatically within the realm of non-commutative geometry, if the indeterminates (the coordinates) are non-commuting.

In any algebra there are always commutative subalgebras, for instance any subalgebra generated by any single element. If the algebra contains for instance a nonzero nilpotent element ($a^n = 0$ for some $n > 2$), then the subalgebra generated by this element satisfies $(C1)$. Indeed, any two elements $f_1 = ap(a) + p_0$ and $f_2 = aq(a) + q_0$ in this commutative subalgebra are annihilated by $F(s,t) = (q_0 s - p_0 t)^n$, since $F(f_1, f_2) = (q_0 f_1 - p_0 f_2)^n = a^n (q_0 p(a) - p_0 q(a))^n = 0$.

For the non-commutative algebras the problems inspired by $C1$ and $C2$ are to describe, analyze and classify:

(NC1) Algebras over a field K such that in all their commutative subalgebras any pair of elements is algebraically dependent over the field.

(NC2) Pairs (A,B) of an algebra A and a subalgebra B over a field K such that any pair of commuting elements in A is algebraically dependent over B.

Problem (NC1) is of course a special case of (NC2).

From the situation in commutative algebras as described above, one might intuitively expect that finding non-commutative algebras satisfying (NC1) is difficult if not impossible task. However, this perception is not quite correct. One of the purposes of the present article is to illustrate this phenomenon.

23.3 Burchnall–Chaundy Construction for Differential Operators

Among the longest known, constantly studied and used all over Mathematics non-commutative algebras is the so called Heisenberg algebra, also called Weyl algebra or Heisenberg–Weyl algebra. It is defined as an algebra over a field K with two generators A and B and defining relations $AB - BA = I$, or equivalently as $\mathscr{H}_1 = K\langle A,B\rangle / \langle AB - BA - I\rangle$, the quotient algebra of a free algebra on two generators A and B by the two-sided ideal generated by $AB - BA - I$ where I is the unit element.

Its ubiquity is due to the fact that the basic operators of differential calculus $A = D = \frac{d}{dx}$ and $M_x : f(x) \mapsto x f(x)$ satisfy the Heisenberg algebra defining relation $DM_x - M_x D = I$ on polynomials, formal power series, differentiable functions or any linear spaces of functions invariant under these operators due to the Leibniz rule. On any such space, the representation (D, M_x) of the commutation relation $AB - BA = I$ defines a representation of the algebra $\mathscr{H}_1$, called sometimes, especially in Physics, the (Heisenberg) canonical representation. We will use sometimes this term as well, for shortage of presentation.

If K is a field of characteristic zero, then the algebra

$$\mathscr{H}_1 = K\langle A,B\rangle / \langle AB - BA - I\rangle$$

is simple, meaning that it does not contain any two-sided ideals different from zero ideal and the whole algebra. The kernel $\mathrm{Ker}(\pi) = \{a \in A \mid \pi(a) = 0\}$ of any representation π of an algebra is a two-sided ideal in the algebra. Thus any non-zero representation of $\mathscr{H}_1$ is faithful, which is in particular holds also for a canonical representation. Thus the algebra $\mathscr{H}_1$ is actually isomorphic to the ring of differential operators with polynomial coefficients acting for instance on the linear space of all polynomials or on the space of formal power series in a single variable.

In the literature on algebraic dependence of commuting elements in the Heisenberg algebra and its generalizations – a result which is fundamental for the algebro-geometric method of constructing and solving certain important non-linear partial differential equations – one can find several different proofs of this fact, each with its own advantages and disadvantages. The first proof of such result utilizes analytical and operator theoretical methods. It was first discovered by Burchnall and Chaundy [2] in the 1920s. Their articles [2–4] contain also pioneering results in the direction of in-depth connections to algebraic geometry. These fundamental papers were largely forgotten for almost fifty years when the main results and the method of the proofs of Burchnall and Chaundy were rediscovered in the context of integrable systems and non-linear differential equations [9, 10, 12]. Since the 1970s, deep connections between algebraic geometry and solutions of non-linear differential equations have been revealed, indicating an enormous richness largely yet to be explored, in the intersection where non-linear differential equations and algebraic geometry meet. This connection is of interest both for its own theoretical beauty and because non-linear differential equations appear naturally in a large variety of applications, thus providing further external motivation and a source of inspiration for further research into commutative subalgebras and their interplay with algebraic geometry.

A second, more algebraic approach to proving the algebraic dependence of commuting differential operators was obtained in a different context by Amitsur [1] in the 1950s. Amitsur's approach is more in the direction of the classical connections with field extensions we have already mentioned. Recently in the 1990s a more algorithmic combinatorial method of proof based on dimension growth considerations has been found in [5, 6]. The main motivating problem for these developments was to describe, as detailed as possible, algebras of commuting differential operators and their properties. The solution of this problem is where the interplay with algebraic geometry enters the scene. The Burchnall–Chaundy result is responsible for this connection as it states that commuting differential operators satisfy equations for certain algebraic curves, which can be explicitly calculated for each pair of commuting operators by the so called eliminant method. The formulas for these curves, obtained from this method by using the corresponding determinants, are important for their further analysis and for applications and further development of the general method and interplay with algebraic geometry.

In the rest of this section we will briefly review the basic steps of the Burchnall–Chaundy construction. For simplicity of exposition until the end of this section we assume that the field of scalars is $K = \mathbb{C}$.

Commutativity of a pair of differential operators

$$P = \sum_{i=0}^{m} p_i(t)\partial^i, \qquad Q = \sum_{i=0}^{n} q_i(t)\partial^i$$

where p_i, q_i are analytic functions in t and $\partial := \frac{\mathrm{d}}{\mathrm{d}t}$, puts severe restrictive conditions on the functions p_i and q_i. In its original formulation the result of Burchnall and Chaundy can be stated as follows.

Theorem 23.1 (Burchnall–Chaundy, 1922). *For any two commuting differential operators P and Q, there is a polynomial $F(x,y) \in \mathbb{C}[x,y]$ such that $F(P,Q) = 0$.*

The polynomial appearing in this theorem is often referred to as the *Burchnall–Chaundy polynomial*.

It is worth mentioning that in their papers Burchnall and Chaundy have neither specified any conditions on what kind of functions the coefficients in the differential operators are, nor the spaces on which these operators act. Thus more precise formulations of the result should contain such a specification. Any space of functions where the construction is valid would be fine (e.g., polynomials or analytic functions in the complex domain). Algebraic steps in the construction are generic. However in order to reach the main conclusions on the existence and annihilating property of the curves, the Burchnall–Chaundy considerations use existence of solutions of an eigenvalue problem for ordinary differential operators and the property that the dimension of the solution space of a homogeneous differential equation does not exceed the order of the operator. To ensure these properties, coefficients of differential operators must be required to belong to not very restrictive but nevertheless specific classes of functions.

We will now sketch Burchnall–Chaundy construction for the convenience of the reader and in connection with further considerations in the present article. In spite of the fact that the Burchnall–Chaundy arguments actually do not constitute a complete proof due to some serious gaps, they provide important insight for building annihilating curves which can be developed into a complete proof and a well functioning construction after appropriate adjustments and restructuring.

If differential operators P and Q of orders m and n, respectively, commute then $P-h$ and Q commute for any constant $h \in \mathbb{C}$. Thus $Q(\mathrm{Ker}(P-h)) \subseteq \mathrm{Ker}(P-h)$. Consequently, if $y_1, \dots, y_m$ is a basis of $\mathrm{Ker}(P-h)$, the fundamental set of solutions of the eigenvalue problem for the differential equation $P(y) - hy = 0$ (note that the existence and the dimension properties of the solution space are assumed to hold here), then $Q(y_1), \dots, Q(y_m)$ are also elements of $\mathrm{Ker}(P-h)$ and hence $Q(\mathbf{y}) = A\mathbf{y}$ for some matrix $A = (a_{i,j})_{i,j=1}^{m}$ with entries from $\mathbb{C}$. Let $k \in \mathbb{C}$ be another arbitrary constant. A common nonzero solution, $Y = \mathbf{c}^T\mathbf{y}, \mathbf{c} \in \mathbb{C}^n$, of eigenvalue problems $PY = hY$, $QY = hY$, or equivalently a nonzero $Y \in \mathrm{Ker}(P-h) \cap \mathrm{Ker}(Q-k)$, exists if and only if $(Q-k)Y = \mathbf{c}^T(A-k)\mathbf{y} = 0$ has a nonzero solution $\mathbf{c}$, which happens only if $\det(A-k) = 0$. This is a polynomial in k of order m and hence corresponding to each h there exists only m values of the constant k (not necessarily all distinct) such that there exists nonzero $Y \in \mathrm{Ker}(P-h) \cap \mathrm{Ker}(Q-k)$ Note that here it was used that the scalar field is algebraically closed. Similarly, corresponding to each k there exists n values of h such that $\mathrm{Ker}(P-h) \cap \mathrm{Ker}(Q-k) \neq \{0\}$. From this Burchnall and Chaundy conclude that if $\mathrm{Ker}(P-h) \cap \mathrm{Ker}(Q-k) \neq \{0\}$, then h and k satisfy some polynomial equation $F(h,k) = 0$, where F is a polynomial of degree n in h and m in k with coefficients from $\mathbb{C}$. There is however a problem with this key conclusion. Surely, the equation $h = 2k + \sin(k)$ gives a bijection between the h's and the k's, but the resulting curve is not algebraic. Thus as we already pointed out, substantial adjustments are necessary in order to rearrange these arguments into a functioning construction and a complete proof. Suppose, however, that the proper adjustments have been made, and that a polynomial F with the above annihilating properties has been found. Then, any $Y \in \mathrm{Ker}(P-h) \cap \mathrm{Ker}(Q-k)$ is also a solution of the differential equation $F(P,Q)Y = F(h,k)Y = 0$ which is of order mn unless it happens that $F(P,Q) = 0$. Thus there can be at most mn linearly independent $Y \in \mathrm{Ker}(P-h) \cap \mathrm{Ker}(Q-k)$. Note that here again a specific property of the dimension of the solution space of a differential equation is assumed to hold. For each h there exists k such that $\mathrm{Ker}(P-h) \cap \mathrm{Ker}(Q-k) \neq \{0\}$. Since the field $\mathbb{C}$ is infinite (note that it is another special property of the field), one can choose infinitely many pairwise distinct numbers h with corresponding k and nonzero functions $Y_{h,k} \in \mathrm{Ker}(P-h) \cap \mathrm{Ker}(Q-k)$. But any nonempty set of eigenfunctions with pairwise distinct eigenvalues for a linear operator is always linearly independent. Thus the dimension of the solution space $\mathrm{Ker}(F(P,Q))$ is infinite. But this contradicts to the already proved $\mathrm{Ker}(F(P,Q)) \leq mn$ unless $F(P,Q) = 0$. Therefore, indeed $F(P,Q) = 0$ which is exactly what was claimed.

A beautiful feature of the Burchnall–Chaundy arguments in the differential operator case, however, is that they are almost constructive in the sense that they actually tell us, after taking a closer look, how to compute such annihilating curves, given the commuting operators. This is done by constructing the *resultant* (or *eliminant*) of operators P and Q. We sketch this construction, as it is important to have in mind for this article. To this end, for complex variables h and k, one writes:

$$\partial^r(P - h\mathbf{1}) = \sum_{i=0}^{m+r} \theta_{i,r}\partial^i - h\partial^r, \qquad r = 0,1,\ldots,n-1 \tag{23.1}$$

$$\partial^r(Q - k\mathbf{1}) = \sum_{i=0}^{n+r} \omega_{i,r}\partial^i - k\partial^r, \qquad r = 0,1,\ldots,m-1 \tag{23.2}$$

where $\theta_{i,r}$ and $\omega_{i,r}$ are certain functions built from the coefficients of P and Q respectively, whose exact form is calculated by moving ∂^r through to the right of the coefficients, using the Leibniz rule. The coefficients of the powers of ∂ on the right-hand side in (23.1) and (23.2) build up the rows of a matrix exactly as written. That is, as the first row we take the coefficients in $\sum_{i=0}^{m} \theta_{i,0}\partial^i - h\partial^0$, and as the second row the coefficients in $\sum_{i=0}^{m+1} \theta_{i,1}\partial^i - h\partial$, continuing this until $k = n-1$. As the nth row we take the coefficients in $\sum_{i=0}^{n} \omega_{i,0}\partial^i - k\partial^0$, and as the $(n+1)^{\text{th}}$ row we take the coefficients in $\sum_{i=0}^{n+1} \omega_{i,1}\partial^i - k\partial$ and so on. In this manner we get a $(n+m)\times(n+m)$-matrix using (23.1) and (23.2). The determinant of this matrix yields a trivariate polynomial $F(x,h,k)$ over $\mathbb{C}$. When written as $F(x,h,k) = \sum_i \delta_i(h,k)x^i$, it can be proved, using existence and uniqueness results for ordinary differential equations, that $\delta_i(P,Q) = 0$ for all i. It is not difficult to see that the δ_i are not all zero.

For clarity, we include the following example.

Example 23.1. Let P and Q be as above, with $m = 3$ and $n = 2$. We then have $F(x,h,k) =$

$$\begin{vmatrix} p_{0,0}(x)-h & p_{0,1}(x) & p_{0,2}(x) & p_{0,3}(x) & 0 \\ p_{1,0}(x) & p_{1,1}(x)-h & p_{1,2}(x) & p_{1,3}(x) & p_{1,4}(x) \\ q_{0,0}(x)-k & q_{0,1}(x) & q_{0,2}(x) & 0 & 0 \\ q_{1,0}(x) & q_{1,1}(x)-k & q_{1,2}(x) & q_{1,3}(x) & 0 \\ q_{2,0}(x) & q_{2,1}(x) & q_{2,2}(x)-k & q_{2,3}(x) & q_{2,4}(x) \end{vmatrix}.$$

Remark 23.1. Thus the pairs of eigenvalues $(h,k) \in \mathbb{C}^2$ corresponding to the same joint eigenfunction, i.e., corresponding to the same non-zero y such that

$$Py = hy \qquad \text{and} \qquad Qy = ky,$$

lie on curves Z_i defined by the δ_i. There is a "dictionary" between certain geometric and analytic data [12] allowing in particular construction of common eigenfunctions for commuting operators using geometric data associated to the compactification of the annihilating algebraic curve.

Burchnall–Chaundy result for operators with polynomial coefficients can be reformulated in more general terms for the abstract Heisenberg algebra $\mathscr{H}_1$ rather than in terms of its specific canonical representation by differential operators. This specialization of coefficients to polynomials does not influence the Burchnall–Chaundy construction of the annihilating algebraic curves. Thus restricting to this context does not affect the main ingredients needed for building the interplay with algebraic geometry. At the same time, when moreover reformulated entirely in the general terms of the algebra $\mathscr{H}_1$, this specialization becomes a generalization in another way, because establishing algebraic dependence of commuting elements directly in the Heisenberg algebra, without passing to the canonical representation, makes the result valid not just for differential operators, but also for any other representations of $\mathscr{H}_1$ by other kinds of operators.

Theorem 23.2 ("Burchnall–Chaundy", algebraic version). *Let P and Q be two commuting elements in $\mathscr{H}_1$, the Heisenberg algebra. Then there is a bivariate polynomial $F(x,y) \in \mathbb{C}[x,y]$ such that $F(P,Q) = 0$.*

23.4 Burchnall–Chaundy Theory for the q-Deformed Heisenberg Algebra

Let K be a field. If $q \in K$ then $\mathscr{H}_q = \mathscr{H}_K(q)$, the q-deformed Heisenberg algebra over K, is the unital associative K-algebra which is generated by two elements A and B, subject to the commutation relation $AB - qBA = I$. Though this algebra sometimes is also called the q-deformed Weyl or q-deformed Heisenberg–Weyl algebra in various contexts, we will follow systematically the terminology in [6]. We will indicate how one can prove that – under a condition on q – for any commuting $P, Q \in \mathscr{H}_q$ of order at least one (where "order" will be defined below), there exist finitely many explicitly calculable polynomials $p_i \in K[X,Y]$ such that $p_i(P,Q) = 0$ for all i, and at least one of the p_i is non-zero. The number of polynomials depends on the coefficients of P and Q, as well as on their orders. The polynomials p_i can be obtained by the mentioned before so-called eliminant construction which was introduced for differential operators (the case of $q = 1$) by Burchnall and Chaundy in [2–4] in 1920s, and which we will employ and extend to the context of general q-deformed

Heisenberg algebras showing that analogous determinant (resultant) construction of the annihilating algebraic curves works for q-deformed Heisenberg algebras well.

We assume $q \neq 0$ throughout (our method breaks down when $q = 0$) and define the q-integer $\{n\}_q$, for $n \in \mathbb{Z}$, by

$$\{n\}_q = \begin{cases} \frac{q^n - 1}{q-1} & q \neq 1; \\ n & q = 1. \end{cases}$$

Following [6, Definition 5.2] we will say that $q \in K^* = K \setminus \{0, 1\}$ is of *torsion type* if there is a positive integer solution p to $q^p = 1$, and of *free type* if the only integer solution to $q^p = 1$ is $p = 0$. In the torsion type case the least such positive integer p is called the order of q and in free type case the order of q is said to be zero. If $q = 1$ then it is said to be of a torsion type if K is a field of non-zero characteristic, and of free type if the field is of characteristic zero.

Remark 23.2. The following are equivalent for $q \neq 0$:

1. For $n \in \mathbb{Z}$, $\{n\}_q = 0$ if and only if $n = 0$
2. For $n_1, n_2 \in \mathbb{Z}$, $\{n_1\}_q = \{n_2\}_q$ if and only if $n_1 = n_2$
3. $\begin{cases} q \text{ is not a root of unity other than } 1, & \text{if } \operatorname{char} k = 0 \\ q \text{ is not a root of unity,} & \text{if } \operatorname{char} k \neq 0 \end{cases}$

Part (2) of this remark is essential when one considers the dimension of eigenspaces for q-difference operators which is important for our extension of the Burchnall–Chaundy construction.

Note also that under our assumptions K is infinite.

Let $\mathscr{L}$ be the K-vector space of all formal Laurent series in a single variable t with coefficients in K. Define

$$M\Big(\sum_{n=-\infty}^{\infty} a_n t^n\Big) = \sum_{n=-\infty}^{\infty} a_n t^{n+1} = \sum_{n=-\infty}^{\infty} a_{n-1} t^n,$$

$$D_q\Big(\sum_{n=-\infty}^{\infty} a_n t^n\Big) = \sum_{n=-\infty}^{\infty} a_n \{n\}_q t^{n-1} = \sum_{n=-\infty}^{\infty} a_{n+1} \{n+1\}_q t^n.$$

Alternatively, one could introduce $\mathscr{L}$ as the vector space of all functions from $\mathbb{Z}$ to K and let M act as the right shift and D_q as a weighted left shift, but the Laurent series model is more appealing.

The algebra $\mathscr{H}_q$ has $\{I, A, A^2, \ldots\}$ as a free basis in its natural structure as a left $K[X]$-module with X mapped to B.

So an arbitrary non-zero element P of $\mathscr{H}_q$ can be written as

$$P = \sum_{j=0}^{m} p_j(B) A^j, p_m \neq 0,$$

for uniquely determined $p_j \in K[X]$ and $m \geq 0$. The integer m is called the *order* of P (or degree with respect to A) [6].

By sending A to D_q and B to M, $\mathscr{L}$ becomes an $\mathscr{H}_q$-module. If we make the additional assumption that $\{n\}_q \neq 0$ or equivalently $q^n \neq 1$ for all non-zero $n \in \mathbb{Z}$, then this representation is faithful [6, Theorem 8.3].

We will assume that $q \neq 0$ and $\{n\}_q \neq 0$ for $n \neq 0$ throughout this paper and identify $\mathscr{H}_q$ with its image in $\mathrm{End}_K(\mathscr{L})$ under the previously defined representation without further notice. In the image of $\mathscr{H}_q$ under the representation, $\{1, D_q, D_q^2, \dots\}$ is a free basis in its natural structure as a left $K[X]$-module with X being mapped to M, and so any P can be written as

$$P = \sum_{j=0}^{m} p_j(M) D_q^j, \; p_m \neq 0,$$

for uniquely determined $p_j \in K[X]$ and $m \geq 0$. The integer m is called the *order* of P.

The important result in the context of this article is an extension of the Burchnall–Chaundy theorem in algebraic version for the q-deformed Heisenberg algebra $\mathscr{H}_q$, due to Hellström and Silvestrov [6, Therem 7.5].

Theorem 23.3 (Hellström–Silvestrov [6]). *Let $q \in K^* = K \setminus \{0\}$, and let P and Q be two commuting elements in $\mathscr{H}_q$. Then there is a bivariate polynomial $F(x,y) \in \mathrm{Z}(\mathscr{H}_q)[x,y]$, with coefficients in the center $\mathrm{Z}(\mathscr{H}_q)$ of $\mathscr{H}_q$, such that $F(P,Q) = 0$.*

If q is not a root of unity ($\{n\}_q$ are different for different n), the center of $\mathscr{H}_q$ is trivial, i.e., consists of scalar multiples of the identity $\mathrm{Z}(\mathscr{H}_q) = KI$. Thus in this case there exists a "genuine" algebraic curve over the scalar field K as Theorem 23.3 takes the following form [6, Theorem 7.4].

Theorem 23.4 (Hellström–Silvestrov [6]). *Let $q \in K^* = K \setminus \{0\}$ be of free type, and let P and Q be two commuting elements in $\mathscr{H}_q$. Then there is a bivariate polynomial $F(x,y) \in K[x,y]$, with coefficients in K, such that*

$$F(P,Q) = 0.$$

Note however, that when q is of torsion type and order d (i.e., d is the smallest positive integer such that $q^d = 1$), then the center is

$$\mathrm{Z}(\mathscr{H}_q) = K[A^d, B^d],$$

the subalgebra spanned by $\{A^d, B^d\}$, where d is the order of q, the minimal positive integer such that $q^d = 1$ [6, Corollary 6.12]. The conclusion of Theorem 23.4, that is the algebraic dependence of commuting elements over the field K, does not hold for q of torsion type, since if p is the order of q, then $\alpha = A^p$ and $\beta = B^p$ commute, but do not satisfy any commutative polynomial relation. Thus in this case Theorem 23.3 has indeed to be invoked.

The proof of Theorem 23.3 as given in [6] is purely existential. However, while it says essentially nothing theoretically on the form or properties of the annihilating curves, the construction used in the proof actually provides an explicit computationally implementable algorithm for producing annihilating polynomials.

A specialization of a part of the general conjecture made by S. Silvestrov in 1994 is that the determinant scheme devised by Burchnall and Chaundy could be used to calculate the polynomial even in the case of $\mathscr{H}_q$. The Burchnall–Chaundy eliminant construction adaptation to $\mathscr{H}_q$ and a series of examples, indicating that the construction indeed yields annihilating algebraic curves for commuting elements in $\mathscr{H}_q$, was first presented in [11]. But no full proof of the conjecture that this adaption is possible for all commuting elements of $\mathscr{H}_q$ was given. Even though a direct generalization of the classical arguments of Burchnall and Chaundy was attempted some important technical steps were missing. The main reason why an analogous proof for q-difference operators (i.e., elements in $\mathscr{H}_q$) is problematic is that the solution space for q-difference equations may not in general be as well-behaved as for ordinary differential operators, and the proof of the classical Burchnall–Chaundy theorem relies heavily on properties of the solution spaces to the eigenvalue-problems for the differential operators P and Q.

Nevertheless, in the case of q of free type we have succeeded now to extend the original Burchnall–Chaundy eliminant method to the q-difference operators and hence to $\mathscr{H}_q$. Roughly speaking, the key technical idea is to choose an appropriate representation space for the canonical representation of $\mathscr{H}_q$ generated by M_x, D_q, so that the necessary key ingredients about dimensions of eigenspaces of q-difference operators in the Burchnall–Chaundy type proof are still available. The module $\mathscr{L}$ introduced above has all these properties. It is thus that, for q of free type, we again have a determinant based construction providing annihilating curves. The complete proofs will be presented in [8]; they are considerably more involved than the original work by Burchnall and Chaundy.

An extension of the result and the Burchnall–Chaundy type construction to the case of q of torsion type, i.e., when $\mathscr{H}_q$ has non-trivial center, is still not available. This is just one of many reasons why another proof of the possibility of adapting to $\mathscr{H}_q$ the determinant construction of the annihilating curves, relying on purely algebraic methods, i.e., without passing to a specific representation of $\mathscr{H}_q$, is desirable. The existence of such proof is a specialization to $\mathscr{H}_q$ of the third part of the aforementioned general conjecture by S. Silvestrov.

23.4.1 Eliminant Determinant Construction for q-Deformed Heisenberg Algebra

In this section we will briefly outline our extension of the Burchnall–Chaundy result and constructions to the q-difference operators, or to the abstract algebra context of the q-deformed Heisenberg algebra. The complete details of the proofs will be presented in [8].

Let $P, Q \in \mathscr{H}_q$ be of order $m \geq 1$ and $n \geq 1$ respectively. Write for $k = 0, \ldots, n-1$,

$$D_q^k P = \sum_{j=0}^{m+k} p_{k,j}(M) D_q^j, \text{ with } p_{k,j} \in K[X],$$

and, for $l = 0, \ldots, m-1$, write

$$D_q^l Q = \sum_{j=0}^{n+l} q_{l,j}(M) D_q^j, \text{ with } q_{l,j} \in K[X].$$

By analogy with the Burchnall–Chaundy method for differential operators, we build up an $(m+n) \times (m+n)$-matrix as follows. For $k = 1, \ldots, n$, the k-th row is given by the coefficients of powers of D_q in the expression

$$D_q^{k-1} P - \lambda D_q^{k-1} = \sum_{j=0}^{m+k-1} p_{k-1,j}(M) D_q^j - \lambda D_q^{k-1},$$

where λ is a formal variable. For $k \in \{n+1, \ldots, m+n\}$, the k-th row is given by the coefficients of D_q in

$$D_q^{k-n-1} Q - \mu D_q^{k-n-1} = \sum_{j=0}^{k-1} p_{k-n-1,j}(M) D_q^j - \mu D_q^{k-n-1},$$

where μ is a formal variable different from λ. The determinant of this matrix is called the *eliminant* of P and Q. We denote it $\Delta_{(P,Q)}(M, \lambda, \mu)$. It is a polynomial with coefficients in K.

Theorem 23.5 ([8]). *Let K be a field and $0 \neq q \in K$ be such that $\{n\}_q = 0$ if and only if $n = 0$. Suppose*

$$P = \sum_{j=0}^{m} p_j(M) D_q^j \ (m \geq 1, p_m \neq 0) \quad \text{and} \quad Q = \sum_{j=0}^{n} q_j(M) D_q^j \ (n \geq 1, q_n \neq 0)$$

are commuting elements of $\mathscr{H}_q$, and let $\Delta_{P,Q}(M, \lambda, \mu)$ be the eliminant constructed as above. Then $\Delta_{P,Q} \neq 0$. In fact, if $q_n(M) = \sum_i a_i M^i$ $(a_i \in K)$ then $\Delta_{P,Q}$ has degree n as a polynomial in λ, and its non-zero coefficient of λ^n is equal to $(-1)^n \prod_{k=0}^{m-1} (\sum_i a_i q^{k \cdot i} M^i)$. Likewise, if $p_m(M) = \sum_i b_i M^i$, then $\Delta_{P,Q}$ has degree m as a polynomial in μ, and its non-zero coefficient of μ^m is $(-1)^m \prod_{k=0}^{n-1} (\sum_i b_i q^{k \cdot i} M^i)$.

Let $s = n \cdot \max_j \deg(p_j) + m \cdot \max_j \deg(q_j)$, and

$$t = \frac{1}{2} \cdot n \cdot (n-1) \cdot \max_j \deg(p_j) + \frac{1}{2} \cdot m \cdot (m-1) \cdot \max_j \deg(q_j),$$

and define the polynomials δ_i $(i = 1, \ldots, s)$ by $\Delta_{P,Q}(M, \lambda, \mu) = \sum_{i=0}^{s} \delta_i(\lambda, \mu) M^i$.

Then:

1. *Each of the coefficients of the δ_i can be expressed as $\sum_{l=0}^{t} r_l q^l$ with the r_l in the subring of K which is generated by the coefficients of all the p_j and the q_j.*
2. *At least one of the δ_i is non-zero.*
3. *$\delta_i(P,Q) = 0$ for all $i = 0,\dots,s$.*

The reader will easily convince himself of all statements in the theorem other than (3). We will now sketch the main idea of the proof of (3) as given in [8].

The idea is as follows. Suppose $\lambda_0, \mu_0 \in K$ and $0 \neq v_{(\lambda_0,\mu_0)} \in \mathscr{L}$ is a common eigenvector of P and Q:

$$Pv_{(\lambda_0,\mu_0)} = \lambda_0 v_{(\lambda_0,\mu_0)},$$

$$Qv_{(\lambda_0,\mu_0)} = \mu_0 v_{(\lambda_0,\mu_0)}.$$

Then the specialization $\lambda = \lambda_0, \mu = \mu_0$ of the matrix of endomorphisms of $\mathscr{L}$ that defines the eliminant has the column vector $(v_{(\lambda_0,\mu_0)},\dots,D_q^{m+n-1}v_{(\lambda_0,\mu_0)})$ in its kernel. Hence $\Delta_{P,Q}(M,\lambda_0,\mu_0)\,v_{(\lambda_0,\mu_0)} = 0$. Now it does not follow automatically from this that $\Delta_{P,Q}(M,\lambda_0,\mu_0) = 0$ in $\mathscr{H}_q$ since a polynomial in M might have non-trivial kernel, as the example $(M-1)\sum_n t^n = 0$ shows. However, embedding K in an algebraically closed field if necessary, we are able to show that there exist infinitely many such (λ_0,μ_0) where we *can* conclude that $\Delta_{P,Q}(M,\lambda_0,\mu_0) = 0$ in $\mathscr{H}_q$. Thus the operators $\delta_i(P,Q)$ have an infinite-dimensional kernel, and it is possible to show that this implies that $\delta_i(P,Q) = 0$ in $\mathscr{H}_q$.

It is more complicated to show that there exist infinitely many (λ_0,μ_0) with the required property. The idea is to exploit the fact that $v_{(\lambda_0,\mu_0)}$ is both in the kernel of the operator $P-\lambda_0$ of order $m \geq 1$ and of the operator $\Delta_{P,Q}(M,\lambda_0,\mu_0)$ which, if it is not zero, is not constant. We can describe the kernel of a non-constant polynomial element $p(M)$ of $\mathscr{H}_q$ and the action of $P-\lambda_0$ on it explicitly enough to show that any such $v_{(\lambda_0,\mu_0)}$ is in a subspace of *finite* dimension which (for fixed q) depends only on the leading coefficient of P and the degree of P, but not on λ_0 or μ_0. Hence for the infinity of pairs (λ_0,μ_0) that can be shown to exist, it can, by linear independence, only for finitely many pairs be the case that $\Delta_{P,Q}(M,\lambda_0,\mu_0)$ is not zero. For the remaining pairs, the specialized eliminant must be zero.

23.4.2 Hellström–Silvestrov Proof of Algebraic Dependence in $\mathscr{H}_q$

In this section we review some ideas of the proof of Theorems 23.3 and 23.4 obtained by Hellström and Silvestrov in [6].

Let us first q be of free type. Then $\mathrm{Z}(\mathscr{H}_q)$ is isomorphic to K, and hence our aim is to prove Theorem 23.4. There are three cases to consider. In the simplest case, when α is of the form cI for some $c \in K$, the polynomial $P(x,y) = x - c$ satisfies is annihilating for (α,β) since $\alpha - cI = 0$.

In the second case assume that α,β are linear combinations of monomials with equal degrees in A and B (denoted by $\alpha,\beta \sqsubseteq K_0$ as in [6]), and that there is no

$c \in K$ such that $\alpha = cI$. Let $a = \deg \alpha > 0$ and $b = \deg \beta$. A general expression for $P(\alpha, \beta)$, where P has at most degree b in the first variable and at most degree a in the second is

$$\sum_{\substack{0 \leqslant i \leqslant b \\ 0 \leqslant j \leqslant a}} p_{ij} \alpha^i \beta^j. \tag{23.3}$$

This sum is a linear combination of the $(a+1)(b+1)$ vectors $\{\alpha^i \beta^j\}_{i=0;j=0}^{b;a}$, and all these vectors belong to the vector space $\mathrm{Cen}(2ab, 0, \alpha) = \{\beta \sqsubseteq K_0 \mid [\alpha, \beta] = 0 \text{ and } \deg \beta \leq 2ab\}$. Now the point is that it can be shown that the dimension of this space is strictly smaller than $(a+1)(b+1)$. Hence there exist numbers $\{p_{ij}\}_{i,j}$, not all zero, which make (23.3) zero. Thus there exists a P as required.

The similar algorithmic dimension growth type proof in the case when $\alpha \not\sqsubseteq K_0$ or $\beta \not\sqsubseteq K_0$ is the most technical part. It is based on application of [6, Theorem 7.3] and [6, Corollary 6.9] which are proved using other concepts and results in the book. Thus we refer the reader to [6] for details.

Let us now assume that q is of torsion type and that p is its order. Observe that $\mathscr{H}_q$, seen as a $\mathrm{Z}(\mathscr{H}_q)$-module, contains a spanning set of p^2 elements, namely $\{B^i A^j\}_{0 \leqslant i,j < p}$. Also observe that $\mathrm{Z}(\mathscr{H}_q)$, by [6, Theorem 4.9] and [6, Theorem 5.7], is an integral domain. Hence any subset of $\mathscr{H}_q$ that contains more elements than a spanning set will be linearly dependent. Consider the polynomial

$$P(x,y) = \sum_{i=0}^{p} \sum_{j=0}^{p} c_{ij} x^i y^j$$

where $c_{ij} \in \mathrm{Z}(\mathscr{H}_q)$. Clearly $P(\alpha, \beta)$ will be a linear combination of $(p+1)^2 > p^2$ elements in $\mathscr{H}_q$ and these elements are linearly dependent. Thus there are coefficients $\{c_{ij}\}_{0 \leqslant i,j \leqslant p} \subset \mathrm{Z}(\mathscr{H}_q)$, not all zero, such that $P(\alpha, \beta) = 0$. This is exactly what the theorem claims.

Acknowledgements This work was supported by a visitor's grant of the Netherlands Organisation for Scientific Research (NWO), the Swedish Foundation for International Cooperation in Research and Higher Education (STINT), the Crafoord Foundation, the Royal Physiographic Society in Lund, the Royal Swedish Academy of Sciences and the Swedish Research Council. We are also grateful to Lars Hellström and Daniel Larsson for helpful comments and discussions.

References

1. Amitsur, S.A.: Commutative linear differential operators. Pacific J. Math. **8**, 1–10 (1958)
2. Burchnall, J.L., Chaundy, T.W.: Commutative ordinary differential operators. Proc. London Math. Soc. (Ser. 2) **21**, 420–440 (1922)
3. Burchnall, J.L., Chaundy, T.W.: Commutative ordinary differential operators. Proc. Roy. Soc. London A **118**, 557–583 (1928)
4. Burchnall, J.L., Chaundy, T.W.: Commutative ordinary differential operators. II. – The Identity $P^n = Q^m$. Proc. Roy. Soc. London A **134**, 471–485 (1932)

5. Hellström, L.: Algebraic dependence of commuting differential operators. Disc. Math. **231**, no. 1–3, 246–252 (2001)
6. Hellström, L., Silvestrov, S.D.: Commuting elements in q-deformed Heisenberg algebras. World Scientific, New Jersey (2000)
7. Hellström, L., Silvestrov, S.: Ergodipotent maps and commutativity of elements in non-commutative rings and algebras with twisted intertwining. J. Algebra **314**, 17–41 (2007)
8. de Jeu, M., Svensson, P.C., Silvestrov, S.: Algebraic curves for commuting elements in the q-deformed Heisenberg algebra, arXiv:0710.2748v1 [math.RA], 17pp., to appear (2007)
9. Krichever, I.M.: Integration of non-linear equations by the methods of algebraic geometry. Funktz. Anal. Priloz. **11**, no. 1, 15–31 (1977)
10. Krichever, I.M.: Methods of algebraic geometry in the theory of nonlinear equations. Uspekhi Mat. Nauk, **32**, no. 6, 183–208 (1977)
11. Larsson, D., Silvestrov, S.D.: Burchnall–Chaundy theory for q-difference operators and q-deformed Heisenberg algebras. J. Nonlin. Math. Phys. **10**, suppl. 2, 95–106 (2003)
12. Mumford, D.: An algebro-geometric construction of commuting operators and of solutions to the Toda lattice equation, Korteweg–de Vries equation and related non-linear equations. Proc. Int. Symp. on Algebraic Geometry, Kyoto, 115–153 (1978)
13. Nesterenko, Yu.: Modular functions and transcendence problems. C.R. Acad. Sci. Paris Ser. I Math. **322**, no. 10, 909–914 (1996)
14. Nesterenko, Yu., Philippon, P.: Introduction to algebraic independence theory, Lecture Notes in Mathematics 1752. Springer, Berlin (2001)

Chapter 24
Crossed Product-Like and Pre-Crystalline Graded Rings

Johan Öinert and Sergei D. Silvestrov

Abstract We introduce crossed product-like rings, as a natural generalization of crystalline graded rings, and describe their basic properties. Furthermore, we prove that for certain pre-crystalline graded rings and every crystalline graded ring $\mathcal{A}$, for which the base subring $\mathcal{A}_0$ is commutative, each non-zero two-sided ideal has a non-zero intersection with $C_{\mathcal{A}}(\mathcal{A}_0)$, i.e. the commutant of $\mathcal{A}_0$ in $\mathcal{A}$. We also show that in general this property need not hold for crossed product-like rings.

24.1 Introduction

In the recent paper [3], by E. Nauwelaerts and F. Van Oystaeyen, so called crystalline graded rings were introduced, as a general class of group graded rings containing as special examples the algebraic crossed products, the Weyl algebras, the generalized Weyl algebras and generalizations of Clifford algebras. In the paper [4], by T. Neijens, F. Van Oystaeyen and W.W. Yu, the structure of the center of special classes of crystalline graded rings and generalized Clifford algebras was studied.

In this paper we prove that if $\mathcal{A}$ is a crystalline graded ring, where the base subring $\mathcal{A}_0$ is commutative, then each non-zero two-sided ideal has a non-zero intersection with $C_{\mathcal{A}}(\mathcal{A}_0) = \{b \in \mathcal{A} \mid ab = ba, \quad \forall a \in \mathcal{A}_0\}$, i.e. the commutant of $\mathcal{A}_0$ in $\mathcal{A}$ (Corollary 24.10). Furthermore, we define *pre-crystalline graded rings* as the obvious generalization of crystalline graded rings, and show that under certain conditions the previously mentioned property also holds for pre-crystalline group graded rings (Theorem 24.2).

J. Öinert
Centre for Mathematical Sciences, Lund University, Box 118, 22100 Lund, Sweden
e-mail: Johan.Oinert@math.lth.se

S.D. Silvestrov
Centre for Mathematical Sciences, Lund University, Box 118, 22100 Lund, Sweden
e-mail: Sergei.Silvestrov@math.lth.se

S. Silvestrov et al. (eds.), *Generalized Lie Theory in Mathematics, Physics and Beyond,*

We also introduce *crossed product-like rings* which is as a broad class of rings, containing as special cases the pre-crystalline graded rings and hence also the crystalline graded rings. Crossed product-like rings are in general only monoid graded in contrast to crystalline graded rings and algebraic crossed products, which are group graded. A crossed product-like ring graded by the monoid M and with base ring $\mathcal{A}_0$ will be denoted by $\mathcal{A}_0 \Diamond_\sigma^\alpha M$, where $\sigma_s : \mathcal{A}_0 \to \mathcal{A}_0$ is an additive and multiplicative set map for each $s \in M$ and $\alpha : M \times M \to \mathcal{A}_0 \setminus \{0\}$ is another set map (see Lemma 24.1 for details). The product in $\mathcal{A}_0 \Diamond_\sigma^\alpha M$ is given by the bilinear extension of the rule

$$(a u_s)(b u_t) = a \, \sigma_s(b) \, \alpha(s,t) \, u_{st}$$

for $a, b \in \mathcal{A}_0$ and $s, t \in M$. In contrary to the case of algebraic crossed products or more generally crystalline graded rings, for crossed product-like rings the maps σ_s, $s \in M$, may be non-surjective, thus allowing non-standard examples of rings to fit in (Example 24.2).

The center $Z(\mathcal{A})$ (Proposition 24.2) and the commutant $C_{\mathcal{A}}(\mathcal{A}_0)$ of the base subring in a general crossed product-like ring $\mathcal{A} = \mathcal{A}_0 \Diamond_\sigma^\alpha M$ (Theorem 24.1) will be described and we will give an example of a (possibly) non-invertible dynamical system, to which we associate a monoid graded crossed product-like ring and show that it contains a non-zero two-sided ideal which has zero intersection with the commutant of the base subring (Proposition 24.3). This displays a difference between monoid graded and group graded rings with regards to the intersection property that we are investigating.

24.2 Preliminaries and Definitions

We shall begin by giving the definitions of the rings that we are investigating and also describe some of their basic properties.

Definition 24.1 (Graded Ring). Let M be a monoid with neutral element e. A ring $\mathcal{R}$ is said to be *graded by M* if $\mathcal{R} = \oplus_{s \in M} \mathcal{R}_s$ for additive subgroups $\mathcal{R}_s$, $s \in M$, satisfying $\mathcal{R}_s \mathcal{R}_t \subseteq \mathcal{R}_{st}$ for all $s, t \in M$. Moreover, if $\mathcal{R}_s \mathcal{R}_t = \mathcal{R}_{st}$ holds for all $s, t \in M$, then $\mathcal{R}$ is said to be *strongly graded by M*.

It may happen that M is in fact a group and when we want to emphasize that a ring $\mathcal{R}$ is graded by a monoid respectively a group we shall say that it is *monoid graded* respectively *group graded*. It is worth noting that in an associative and graded ring $\mathcal{R} = \oplus_{s \in M} \mathcal{R}_s$, by the gradation property, $\mathcal{R}_e$ will always be a subring and furthermore for every $s \in M$, $\mathcal{R}_s$ is an $\mathcal{R}_e$-bimodule. For a thorough exposition of the theory of graded rings we refer to [1, 2].

So called *crystalline graded rings* were defined in [3] and further investigated in [4]. Right now we wish to weaken some of the conditions in the definition of those rings, in order to allow more general rings to fit in.

Definition 24.2 (Crossed Product-Like Ring). An associative and unital ring $\mathcal{A}$ is said to be *crossed product-like* if:

- There is a monoid M (with neutral element e).
- There is a map $u : M \to \mathcal{A}, \quad s \mapsto u_s$ such that $u_e = 1_{\mathcal{A}}$ and $u_s \neq 0_{\mathcal{A}}$ for every $s \in M$.
- There is a subring $\mathcal{A}_0 \subseteq \mathcal{A}$ containing $1_{\mathcal{A}}$.

such that the following conditions are satisfied:

(P1) $\mathcal{A} = \bigoplus_{s\in M} \mathcal{A}_0 u_s$.
(P2) For every $s \in M$, $u_s \mathcal{A}_0 \subseteq \mathcal{A}_0 u_s$ and $\mathcal{A}_0 u_s$ is a free left $\mathcal{A}_0$-module of rank one.
(P3) The decomposition in P1 makes $\mathcal{A}$ into an M-graded ring with $\mathcal{A}_0 = \mathcal{A}_e$.

If M is a group and we want to emphasize that, then we shall say that $\mathcal{A}$ is a *crossed product-like group graded ring* . If we want to emphasize that M is only a monoid, we shall say that $\mathcal{A}$ is a *crossed product-like monoid graded ring* .

Similarly to the case of algebraic crossed products, one is able to find maps that can be used to describe the formation of arbitrary products in the ring. Some key properties of these maps are highlighted in the following lemma.

Lemma 24.1. *With notation and definitions as above:*

(1) For every $s \in M$, there is a set map $\sigma_s : \mathcal{A}_0 \to \mathcal{A}_0$ defined by $u_s a = \sigma_s(a) u_s$ for $a \in \mathcal{A}_0$. The map σ_s is additive and multiplicative. Moreover, $\sigma_e = \mathrm{id}_{\mathcal{A}_0}$.
(2) There is a set map $\alpha : M \times M \to \mathcal{A}_0$ defined by $u_s u_t = \alpha(s,t) u_{st}$ for $s,t \in M$. For any triple $s,t,w \in M$ and $a \in \mathcal{A}_0$ the following equalities hold:

$$\alpha(s,t)\alpha(st,w) = \sigma_s(\alpha(t,w))\alpha(s,tw) \tag{24.1}$$

$$\sigma_s(\sigma_t(a))\alpha(s,t) = \alpha(s,t)\sigma_{st}(a) \tag{24.2}$$

(3) For every $s \in M$ we have $\alpha(s,e) = \alpha(e,s) = 1_{\mathcal{A}}$.

Proof. The proof is analogous to the proof of Lemma 2.1 in [3]. □

By the foregoing lemma we see that, for arbitrary $a,b \in \mathcal{A}_0$ and $s,t \in M$, the product of $a u_s$ and $b u_t$ in the crossed product-like ring $\mathcal{A}$ may be written as

$$(a u_s)(b u_t) = a\, \sigma_s(b)\, \alpha(s,t) u_{st}$$

and this is the motivation for the name *crossed product-like*. A crossed product-like ring $\mathcal{A}$ with the above properties will be denoted by $\mathcal{A}_0 \diamond_\sigma^\alpha M$, indicating the maps σ and α.

Remark 24.1. Note that for $s \in M \setminus \{e\}$ we need not necessarily have $\sigma_s(1_{\mathcal{A}_0}) = 1_{\mathcal{A}_0}$ and hence σ_s need not be a ring morphism.

Definition 24.3 (Pre-Crystalline Graded Ring). A crossed product-like ring $\mathcal{A}_0 \diamond_\sigma^\alpha M$ where for each $s \in M$, $\mathcal{A}_0 u_s = u_s \mathcal{A}_0$, is said to be a *pre-crystalline graded ring* .

For a pre-crystalline graded ring $\mathcal{A}_0 \diamond_\sigma^\alpha M$, the following lemma gives us additional information about the maps σ and α defined in Lemma 24.1.

Lemma 24.2. *If $\mathcal{A}_0\diamond_\sigma^\alpha M$ is a pre-crystalline graded ring, then the following holds:*

(1) For every $s \in M$, the map $\sigma_s : \mathcal{A}_0 \to \mathcal{A}_0$ is a surjective ring morphism.
(2) If M is a group, then

$$\alpha(g,g^{-1}) = \sigma_g(\alpha(g^{-1},g))$$

for each $g \in M$.

Proof. The proof is the same as for Lemma 2.1 in [3]. □

In a pre-crystalline graded ring, one may show that for $s,t \in M$, the $\alpha(s,t)$ are normalizing elements of $\mathcal{A}_0$ in the sense that $\mathcal{A}_0\alpha(s,t) = \alpha(s,t)\mathcal{A}_0$ (see [3, Proposition 2.3]). If we in addition assume that $\mathcal{A}_0$ is commutative, then we see by Lemma 24.1 that the map $\sigma : M \to \mathrm{End}(\mathcal{A}_0)$ is a monoid morphism.

For a pre-crystalline group graded ring $\mathcal{A}_0\diamond_\sigma^\alpha G$, we let $S(G)$ denote the multiplicative set in $\mathcal{A}_0$ generated by $\{\alpha(g,g^{-1}) \mid g \in G\}$ and let $S(G \times G)$ be the multiplicative set generated by $\{\alpha(g,h) \mid g,h \in G\}$.

Lemma 24.3 (Corollary 2.7 in [3]). *If $\mathcal{A} = \mathcal{A}_0\diamond_\sigma^\alpha G$ is a pre-crystalline group graded ring, then the following are equivalent:*

- *$\mathcal{A}_0$ is $S(G)$-torsion free.*
- *$\mathcal{A}$ is $S(G)$-torsion free.*
- *$\alpha(g,g^{-1})a_0 = 0$ for some $g \in G$ implies $a_0 = 0$.*
- *$\alpha(g,h)a_0 = 0$ for some $g,h \in G$ implies $a_0 = 0$.*
- *$\mathcal{A}_0 u_g = u_g \mathcal{A}_0$ is also free as a right $\mathcal{A}_0$-module, with basis u_g, for every $g \in G$.*
- *For every $g \in G$, σ_g is bijective and hence a ring automorphism of $\mathcal{A}_0$.*

From Lemma 24.3 we see that when $\mathcal{A}_0$ is $S(G)$-torsion free in a pre-crystalline group graded ring $\mathcal{A}_0\diamond_\sigma^\alpha G$, we have $\mathrm{im}(\sigma) \subseteq \mathrm{Aut}(\mathcal{A}_0)$. We shall now state the definition of a crystalline graded ring.

Definition 24.4 (Crystalline Graded Ring). A pre-crystalline group graded ring $\mathcal{A}_0\diamond_\sigma^\alpha G$ which is $S(G)$-torsion free is said to be a *crystalline graded ring*.

24.3 The Commutant of $\mathcal{A}_0$ in a Crossed Product-Like Ring

The commutant of the subring $\mathcal{A}_0$ in the crossed product-like ring $\mathcal{A} = \mathcal{A}_0\diamond_\sigma^\alpha M$ is defined by

$$C_{\mathcal{A}}(\mathcal{A}_0) = \{b \in \mathcal{A} \mid ab = ba, \quad \forall a \in \mathcal{A}_0\}.$$

In this section we will describe $C_{\mathcal{A}}(\mathcal{A}_0)$ in various crossed product-like rings. Theorem 24.1 tells us exactly when an element of a crossed product-like ring $\mathcal{A} = \mathcal{A}_0\diamond_\sigma^\alpha M$ lies in $C_{\mathcal{A}}(\mathcal{A}_0)$.

Theorem 24.1. *In a crossed product-like ring $\mathcal{A} = \mathcal{A}_0 \diamond_\sigma^\alpha M$, we have*

$$C_{\mathcal{A}}(\mathcal{A}_0) = \left\{ \sum_{s\in M} r_s u_s \in \mathcal{A}_0 \diamond_\sigma^\alpha M \;\middle|\; r_s \sigma_s(a) = a r_s, \quad \forall a \in \mathcal{A}_0, s \in M \right\}.$$

Proof. The proof is established through the following sequence of equivalences:

$$\begin{aligned}\sum_{s\in M} r_s u_s \in C_{\mathcal{A}}(\mathcal{A}_0) &\iff \left(\sum_{s\in M} r_s u_s\right) a = a \left(\sum_{s\in M} r_s u_s\right), \quad \forall a \in \mathcal{A}_0 \\ &\iff \sum_{s\in M} r_s \sigma_s(a) u_s = \sum_{s\in M} a r_s u_s, \quad \forall a \in \mathcal{A}_0 \\ &\iff \text{For each } s \in M: \; r_s \sigma_s(a) = a r_s, \quad \forall a \in \mathcal{A}_0.\end{aligned}$$

□

If $\mathcal{A}_0$ is commutative, then for each $r \in \mathcal{A}_0$ we denote its annihilator ideal in $\mathcal{A}_0$ by $\mathrm{Ann}(r) = \{a \in \mathcal{A}_0 \mid ar = 0\}$ and get a simplified description of $C_{\mathcal{A}}(\mathcal{A}_0)$.

Corollary 24.1. *If $\mathcal{A} = \mathcal{A}_0 \diamond_\sigma^\alpha M$ is a crossed product-like ring and $\mathcal{A}_0$ is commutative, then*

$$C_{\mathcal{A}}(\mathcal{A}_0) = \left\{ \sum_{s\in M} r_s u_s \in \mathcal{A}_0 \diamond_\sigma^\alpha M \;\middle|\; \sigma_s(a) - a \in \mathrm{Ann}(r_s), \quad \forall a \in \mathcal{A}_0, s \in M \right\}.$$

When $\mathcal{A}_0$ is commutative it is clear that $\mathcal{A}_0 \subseteq C_{\mathcal{A}}(\mathcal{A}_0)$. Using the explicit description of $C_{\mathcal{A}}(\mathcal{A}_0)$ in Corollary 24.1, we immediately get necessary and sufficient conditions for $\mathcal{A}_0$ to be maximal commutative in $\mathcal{A}$, i.e. $\mathcal{A}_0 = C_{\mathcal{A}}(\mathcal{A}_0)$.

Corollary 24.2. *If $\mathcal{A}_0 \diamond_\sigma^\alpha M$ is a crossed product-like ring where $\mathcal{A}_0$ is commutative, then $\mathcal{A}_0$ is maximal commutative in $\mathcal{A}_0 \diamond_\sigma^\alpha M$ if and only if, for each pair $(s, r_s) \in (M \setminus \{e\}) \times (\mathcal{A}_0 \setminus \{0_{\mathcal{A}_0}\})$, there exists $a \in \mathcal{A}_0$ such that $\sigma_s(a) - a \notin \mathrm{Ann}(r_s)$.*

Corollary 24.3. *Let $\mathcal{A}_0 \diamond_\sigma^\alpha M$ be a crossed product-like ring where $\mathcal{A}_0$ is commutative. If for each $s \in M \setminus \{e\}$ it is always possible to find some $a \in \mathcal{A}_0$ such that $\sigma_s(a) - a$ is not a zero-divisor in $\mathcal{A}_0$, then $\mathcal{A}_0$ is maximal commutative in $\mathcal{A}_0 \diamond_\sigma^\alpha M$.*

The next corollary is a consequence of Corollary 24.2.

Corollary 24.4. *If the subring $\mathcal{A}_0$ in the crossed product-like ring $\mathcal{A}_0 \diamond_\sigma^\alpha M$ is maximal commutative, then $\sigma_s \neq \mathrm{id}_{\mathcal{A}_0}$ for each $s \in M \setminus \{e\}$.*

The description of the commutant $C_{\mathcal{A}}(\mathcal{A}_0)$ from Corollary 24.1 can be further refined in the case when $\mathcal{A}_0$ is an integral domain.

Corollary 24.5. *If $\mathcal{A}_0$ is an integral domain, then the commutant of $\mathcal{A}_0$ in the crossed product-like ring $\mathcal{A}_0 \diamond_\sigma^\alpha M$ is*

$$C_{\mathcal{A}}(\mathcal{A}_0) = \left\{ \sum_{s \in \sigma^{-1}(\mathrm{id}_{\mathcal{A}_0})} r_s u_s \in \mathcal{A}_0 \diamond_\sigma^\alpha M \;\middle|\; r_s \in \mathcal{A}_0 \right\}$$

where $\sigma^{-1}(\mathrm{id}_{\mathcal{A}_0}) = \{s \in M \mid \sigma_s = \mathrm{id}_{\mathcal{A}_0}\}$.

The following corollary can be derived directly from Corollary 24.4 together with either Corollary 24.3 or Corollary 24.5.

Corollary 24.6. *If* $\mathcal{A}_0 \diamond_\sigma^\alpha M$ *is a crossed product-like ring where* $\mathcal{A}_0$ *is an integral domain, then* $\sigma_s \neq \mathrm{id}_{\mathcal{A}_0}$ *for all* $s \in M \setminus \{e\}$ *if and only if* $\mathcal{A}_0$ *is maximal commutative in* $\mathcal{A}_0 \diamond_\sigma^\alpha M$.

We will now give a sufficient condition for $C_{\mathcal{A}}(\mathcal{A}_0)$ to be commutative.

Proposition 24.1. *If* $\mathcal{A}_0 \diamond_\sigma^\alpha M$ *is a crossed product-like ring where* $\mathcal{A}_0$ *is commutative,* M *is abelian and* $\alpha(s,t) = \alpha(t,s)$ *for all* $s,t \in M$*, then* $C_{\mathcal{A}}(\mathcal{A}_0)$ *is commutative.*

Proof. Let $\sum_{s \in M} r_s u_s$ and $\sum_{t \in M} p_t u_t$ be arbitrary elements in $C_{\mathcal{A}}(\mathcal{A}_0)$. By our assumptions and Corollary 24.1 we get

$$\begin{aligned}
\left(\sum_{s \in M} r_s u_s\right)\left(\sum_{t \in M} p_t u_t\right) &= \sum_{(s,t) \in M \times M} r_s \sigma_s(p_t) \alpha(s,t) u_{st} \\
&= \sum_{(s,t) \in M \times M} r_s p_t \alpha(s,t) u_{st} \\
&= \sum_{(s,t) \in M \times M} p_t \sigma_t(r_s) \alpha(t,s) u_{ts} \\
&= \left(\sum_{t \in M} p_t u_t\right)\left(\sum_{s \in M} r_s u_s\right).
\end{aligned}$$

□

24.4 The Center of a Crossed Product-Like Ring $\mathcal{A}_0 \diamond_\sigma^\alpha M$

In this section we will describe the center $Z(\mathcal{A}) = \{b \in \mathcal{A} \mid ab = ba, \forall a \in \mathcal{A}\}$ of a crossed product-like ring $\mathcal{A} = \mathcal{A}_0 \diamond_\sigma^\alpha M$. Note that $Z(\mathcal{A}_0 \diamond_\sigma^\alpha M) \subseteq C_{\mathcal{A}}(\mathcal{A}_0)$.

Proposition 24.2. *The center of a crossed product-like ring* $\mathcal{A} = \mathcal{A}_0 \diamond_\sigma^\alpha M$ *is*

$$Z(\mathcal{A}) = \left\{ \sum_{g \in M} r_g u_g \;\middle|\; \sum_{g \in M} r_g \alpha(g,s) u_{gs} = \sum_{g \in M} \sigma_s(r_g) \alpha(s,g) u_{sg} \right.$$

$$\left. r_s \sigma_s(a) = a r_s, \quad \forall a \in \mathcal{A}_0,\ s \in M \right\}.$$

Proof. Let $\sum_{g\in M} r_g u_g \in \mathcal{A}_0\diamond_\sigma^\alpha M$ be an element which commutes with every element in $\mathcal{A}_0\diamond_\sigma^\alpha M$. Then, in particular $\sum_{g\in M} r_g u_g$ must commute with every $a \in \mathcal{A}_0$. From Theorem 24.1 we immediately see that this implies $r_s\,\sigma_s(a) = a\,r_s$ for every $a \in \mathcal{A}_0$ and $s \in M$. Furthermore, $\sum_{g\in M} r_g u_g$ must commute with u_s for every $s \in M$. This yields

$$\sum_{g\in M} r_g\,\alpha(g,s)\,u_{gs} = \left(\sum_{g\in M} r_g u_g\right) u_s = u_s \left(\sum_{g\in M} r_g u_g\right) = \sum_{g\in M} \sigma_s(r_g)\,\alpha(s,g)\,u_{sg}$$

for each $s \in M$.

Conversely, suppose that $\sum_{g\in M} r_g u_g \in \mathcal{A}_0\diamond_\sigma^\alpha M$ is an element satisfying $r_s\,\sigma_s(a) = a\,r_s$ and $\sum_{g\in M} r_g\,\alpha(g,s)\,u_{gs} = \sum_{g\in M} \sigma_s(r_g)\,\alpha(s,g)\,u_{sg}$ for every $a \in \mathcal{A}_0$ and $s \in M$. Let $\sum_{s\in M} a_s u_s \in \mathcal{A}_0\diamond_\sigma^\alpha M$ be arbitrary. Then

$$\begin{aligned}
\left(\sum_{g\in M} r_g u_g\right)\left(\sum_{s\in M} a_s u_s\right) &= \sum_{(g,s)\in M\times M} r_g\,\sigma_g(a_s)\,\alpha(g,s)\,u_{gs} \\
&= \sum_{(g,s)\in M\times M} a_s\,r_g\,\alpha(g,s)\,u_{gs} \\
&= \sum_{s\in M} a_s \left(\sum_{g\in M} r_g\,\alpha(g,s)\,u_{gs}\right) \\
&= \sum_{s\in M} a_s \left(\sum_{g\in M} \sigma_s(r_g)\,\alpha(s,g)\,u_{sg}\right) \\
&= \sum_{(s,g)\in M\times M} a_s\,\sigma_s(r_g)\,\alpha(s,g)\,u_{sg} \\
&= \left(\sum_{s\in M} a_s u_s\right)\left(\sum_{g\in M} r_g u_g\right)
\end{aligned}$$

and hence $\sum_{g\in M} r_g u_g$ commutes with every element in $\mathcal{A}_0\diamond_\sigma^\alpha M$. □

A crossed product-like group graded ring offers a more simple description of its center.

Corollary 24.7. *The center of a crossed product-like group graded ring* $\mathcal{A} = \mathcal{A}_0\diamond_\sigma^\alpha G$ *is*

$$Z(\mathcal{A}) = \left\{ \sum_{g\in G} r_g u_g \;\middle|\; r_{ts^{-1}}\,\alpha(ts^{-1},s) = \sigma_s(r_{s^{-1}t})\,\alpha(s,s^{-1}t), \right.$$
$$\left. r_s\,\sigma_s(a) = a\,r_s, \quad \forall a \in \mathcal{A}_0,\ (s,t) \in G\times G \right\}.$$

Corollary 24.8. *The center of a crossed product-like ring $\mathcal{A} = \mathcal{A}_0 \Diamond_\sigma^\alpha G$ graded by an abelian group G is*

$$Z(\mathcal{A}) = \left\{ \sum_{g\in G} r_g u_g \,\middle|\, r_g\, \alpha(g,s) = \sigma_s(r_g)\, \alpha(s,g), \right.$$
$$\left. r_s\, \sigma_s(a) = a\, r_s, \quad \forall a \in \mathcal{A}_0,\ s \in G \right\}.$$

24.5 Intersection Theorems

For any group graded algebraic crossed product $\mathcal{A}$, where the base ring $\mathcal{A}_0$ is commutative, it was shown in [5] that

$$I \cap C_{\mathcal{A}}(\mathcal{A}_0) \neq \{0\}$$

for each non-zero two-sided ideal I in $\mathcal{A}$ (see also [6]). In this section we will investigate if this property holds for more general classes of rings. It turns out that it need not hold in a crossed product-like monoid graded ring. However, we shall see that under certain conditions it will hold for pre-crystalline graded rings and that it always holds for crystalline graded rings if $\mathcal{A}_0$ is assumed to be commutative.

24.5.1 Crossed Product-Like Monoid Graded Rings

Let X be a non-empty set and $\gamma : X \to X$ a map (not necessarily injective nor surjective). We define the following sets:

$$\begin{aligned}
\mathrm{Per}^n(\gamma) &= \{x \in X \mid \gamma^{\circ(n)}(x) = x\}, \quad n \in \mathbb{Z}_{\geq 0} \\
\mathrm{Per}(\gamma) &= \bigcup_{n\in\mathbb{Z}_{\geq 0}} \mathrm{Per}^n(\gamma) \\
\mathrm{Aper}(\gamma) &= X \setminus \mathrm{Per}(\gamma)
\end{aligned}$$

We let $\mathbb{C}^X$ denote the algebra of complex-valued functions on X, with addition and multiplication being the usual pointwise operations. We define a map

$$\sigma : \mathbb{Z}_{\geq 0} \to \mathrm{End}(\mathbb{C}^X)$$

by

$$\sigma_n(f) = f \circ \gamma^{\circ(n)}, \quad f \in \mathbb{C}^X$$

for $n \in \mathbb{Z}_{\geq 0}$. Let $\mathcal{A}$ be the ring consisting of finite sums of the form $\sum_{s\in\mathbb{Z}_{\geq 0}} f_s u_s$ where $f_s \in \mathbb{C}^X$ for $s \in \mathbb{Z}_{\geq 0}$, and where u_i are non-zero elements satisfying $u_0 = 1$, $u_s u_t = u_{s+t}$ for all $s,t \in \mathbb{Z}_{\geq 0}$ and

$$u_s f = \sigma_s(f) u_s , \quad \text{for } f \in \mathbb{C}^X .$$

It is easy to see that $\mathcal{A} = \mathbb{C}^X \diamond_\sigma \mathbb{Z}_{\geq 0}$ is a crossed product-like ring. We omit to indicate the map $\alpha : \mathbb{Z}_{\geq 0} \times \mathbb{Z}_{\geq 0} \to \mathbb{C}^X$ since it is mapped onto the identity in $\mathbb{C}^X$ everywhere.

Proposition 24.3. *With conventions and notation as above, if* $\mathrm{Aper}(\gamma) \neq \emptyset$*, then there exists a non-zero two-sided ideal* I *in* $\mathcal{A} = \mathbb{C}^X \diamond_\sigma \mathbb{Z}_{\geq 0}$ *such that*

$$I \cap C_{\mathcal{A}}(\mathbb{C}^X) = \{0\}.$$

Proof. Suppose that $\mathrm{Aper}(\gamma) \neq \emptyset$. Clearly, $\gamma(\mathrm{Per}(\gamma)) \subseteq \mathrm{Per}(\gamma)$ and $\gamma(\mathrm{Aper}(\gamma)) \subseteq \mathrm{Aper}(\gamma)$. For the pre-images we have $\gamma^{-1}(\mathrm{Per}(\gamma)) \subseteq \mathrm{Per}(\gamma)$ and $\gamma^{-1}(\mathrm{Aper}(\gamma)) \subseteq \mathrm{Aper}(\gamma)$. Choose some $f \in \mathbb{C}^X$ such that $\mathrm{supp}(f) \subseteq \mathrm{Aper}(\gamma)$ and let I be the ideal in $\mathcal{A}$ generated by $f u_1$. An arbitrary element $c \in I$ may be written as

$$c = \sum_{k \in \mathbb{Z}_{\geq 0}} \left(\sum_{r_1, r_2 \in \mathbb{Z}_{\geq 0}} g_{r_1,k} u_{r_1} f u_1 h_{r_2,k} u_{r_2} \right)$$

where $g_{i,j}, h_{i,j} \in \mathbb{C}^X$ for $i, j \in \mathbb{Z}_{\geq 0}$. This may be rewritten as

$$c = \sum_{t \in \mathbb{Z}_{>0}} \underbrace{\left(\sum_{\substack{r_1, r_2 \in \mathbb{Z}_{\geq 0} \\ r_1 + r_2 + 1 = t}} \sum_{k \in \mathbb{Z}_{\geq 0}} g_{r_1,k} \sigma_{r_1}(f) \sigma_{r_1+1}(h_{r_2,k}) \right)}_{=a_t} u_t .$$

By elementary properties of the support of a function, we get

$$\mathrm{supp}(a_t) \subseteq \bigcup_{\substack{r_1, r_2 \in \mathbb{Z}_{\geq 0} \\ r_1 + r_2 + 1 = t}} \bigcup_{k \in \mathbb{Z}_{\geq 0}} \left(\mathrm{supp}(g_{r_1,k}) \cap (\gamma^{\circ(r_1)})^{-1}(\mathrm{supp}(f)) \cap (\gamma^{\circ(r_1+1)})^{-1}(\mathrm{supp}(h_{r_2,k})) \right).$$

from which we conclude that

$$\mathrm{supp}(a_t) \subseteq \bigcup_{\substack{r_1 \in \mathbb{Z}_{\geq 0} \\ r_1 + 1 = t}} (\gamma^{\circ(r_1)})^{-1}(\mathrm{supp}(f)). \tag{24.3}$$

It follows from Corollary 24.1 that $\sum_{t \in \mathbb{Z}_{\geq 0}} a_t u_t$ lies in $C_{\mathcal{A}}(\mathbb{C}^X)$ if and only if $\mathrm{supp}(a_t) \subseteq \mathrm{Per}^t(\gamma)$ for each $t > 0$. If we can show that $\mathrm{Per}^t(\gamma) \bigcap \mathrm{supp}(a_t) = \emptyset$ for every t, then we are finished. Suppose that there exists some t for which $\mathrm{Per}^t(\gamma) \bigcap \mathrm{supp}(a_t) \neq \emptyset$. It now follows by (24.3) that

$$\mathrm{Per}^t(\gamma) \bigcap \bigcup_{\substack{r_1 \in \mathbb{Z}_{\geq 0} \\ r_1 + 1 = t}} (\gamma^{\circ(r_1)})^{-1}(\mathrm{supp}(f)) \neq \emptyset$$

which is absurd, after noting that $\bigcup_{r_1 \in \mathbb{Z}_{\geq 0}, r_1+1=t} (\gamma^{\rho(r_1)})^{-1}(\mathrm{supp}(f)) \subseteq \mathrm{Aper}(\gamma)$, $\mathrm{Per}^t(\gamma) \subseteq \mathrm{Per}(\gamma)$ for each $t \in \mathbb{Z}_{>0}$ and that $\mathrm{Per}(\gamma) \cap \mathrm{Aper}(\gamma) = \emptyset$. This concludes the proof. □

Remark 24.2. Let X be a non-empty set and $\gamma : X \to X$ a bijection. The bijection gives rise to a map $\tilde{\gamma} : \mathbb{Z} \to \mathrm{Aut}(\mathbb{C}^X)$ given by $\tilde{\gamma}_n(f) = f \circ \gamma^{\rho(n)}$, for $f \in \mathbb{C}^X$, and we may define the $\mathbb{Z}$-graded crossed product $\mathcal{A} = \mathbb{C}^X \rtimes_{\tilde{\gamma}} \mathbb{Z}$ (see for example [7–9]). It follows from [9, Theorem 3.1] or more generally from [5, Theorem 2], that

$$I \cap C_{\mathcal{A}}(\mathbb{C}^X) \neq \{0\}$$

for each non-zero two-sided ideal I in $\mathbb{C}^X \rtimes_{\tilde{\gamma}} \mathbb{Z}$.

Proposition 24.3 and Remark 24.2 display a difference between monoid graded crossed product-like rings and group graded crossed products.

24.5.2 Group Graded Pre-Crystalline and Crystalline Graded Rings

Given a pre-crystalline graded ring $\mathcal{A} = \mathcal{A}_0 \Diamond_\sigma^\alpha G$, for each $b \in \mathcal{A}_0$ we define the commutator to be

$$D_b : \mathcal{A} \to \mathcal{A}, \quad \sum_{s \in G} a_s u_s \mapsto b\left(\sum_{s \in G} a_s u_s\right) - \left(\sum_{s \in G} a_s u_s\right) b.$$

From the definition of the multiplication we have

$$\begin{aligned} D_b\left(\sum_{s \in G} a_s u_s\right) &= b\left(\sum_{s \in G} a_s u_s\right) - \left(\sum_{s \in G} a_s u_s\right) b \\ &= \left(\sum_{s \in G} b a_s u_s\right) - \left(\sum_{s \in G} a_s \sigma_s(b) u_s\right) \\ &= \sum_{s \in G} \big(b a_s - a_s \sigma_s(b)\big) u_s \end{aligned}$$

for each $b \in \mathcal{A}_0$.

Theorem 24.2. *If $\mathcal{A} = \mathcal{A}_0 \Diamond_\sigma^\alpha G$ is a pre-crystalline group graded ring where $\mathcal{A}_0$ is commutative and for each $\sum_{s \in G} a_s u_s \in \mathcal{A} \setminus C_{\mathcal{A}}(\mathcal{A}_0)$ there exists $s \in G$ such that $a_s \notin \mathrm{Ker}(\sigma_s \circ \sigma_{s^{-1}})$, then*

$$I \cap C_{\mathcal{A}}(\mathcal{A}_0) \neq \{0\}$$

for every non-zero two-sided ideal I in $\mathcal{A}$.

Proof. Let I be an arbitrary non-zero two-sided ideal in $\mathcal{A}$ and assume that $\mathcal{A}_0$ is commutative and that for each $\sum_{s \in G} a_s u_s \in \mathcal{A} \setminus C_{\mathcal{A}}(\mathcal{A}_0)$ there exists $s \in G$ such that $a_s \notin \mathrm{Ker}(\sigma_s \circ \sigma_{s^{-1}})$. For each $g \in G$ we may define a *translation operator*

$$T_g : \mathcal{A} \to \mathcal{A}, \quad \sum_{s\in G} a_s u_s \mapsto \left(\sum_{s\in G} a_s u_s\right) u_g.$$

Note that, for each $g \in G$, I is invariant under T_g. We have

$$T_g\left(\sum_{s\in G} a_s u_s\right) = \left(\sum_{s\in G} a_s u_s\right) u_g = \sum_{s\in G} a_s \alpha(s,g)\, u_{sg}$$

for every $g \in G$. By the assumptions and together with [3, Corollary 2.4] it is clear that for each element $c \in \mathcal{A} \setminus C_{\mathcal{A}}(\mathcal{A}_0)$ it is always possible to choose some $g \in G$ and let T_g operate on c to end up with an element where the coefficient in front of u_e is non-zero.

Note that, for each $b \in \mathcal{A}_0$, I is invariant under D_b. Furthermore, we have

$$\begin{aligned} D_b\left(\sum_{s\in G} a_s u_s\right) &= \sum_{s\in G} (b a_s - a_s \sigma_s(b))\, u_s \\ &= \sum_{s\neq e} (b a_s - a_s \sigma_s(b))\, u_s = \sum_{s\neq e} d_s u_s \end{aligned}$$

since $(b a_e - a_e \sigma_e(b)) = b a_e - a_e b = 0$. Note that $C_{\mathcal{A}}(\mathcal{A}_0) = \bigcap_{b\in\mathcal{A}_0} \mathrm{Ker}(D_b)$ and hence for any $\sum_{s\in G} a_s u_s \in \mathcal{A} \setminus C_{\mathcal{A}}(\mathcal{A}_0)$ we are always able to choose $b \in \mathcal{A}_0$ and the corresponding D_b and have $\sum_{s\in G} a_s u_s \notin \mathrm{Ker}(D_b)$. Therefore we can always pick an operator D_b which kills the coefficient in front of u_e without killing everything. Hence, if $a_e \neq 0_{\mathcal{A}_0}$, the number of non-zero coefficients of the resulting element will always be reduced by at least one.

The ideal I is assumed to be non-zero, which means that we can pick some non-zero element $\sum_{s\in G} r_s u_s \in I$. If $\sum_{s\in G} r_s u_s \in C_{\mathcal{A}}(\mathcal{A}_0)$, then we are finished, so assume that this is not the case. Note that $r_s \neq 0_{\mathcal{A}_0}$ for finitely many $s \in G$. Recall that the ideal I is invariant under T_g and D_a for all $g \in G$ and $a \in \mathcal{A}_0$. We may now use the operators $\{T_g\}_{g\in G}$ and $\{D_a\}_{a\in\mathcal{A}_0}$ to generate new elements of I. More specifically, we may use the T_g:s to translate our element $\sum_{s\in G} r_s u_s$ into a new element which has a non-zero coefficient in front of u_e (if needed) after which we use the D_a operator to kill this coefficient and end up with yet another new element of I which is non-zero but has a smaller number of non-zero coefficients. We may repeat this procedure and in a finite number of iterations arrive at an element of I which lies in $C_{\mathcal{A}}(\mathcal{A}_0) \setminus \mathcal{A}_0$, and if not we continue the above procedure until we reach an element in $\mathcal{A}_0 \setminus \{0_{\mathcal{A}_0}\}$. In particular $\mathcal{A}_0 \subseteq C_{\mathcal{A}}(\mathcal{A}_0)$ since $\mathcal{A}_0$ is commutative and hence $I \cap C_{\mathcal{A}}(\mathcal{A}_0) \neq \{0\}$. □

Corollary 24.9. *If $\mathcal{A} = \mathcal{A}_0 \diamond_\sigma^\alpha G$ is a pre-crystalline group graded ring where $\mathcal{A}_0$ is maximal commutative and for each $\sum_{s\in G} a_s u_s \in \mathcal{A} \setminus \mathcal{A}_0$ there exists $s \in G$ such that $a_s \notin \mathrm{Ker}(\sigma_s \circ \sigma_{s^{-1}})$, then*

$$I \cap \mathcal{A}_0 \neq \{0\}$$

for every non-zero two-sided ideal I in $\mathcal{A}$.

A crystalline graded ring has no $S(G)$-torsion and hence $\mathrm{Ker}(\sigma_s \circ \sigma_{s^{-1}}) = \{0_{\mathcal{A}_0}\}$ by [3, Corollary 2.4]. Therefore we get the following corollary which is a generalization of a result for algebraic crossed products in [5, Theorem 2].

Corollary 24.10. *If $\mathcal{A} = \mathcal{A}_0 \diamond_\sigma^\alpha G$ is a crystalline graded ring where $\mathcal{A}_0$ is commutative, then*

$$I \cap C_{\mathcal{A}}(\mathcal{A}_0) \neq \{0\}$$

for every non-zero two-sided ideal I in $\mathcal{A}$.

When $\mathcal{A}_0$ is maximal commutative we get the following corollary.

Corollary 24.11. *If $\mathcal{A}_0 \diamond_\sigma^\alpha G$ is a crystalline graded ring where $\mathcal{A}_0$ is maximal commutative, then*

$$I \cap \mathcal{A}_0 \neq \{0\}$$

for every non-zero two-sided ideal I in $\mathcal{A}_0 \diamond_\sigma^\alpha G$.

24.6 Examples of Crossed Product-Like and Crystalline Graded Rings

Example 24.1 (The Quantum Torus). Let $q \in \mathbb{C} \setminus \{0,1\}$. Denote by $\mathbb{C}_q[x,x^{-1},y,y^{-1}]$ the unital ring of polynomials over $\mathbb{C}$ with four generators $\{x,x^{-1},y,y^{-1}\}$ and the defining commutation relations $xx^{-1} = x^{-1}x = 1$, $yy^{-1} = y^{-1}y = 1$ and

$$yx = qxy. \tag{24.4}$$

This means that $\mathbb{C}_q[x,x^{-1},y,y^{-1}] = \frac{\mathbb{C}\langle x,x^{-1},y,y^{-1}\rangle}{\{xx^{-1}=x^{-1}x=1,\ yy^{-1}=y^{-1}y=1,\ yx=qxy\}}$. This ring is sometimes called *the twisted Laurent polynomial ring* or *the quantum torus*. Following the notation of Definition 24.2, let $\mathcal{A}_0 = \mathbb{C}[x,x^{-1}]$, $M = (\mathbb{Z},+)$ and $u_n = y^n$ for $n \in M$. It is not difficult to show that

$$\mathcal{A} = \mathbb{C}_q[x,x^{-1},y,y^{-1}] = \bigoplus_{n\in\mathbb{Z}} \mathbb{C}[x,x^{-1}]y^n = \bigoplus_{n\in M} \mathcal{A}_0\, u_n$$

and that the other conditions in Definition 24.2 are satisfied as well. Therefore, $\mathbb{C}_q[x,x^{-1},y,y^{-1}]$ can be viewed as a crossed product-like group graded ring. From the defining commutation relations and the choice of u_n, it follows that $\sigma_n : P(x) \mapsto P(q^n x)$ for $n \in M$ and $P(x) \in \mathcal{A}_0$, and that $\alpha(s,t) = 1_{\mathcal{A}}$ for all $s,t \in M$, following the notation of Lemma 24.1.

In the current example, $\mathcal{A}_0$ is an integral domain. Thus, by Corollary 24.6, the subring $\mathcal{A}_0 = \mathbb{C}[x,x^{-1}]$ is maximal commutative in $\mathbb{C}_q[x,x^{-1},y,y^{-1}]$ if and only if $q^n x \neq x$ for every $n \neq 0$ or equivalently if and only if q is not root of unity.

Example 24.2 (Twisted Functions in the Complex Plane). Let $\mathbb{C}^{\mathbb{C}}$ be the algebra of functions $\mathbb{C} \to \mathbb{C}$, with addition and multiplication being the usual pointwise operations. Fix a pair of numbers $q \in \mathbb{C} \setminus \{0\}$ and $d \in \mathbb{Z}_{>0}$ and consider the map

$$\gamma : \mathbb{C} \to \mathbb{C}, \quad z \mapsto qz^d.$$

We define

$$\sigma_n : \mathbb{C}^{\mathbb{C}} \to \mathbb{C}^{\mathbb{C}}, \quad f \mapsto f \circ \gamma^{\circ(n)}$$

for $n \in \mathbb{Z}_{>0}$ and set $\sigma_0 = \mathrm{id}_{\mathbb{C}^{\mathbb{C}}}$. If $n \geq 2$, then this yields

$$(\sigma_n(f))(z) = f(q^{1+d+\ldots+d^{n-1}} z^{d^n})$$

for $f \in \mathbb{C}^{\mathbb{C}}$ and $z \in \mathbb{C}$. Let $\mathcal{A}$ be the algebra consisting of finite sums of the form $\sum_{i \in \mathbb{Z}_{\geq 0}} f_i u_i$, where $f_i \in \mathbb{C}^{\mathbb{C}}$ for each $i \in \mathbb{Z}_{\geq 0}$, and u_i are non-zero elements satisfying:

1. $u_0 = 1$
2. $u_n u_m = u_{n+m}, \quad$ for all $n, m \in \mathbb{Z}_{\geq 0}$

and also such that u_i does not in general commute with $\mathbb{C}^{\mathbb{C}}$, but satisfy

$$u_n f = \sigma_n(f) u_n, \quad \text{for } f \in \mathbb{C}^{\mathbb{C}} \text{ and } n \in \mathbb{Z}_{\geq 0}.$$

One may easily verify that this corresponds to the crossed product-like monoid graded ring $\mathbb{C}^{\mathbb{C}} \diamond_\sigma \mathbb{Z}_{\geq 0}$. Consider the set

$$\mathrm{Per}^n(\gamma) = \{z \in \mathbb{C} \mid \sigma_n(f)(z) = f(z), \quad \forall f \in \mathbb{C}^{\mathbb{C}}\}. \tag{24.5}$$

Note that $\mathrm{Per}^0(\gamma) = \mathbb{C}$ and that $\mathrm{Per}^1(\gamma)$ contains the solutions to the equation $qz^d = z$. For $n \in \mathbb{Z}_{\geq 2}$ we have

$$\mathrm{Per}^n(\gamma) = \{z \in \mathbb{C} \mid q^{1+d+\ldots+d^{n-1}} z^{d^n} = z\}. \tag{24.6}$$

By using the formula for a geometric series, it is easy to see that, for each $n \in \mathbb{Z}_{\geq 1}$, Per^n contains the point $z = 0$ and $d^n - 1$ points equally distributed along a circle with radius $r = |q|^{\frac{1}{1-d}}$ and with its center in the origin in the complex plane.

The following proposition follows from Corollary 24.1.

Proposition 24.4. *With notation and definitions as above, for $\mathcal{A} = \mathbb{C}^{\mathbb{C}} \diamond_\sigma \mathbb{Z}_{\geq 0}$ we have*

$$C_{\mathcal{A}}(\mathbb{C}^{\mathbb{C}}) = \left\{ \sum_{i \in \mathbb{Z}_{\geq 0}} f_i u_i \;\middle|\; \mathrm{supp}(f_i) \subseteq \mathrm{Per}^i(\gamma), \quad i \in \mathbb{Z}_{\geq 0} \right\}.$$

It now becomes clear that $\mathbb{C}^{\mathbb{C}}$ is never maximal commutative in $\mathbb{C}^{\mathbb{C}} \diamond_\sigma \mathbb{Z}_{\geq 0}$. In fact Proposition 24.4 makes it possible to explicitly provide the elements in the commutant $C_{\mathcal{A}}(\mathbb{C}^{\mathbb{C}})$ that are not in $\mathbb{C}^{\mathbb{C}}$. If $f_0, f_1, f_2 \in \mathbb{C}^{\mathbb{C}}$, then by using (24.5) and (24.6) and Proposition 24.4, we conclude that $a = f_0 + f_1 u_1 + f_2 u_2$ lies in $C_{\mathcal{A}}(\mathbb{C}^{\mathbb{C}})$ if and only if, $\mathrm{supp}(f_1)$ respectively $\mathrm{supp}(f_2)$ is contained in $\mathrm{Per}^1(\gamma)$ respectively $\mathrm{Per}^2(\gamma)$.

Example 24.3 (The First Weyl Algebra). Following [3], let

$$A_1(\mathbb{C}) = \mathbb{C}\langle x,y\rangle/(yx - xy - 1)$$

be the first Weyl algebra. If we put $\deg(x) = 1$ and $\deg(y) = -1$ and

$$\begin{aligned} A_1(\mathbb{C})_0 &= \mathbb{C}[xy] \\ A_1(\mathbb{C})_n &= \mathbb{C}[xy]x^n, \text{ for } n \geq 0 \\ A_1(\mathbb{C})_m &= \mathbb{C}[xy]y^m, \text{ for } m \leq 0 \end{aligned}$$

then this defines a $\mathbb{Z}$-gradation on $A_1(\mathbb{C})$. We set $u_n = x^n$ if $n \geq 0$ and $u_m = y^{-m}$ if $m \leq 0$. It is clear that $\sigma_1(xy) = xy - 1$ because $x(xy) = (xy-1)x$ and $\sigma_{-1}(xy) = xy + 1$ because $y(xy) = (1+xy)y$. Let us put $t = xy$, then

$$\sigma : \mathbb{Z} \to \mathrm{Aut}_{\mathbb{C}}\, \mathbb{C}[t], \quad n \mapsto (t \mapsto t - n).$$

We can also calculate, for example

$$\begin{aligned} \alpha(n,-n) &= x^n y^n = x^{n-1} t y^{n-1} = (t-(n-1))x^{n-1}y^{n-1} \\ &= (t-n+1)\cdot(t-n+2)\cdot\ldots\cdot(t-2)\cdot(t-1)\cdot t. \end{aligned}$$

Furthermore $\alpha(n,-m)$ with $n > m$ $(n,m \in \mathbb{Z}_{\geq 0})$ can be calculated from

$$x^n y^m = x^{n-m} x^m y^m = x^{n-m}\alpha(m,-m) = \sigma_{n-m}(\alpha(m,-m))x^{n-m}$$

and so

$$\alpha(n,-m) = \sigma_{n-m}(\alpha(m,-m)).$$

It is well-known that $A_1(\mathbb{C})$ is a simple algebra and hence the only non-zero two-sided ideal is $A_1(\mathbb{C})$ itself, which clearly has a non-zero intersection with $\mathbb{C}[xy]$. Furthermore, $\mathbb{C}[xy]$ is an integral domain and $\sigma_n \neq \mathrm{id}_{\mathbb{C}[xy]}$ for each $n \neq 0$ and by Corollary 24.6, the base ring $\mathbb{C}[xy]$ is maximal commutative in $A_1(\mathbb{C})$.

Example 24.4 (Generalized Weyl Algebras). Let $\mathcal{A}_0$ be an associative and unital ring and fix a positive integer n and set $\mathbf{n} = \{1,2,\ldots,n\}$ and let $\sigma = (\sigma_1,\sigma_2,\ldots,\sigma_n)$ be a set of commuting automorphisms of $\mathcal{A}_0$. Let $\mathbf{a} = (a_1,a_2,\ldots,a_n)$ be an n-tuple with nonzero entries in $Z(\mathcal{A}_0)$ such that $\sigma_i(a_j) = a_j$ for $i \neq j$. The *generalized Weyl algebra* $\mathcal{A} = \mathcal{A}_0(\sigma,\mathbf{a})$ is defined as the ring generated by $\mathcal{A}_0$ and $2n$ symbols $X_1,\ldots,X_n,Y_1,\ldots,Y_n$ satisfying the following rules:

1. For each $i \in \{1,\ldots,n\}$:

$$Y_iX_i = a_i, \text{ and } \quad X_iY_i = \sigma_i(a_i)$$

2. For all $d \in \mathcal{A}_0$ and each $i \in \{1,\ldots,n\}$,

$$X_id = \sigma_i(d)X_i, \text{ and } \quad Y_id = \sigma_i^{-1}(d)Y_i$$

3. For $i \neq j$,

$$[Y_i, Y_j] = 0$$
$$[X_i, X_j] = 0$$
$$[X_i, Y_j] = 0$$

where $[x,y] = xy - yx$.

Now for $m \in \mathbb{Z}$ we write $u_m(i) = (X_i)^m$ if $m > 0$ and $u_m(i) = (Y_i)^{-m}$ if $m < 0$. For $\mathbf{k} = (k_1, k_2, \ldots, k_n) \in \mathbb{Z}^n$ we set $u_{\mathbf{k}} = u_{k_1}(1) \cdot \ldots \cdot u_{k_n}(n)$. By putting $\mathcal{A} = \oplus_{\mathbf{k} \in \mathbb{Z}^n} \mathcal{A}_{\mathbf{k}}$ where $\mathcal{A}_{\mathbf{k}} = \mathcal{A}_0 u_{\mathbf{k}}$, we see that $\mathcal{A}$ is a $\mathbb{Z}^n$-graded ring, which is crystalline graded (see [3]). If the base ring $\mathcal{A}_0$ is commutative, just like in many of the examples of this class of rings, we get by Corollary 24.10 that each nonzero two-sided ideal in the generalized Weyl algebra $\mathcal{A}$ contains a non-zero element which commutes with all of $\mathcal{A}_0$.

Acknowledgements We are grateful to Freddy Van Oystaeyen for useful discussions on the topic of this article. This work was partially supported by the Crafoord Foundation, The Royal Physiographic Society in Lund, The Royal Swedish Academy of Sciences, The Swedish Foundation of International Cooperation in Research and Higher Education (STINT), The Swedish Research Council and "LieGrits", a Marie Curie Research Training Network funded by the European Community as project MRTN-CT 2003-505078.

References

1. Caenepeel, S., Van Oystaeyen, F.: Brauer groups and the cohomology of graded rings. Monographs and Textbooks in Pure and Applied Mathematics, 121. Marcel Dekker, New York (1988)
2. Năstăsescu, C., Van Oystaeyen, F.: Methods of Graded Rings. Lecture Notes in Mathematics, 1836. Springer, Berlin (2004)
3. Nauwelaerts, E., Van Oystaeyen, F.: Introducing crystalline graded algebras. Algebr. Represent. Theory **11**, No. 2, 133–148 (2008)
4. Neijens, T., Van Oystaeyen, F., Yu, W.W.: Centers of certain crystalline graded rings. Preprint in preparation (2007)
5. Oinert, J., Silvestrov, S.D.: Commutativity and ideals in algebraic crossed products. To appear in J. Gen. Lie. T. Appl. (2008)
6. Oinert, J., Silvestrov, S.D.: On a correspondence between ideals and commutativity in algebraic crossed products. J. Gen. Lie. T. Appl. **2**, No. 3, 216–220 (2008)
7. Svensson, C., Silvestrov, S., de Jeu, M.: Dynamical systems and commutants in crossed products. Int. J. Math. **18**, no. 4, 455–471 (2007)
8. Svensson, C., Silvestrov, S., de Jeu, M.: Connections between dynamical systems and crossed products of Banach algebras by $\mathbb{Z}$. Preprints in Mathematical Sciences 2007:5, ISSN 1403-9338, LUTFMA-5081-2007, Centre for Mathematical Sciences, Lund University (2007). Available via arXiv: http://arxiv.org/abs/math/0702118v3
9. Svensson, C., Silvestrov, S., de Jeu, M.: Dynamical systems associated with crossed products. Preprints in Mathematical Sciences 2007:22, ISSN 1403-9338, LUTFMA-5088-2007, Centre for Mathematical Sciences, Lund University (2007). Available via arXiv: http://arxiv.org/abs/0707.1881v2

10. Tomiyama, J.: Invitation to C^*-algebras and topological dynamics. World Scientific Advanced Series in Dynamical Systems, 3. World Scientific, Singapore (1987)
11. Tomiyama, J.: The interplay between topological dynamics and theory of C^*-algebras. Lecture Notes Series, 2. Global Anal. Research Center, Seoul (1992)
12. Tomiyama, J.: The interplay between topological dynamics and theory of C^*-algebras. Part 2 (after the Seoul lecture note 1992). Preprint (2000)

Chapter 25
Decomposition of the Enveloping Algebra so(5)

Čestmír Burdík and Ondřej Navrátil

Abstract The adjoint representation of so(5) on its universal enveloping algebra $U(\mathrm{so}(5))$ is explicitly decomposed into irreducible components. It is shown that the commutant of raising operators of so(5) is generated by seven elements. The explicit form of these elements is given.

25.1 Introduction

Throughout the paper we will denote by $\mathfrak{g}$ the Lie algebra and by $U(\mathfrak{g})$ its enveloping algebra [1].

We will study $U(\mathfrak{g})$ as the vector space, on which $\mathfrak{g}$ is represented via the adjoint action ρ, i.e.

$$\rho(x)a = [x, a] \quad \text{for } x \in \mathfrak{g} \quad \text{and } a \in U(\mathfrak{g}) .$$

In order to decompose the representation ρ it is sufficient to determine the space $\mathscr{B}$ of the highest weight vectors. For $\mathfrak{g} = \mathrm{sl}(2)$ this program was very easily solved, see [2]. Let $\{\mathbf{E}_{12}, \mathbf{E}_{21}, \mathbf{H}_1\}$ be a basis in sl(2). Then the vector space of all highest weight vectors has basis $\mathbf{C}_2^m \mathbf{E}_{12}^n$, $n, m = 0, 1, 2, \ldots$, where $\mathbf{C}_2$ is a Casimir operator of sl(2). It is trivial to see that all such elements are commutant of $\{\mathbf{E}_{12}\}$.

The case of algebra sl(3) was solved in [3] and generally in [4]. The commutant $\mathscr{B}$ of $\{\mathbf{E}_{12}, \mathbf{E}_{23}\}$ in $U(\mathrm{sl}(3))$ was explicitly described. It is generated by six elements and each of the highest weight vectors can by written as a polynomial in these elements.

Č. Burdík
Department of Mathematics, Faculty of Nuclear Sciences and Physical Engineering
Czech Technical University, Trojanova 13, 120 00 Prague 2, Czech Republic
e-mail: burdik@kmalpha.fjfi.cvut.cz

O. Navrátil
Department of Mathematics, Faculty of Transportation Sciences, Czech Technical University
Na Florenci 25, 110 00 Prague, Czech Republic
e-mail: navratil@fd.cvut.cz

S. Silvestrov et al. (eds.), *Generalized Lie Theory in Mathematics, Physics and Beyond,*

I this paper, we study the case of algebra $\mathfrak{g} = \mathrm{so}(5)$ and by construction we use the following well know fact:
Let $\mathfrak{g}$ be a Lie algebra with the basis $\mathbf{e}_k$, $k = 1, \ldots, n$, and structure constants c_{ij}^k. Then the mapping

$$\mathbf{e}_i \mapsto \sum_{j,k=1}^{n} c_{ij}^k x_k \frac{\partial}{\partial x_j} \tag{25.1}$$

forms a representation of algebra $\mathfrak{g}$ by differential operators of the first order.

This is the representation of $\mathrm{ad}_{\mathfrak{g}}$ on the enveloping algebra expressed in basis of symmetric polynomials in $U(\mathfrak{g})$. More precisely, if x_k is an image of $\mathbf{e}_k$ by canonical linear isomorphism $U(\mathfrak{g}) \to \mathrm{Sym}(\mathfrak{g})$, than the relation (25.1) gives the action of $\mathrm{ad}_{\mathbf{e}_i}$.

From definition (25.1) it is easy to see that for any N polynomials of the order of N form the invariant subspace of representation $\mathrm{ad}_{\mathfrak{g}}$.

25.2 The Lie Algebra so(5)

The basis of algebra $\mathfrak{g} = \mathrm{so}(5,\mathbb{C})$ consists of the elements $\mathbf{J}_{\alpha\beta} = -\mathbf{J}_{\beta\alpha}$, $\alpha, \beta = 1, 2, \ldots, 5$, which fulfill the commutation relations

$$\left[\mathbf{J}_{\alpha\beta}, \mathbf{J}_{\mu\nu}\right] = \delta_{\beta\mu}\mathbf{J}_{\alpha\nu} + \delta_{\alpha\nu}\mathbf{J}_{\beta\mu} - \delta_{\beta\nu}\mathbf{J}_{\alpha\mu} - \delta_{\alpha\mu}\mathbf{J}_{\beta\nu}\,. \tag{25.2}$$

If we define

$$\begin{array}{ll}
\mathbf{H}_1 = \mathrm{i}\mathbf{J}_{12} & \mathbf{H}_2 = \mathrm{i}\mathbf{J}_{34} \\
\mathbf{E}_1 = \frac{1}{\sqrt{2}}\left(\mathbf{J}_{45} + \mathrm{i}\mathbf{J}_{35}\right), & \mathbf{E}_2 = \frac{1}{2}\left(\mathbf{J}_{23} + \mathrm{i}\mathbf{J}_{13} - \mathbf{J}_{14} + \mathrm{i}\mathbf{J}_{24}\right), \\
\mathbf{E}_3 = \frac{1}{\sqrt{2}}\left(\mathbf{J}_{15} - \mathrm{i}\mathbf{J}_{25}\right), & \mathbf{E}_4 = \frac{1}{2}\left(\mathbf{J}_{23} + \mathrm{i}\mathbf{J}_{13} + \mathbf{J}_{14} - \mathrm{i}\mathbf{J}_{24}\right), \\
\mathbf{F}_1 = \frac{1}{\sqrt{2}}\left(-\mathbf{J}_{45} + \mathrm{i}\mathbf{J}_{35}\right), & \mathbf{F}_2 = \frac{1}{2}\left(-\mathbf{J}_{23} + \mathrm{i}\mathbf{J}_{13} + \mathbf{J}_{14} + \mathrm{i}\mathbf{J}_{24}\right), \\
\mathbf{F}_3 = \frac{1}{\sqrt{2}}\left(-\mathbf{J}_{15} - \mathrm{i}\mathbf{J}_{25}\right), & \mathbf{F}_4 = \frac{1}{2}\left(-\mathbf{J}_{23} + \mathrm{i}\mathbf{J}_{13} - \mathbf{J}_{14} - \mathrm{i}\mathbf{J}_{24}\right),
\end{array}$$

we obtain from (25.2) the commutation relations

$$\begin{array}{llll}
\left[\mathbf{H}_1, \mathbf{E}_1\right] = 0, & \left[\mathbf{H}_1, \mathbf{E}_2\right] = \mathbf{E}_2, & \left[\mathbf{H}_1, \mathbf{E}_3\right] = \mathbf{E}_3, & \left[\mathbf{H}_1, \mathbf{E}_4\right] = \mathbf{E}_4, \\
\left[\mathbf{H}_2, \mathbf{E}_1\right] = \mathbf{E}_1, & \left[\mathbf{H}_2, \mathbf{E}_2\right] = -\mathbf{E}_2, & \left[\mathbf{H}_2, \mathbf{E}_3\right] = 0, & \left[\mathbf{H}_2, \mathbf{E}_4\right] = \mathbf{E}_4 \\
\left[\mathbf{H}_1, \mathbf{F}_1\right] = 0, & \left[\mathbf{H}_1, \mathbf{F}_2\right] = -\mathbf{F}_2, & \left[\mathbf{H}_1, \mathbf{F}_3\right] = -\mathbf{F}_3, & \left[\mathbf{H}_1, \mathbf{F}_4\right] = -\mathbf{F}_4, \\
\left[\mathbf{H}_2, \mathbf{F}_1\right] = -\mathbf{F}_1, & \left[\mathbf{H}_2, \mathbf{F}_2\right] = \mathbf{F}_2, & \left[\mathbf{H}_2, \mathbf{F}_3\right] = 0, & \left[\mathbf{H}_2, \mathbf{F}_4\right] = -\mathbf{F}_4 \\
\left[\mathbf{E}_1, \mathbf{E}_2\right] = \mathbf{E}_3, & \left[\mathbf{E}_1, \mathbf{E}_3\right] = \mathbf{E}_4, & \left[\mathbf{E}_1, \mathbf{E}_4\right] = 0, & \\
\left[\mathbf{E}_2, \mathbf{E}_3\right] = 0, & \left[\mathbf{E}_2, \mathbf{E}_4\right] = 0, & \left[\mathbf{E}_3, \mathbf{E}_4\right] = 0, & \\
\left[\mathbf{F}_1, \mathbf{F}_2\right] = -\mathbf{F}_3, & \left[\mathbf{F}_1, \mathbf{F}_3\right] = -\mathbf{F}_4, & \left[\mathbf{F}_1, \mathbf{F}_4\right] = 0, & \\
\left[\mathbf{F}_2, \mathbf{F}_3\right] = 0, & \left[\mathbf{F}_2, \mathbf{F}_4\right] = 0, & \left[\mathbf{F}_3, \mathbf{F}_4\right] = 0, & \\
\left[\mathbf{E}_1, \mathbf{F}_1\right] = \mathbf{H}_2, & \left[\mathbf{E}_1, \mathbf{F}_2\right] = 0, & \left[\mathbf{E}_1, \mathbf{F}_3\right] = -\mathbf{F}_2, & \left[\mathbf{E}_1, \mathbf{F}_4\right] = -\mathbf{F}_3
\end{array}$$

$$
\begin{array}{llll}
[\mathbf{E}_2,\mathbf{F}_1]=0, & [\mathbf{E}_2,\mathbf{F}_2]=\mathbf{H}_1-\mathbf{H}_2, & [\mathbf{E}_2,\mathbf{F}_3]=\mathbf{F}_1, & [\mathbf{E}_2,\mathbf{F}_4]=0\\
[\mathbf{E}_3,\mathbf{F}_1]=-\mathbf{E}_2, & [\mathbf{E}_3,\mathbf{F}_2]=\mathbf{E}_1, & [\mathbf{E}_3,\mathbf{F}_3]=\mathbf{H}_1, & [\mathbf{E}_3,\mathbf{F}_4]=\mathbf{F}_1,\\
[\mathbf{E}_4,\mathbf{F}_1]=-\mathbf{E}_3, & [\mathbf{E}_4,\mathbf{F}_2]=0, & [\mathbf{E}_4,\mathbf{F}_3]=\mathbf{E}_1, & [\mathbf{E}_4,\mathbf{F}_4]=\mathbf{H}_1+\mathbf{H}_2
\end{array}
$$

which we will use in further calculation.

25.3 The Highest Weight Vectors

The vectors $\mathbf{v}$ with the highest weights, i.e. the commutant $\mathscr{B}$ of the elements $\mathbf{E}_1$ and $\mathbf{E}_2$ in $U(\mathfrak{g})$, are given as solutions of the equations

$$\mathrm{ad}_{\mathbf{E}_1}\mathbf{v}=\mathrm{ad}_{\mathbf{E}_2}\mathbf{v}=0\,. \tag{25.3}$$

If we denote x_k, h_k and y_k the images of elements $\mathbf{E}_k$, $\mathbf{H}_k$ and $\mathbf{F}_k$ by mapping $U(\mathfrak{g})\to\mathrm{Sym}(\mathfrak{g})$, we obtain from (25.1) and (25.3) the system of equations

$$
\begin{aligned}
x_3\frac{\partial f}{\partial x_2}+x_4\frac{\partial f}{\partial x_3}-x_1\frac{\partial f}{\partial h_2}+h_2\frac{\partial f}{\partial y_1}-y_2\frac{\partial f}{\partial y_3}-y_3\frac{\partial f}{\partial y_4}&=0\,,\\
-x_3\frac{\partial f}{\partial x_1}-x_2\left(\frac{\partial f}{\partial h_1}-\frac{\partial f}{\partial h_2}\right)+(h_1-h_2)\frac{\partial f}{\partial y_2}+y_1\frac{\partial f}{\partial y_3}&=0\,.
\end{aligned}
\tag{25.4}
$$

The general solutions of (25.4) are

$$f=F\left(X_1,X_2,X_3,X_4,X_5,C_2\right),$$

where F is an arbitrary differentiable function in variables

$$
\begin{aligned}
X_1&=x_1\\
X_2&=2x_2x_4-x_3^2\\
X_3&=x_1x_2+x_3h_2-x_4y_1\\
X_4&=x_1^2x_2-x_1x_3(h_1-h_2)+\tfrac{1}{2}x_4(h_1-h_2)^2+(2x_2x_4-x_3^3)y_2\\
X_5&=x_1^2x_2x_3+x_1x_2x_4(h_1+h_2)-x_1x_3^2(h_1-h_2)-x_3x_4h_2(h_1-h_2)-\\
&\quad -x_1x_3x_4y_1+(2x_2x_4-x_3^2)(x_3y_2+x_4y_3)+x_4^2y_1(h_1-h_2)\\
C_2&=2(x_1y_1+x_2y_2+x_3y_3+x_4y_4)+h_1^2+h_2^2
\end{aligned}
$$

We can say that f has the weight (N_1,N_2) when

$$
\begin{aligned}
x_2\frac{\partial f}{\partial x_2}+x_3\frac{\partial f}{\partial x_3}+x_4\frac{\partial f}{\partial x_4}-y_2\frac{\partial f}{\partial y_2}-y_3\frac{\partial f}{\partial y_3}-y_4\frac{\partial f}{\partial y_4}&=N_1f\,,\\
x_1\frac{\partial f}{\partial x_1}-x_2\frac{\partial f}{\partial x_2}+x_4\frac{\partial f}{\partial x_4}-y_1\frac{\partial f}{\partial y_1}+y_2\frac{\partial f}{\partial y_2}-y_4\frac{\partial f}{\partial y_4}&=N_2f\,.
\end{aligned}
$$

It is easy to see that the functions X_1 and X_4 have the weight $(1,1)$, the function X_2 has the weight $(2,0)$, X_3 has the weight $(1,0)$, X_5 has the weight $(2,1)$ and finally C_2, which is an element of the center $U(\mathfrak{g})$, has the weight $(0,0)$.

Since the determinant

$$\frac{D(X_1,X_2,X_3,X_4,X_5,C_2)}{D(x_1,x_2,x_4,y_2,y_3,y_4)} = -4x_2x_4^3(2x_2x_4-x_3^2)^2 \neq 0,$$

the functions X_k, $k=1,\ldots,5$ and C_2 are functionally independent. By direct calculation we can show that the relation

$$X_5^2 + X_4^2X_2 - 2X_1X_4X_3^2 - C_2X_1X_4X_2 + C_4X_1^2X_2 = 0 \tag{25.5}$$

is valid, where

$$\begin{aligned} C_4 = {} & \tfrac{1}{4}\left(h_1^2-h_2^2-2x_1y_1+2x_3y_3\right)^2 + x_2y_2(h_1+h_2)^2 + x_4y_4(h_1-h_2)^2 - \\ & -2(x_1x_2y_3+x_3y_1y_2)(h_1+h_2) - 2(x_1x_3y_4+x_4y_1y_3)(h_1-h_2)^2 + \\ & +2(x_1^2x_2y_4+x_4y_1^2y_3-x_2x_4y_3^2-x_3^2y_2y_4+2x_1x_3y_1y_3+2x_2x_4y_2y_4) \end{aligned}$$

is an element of the center of $U(\mathfrak{g})$, and has the weight $(0,0)$.

Now we show that the vector space of all highest weight polynomials, i.e. the polynomials, which fulfill (25.4), has the basis

$$C_2^{k_1}C_4^{k_2}X_1^{n_1}X_2^{n_2}X_3^{n_3}X_4^{n_4}X_5^{n_5}, \tag{25.6}$$

where $k_1, k_2, n_r \geq 0$ for $r=1,\ldots,4$ and $n_5 = 0, 1$.

It is clear that any linear combination of the polynomials (25.6) are solutions of (25.4), thus the are highest weight polynomials.

The linear independence of the functions (25.6) follows from the fact, that C_4 can be expressed by using the condition (25.5) as a function in variables C_2 and X_k, $k=1,\ldots,5$, which is quadratic in variable X_5, and above proved of functional independency of these functions.

Now we will show that the linear combinations of polynomials of the form (25.6) generate the space of all highest weight polynomials. It is clear that the polynomial (25.6) has the weight (N_1,N_2), where

$$N_1 = n_1+2n_2+n_3+n_4+2n_5, \quad N_2 = n_1+n_4+n_5 \tag{25.7}$$

and is of the order

$$M = 2k_1+4k_2+n_1+2n_2+2n_3+3n_4+4n_5. \tag{25.8}$$

By using the differential operators (25.1) corresponding to generators of the Lie algebra so(5) it is possible from any polynomial with the highest weight (N_1,N_2) of an order of M to generate

$$d_{(N_1,N_2)} = \tfrac{1}{6}\left(N_1+N_2+2\right)\left(N_1-N_2+1\right)\left(2N_1+3\right)\left(2N_2+1\right)$$

of the linearly independent polynomials of an order of M. By direct calculation, which is cumbersome, it is possible to show that for any $M \in \mathbb{N}$ we have

$$\sum_{\mathcal{M}_M} d_{(N_1,N_2)} = \binom{M+9}{M},$$

where the summation is with respect to any $k_1, k_2, n_r \geq 0, r = 1, \ldots, 4$, and $n_5 = 0, 1$, for which (25.8) is valid, and N_1, N_2 are equal to (25.7). Now, because the vector space off all polynomials of an order of M in 10 variables has dimension

$$D_M = \binom{M+9}{M},$$

the polynomials of the form (25.6) generate the vectors space of all highest weight polynomials.

To go back to algebra $U(\mathfrak{g})$, we use the canonical isomorphism from $\mathrm{Sym}(\mathfrak{g})$ to $U(\mathfrak{g})$. First, we define the elements

$$\begin{aligned}
\mathbf{X}_1 &= \mathbf{E}_1\\
\mathbf{X}_2 &= \mathrm{sym}\left(2\mathbf{E}_2\mathbf{E}_4 - \mathbf{E}_3^2\right)\\
\mathbf{X}_3 &= \mathrm{sym}\left(\mathbf{E}_1\mathbf{E}_2 + \mathbf{E}_3\mathbf{H}_2 - \mathbf{E}_4\mathbf{F}_1\right)\\
\mathbf{X}_4 &= \mathrm{sym}\Big(\mathbf{E}_1^2\mathbf{E}_2 - \mathbf{E}_1\mathbf{E}_3\left(\mathbf{H}_1-\mathbf{H}_2\right) + \tfrac{1}{2}\mathbf{E}_4\left(\mathbf{H}_1-\mathbf{H}_2\right)^2 + \left(2\mathbf{E}_2\mathbf{E}_4 - \mathbf{E}_3^3\right)\mathbf{F}_2\Big)\\
\mathbf{X}_5 &= \mathrm{sym}\Big(\mathbf{E}_1^2\mathbf{E}_2\mathbf{E}_3 + \mathbf{E}_1\mathbf{E}_2\mathbf{E}_4\left(\mathbf{H}_1+\mathbf{H}_2\right) - \mathbf{E}_1\mathbf{E}_3^2\left(\mathbf{H}_1-\mathbf{H}_2\right) -\\
&\quad -\mathbf{E}_3\mathbf{E}_4\mathbf{H}_2\left(\mathbf{H}_1-\mathbf{H}_2\right) - \mathbf{E}_1\mathbf{E}_3\mathbf{E}_4\mathbf{F}_1 +\\
&\quad +\left(2\mathbf{E}_2\mathbf{E}_4 - \mathbf{E}_3^2\right)\left(\mathbf{E}_3\mathbf{F}_2 + \mathbf{E}_4\mathbf{F}_3\right) + \mathbf{E}_4^2\mathbf{F}_1\left(\mathbf{H}_1-\mathbf{H}_2\right)\Big)\\
\mathbf{C}_2 &= \mathrm{sym}\Big(2\left(\mathbf{E}_1\mathbf{F}_1 + \mathbf{E}_2\mathbf{F}_2 + \mathbf{E}_3\mathbf{F}_3 + \mathbf{E}_4\mathbf{F}_4\right) + \mathbf{H}_1^2 + \mathbf{H}_2^2\Big)\\
\mathbf{C}_4 &= \mathrm{sym}\Big(\tfrac{1}{4}\left(\mathbf{H}_1^2 - \mathbf{H}_2^2 - 2\mathbf{E}_1\mathbf{F}_1 + 2\mathbf{E}_3\mathbf{F}_3\right)^2 +\\
&\quad +\mathbf{E}_2\mathbf{F}_2\left(\mathbf{H}_1+\mathbf{H}_2\right)^2 + \mathbf{E}_4\mathbf{F}_4\left(\mathbf{H}_1-\mathbf{H}_2\right)^2 -\\
&\quad -2\left(\mathbf{E}_1\mathbf{E}_2\mathbf{F}_3 + \mathbf{E}_3\mathbf{F}_1\mathbf{F}_2\right)\left(\mathbf{H}_1+\mathbf{H}_2\right) - 2\left(\mathbf{E}_1\mathbf{E}_3\mathbf{F}_4 + \mathbf{E}_4\mathbf{F}_1\mathbf{F}_3\right)\left(\mathbf{H}_1-\mathbf{H}_2\right)^2 +\\
&\quad +2\left(\mathbf{E}_1^2\mathbf{E}_2\mathbf{F}_4 + \mathbf{E}_4\mathbf{F}_1^2\mathbf{F}_3 - \mathbf{E}_2\mathbf{E}_4\mathbf{F}_3^2 - \mathbf{E}_3^2\mathbf{F}_2\mathbf{F}_4 + 2\mathbf{E}_1\mathbf{E}_3\mathbf{F}_1\mathbf{F}_3 + 2\mathbf{E}_2\mathbf{E}_4\mathbf{F}_2\mathbf{F}_4\right)\Big).
\end{aligned}$$

Since the set of polynomials (25.6) is a basis in the space of the highest weight polynomials, the set

$$\mathrm{sym}\Big(\mathbf{C}_2^{k_1}\mathbf{C}_4^{k_2}\mathbf{X}_1^{n_1}\mathbf{X}_2^{n_2}\mathbf{X}_3^{n_3}\mathbf{X}_4^{n_4}\mathbf{X}_5^{n_5}\Big) \in U(\mathfrak{g}) \tag{25.9}$$

forms the basis in the space of the highest weight vectors.

We denote $U_n(\mathfrak{g})$ as a subspace of $U(\mathfrak{g})$ with elements of an order less or equal to n. It is a well know fact that for n_1 and n_2 one has

$$\left[U_{n_1}(\mathfrak{g}), U_{n_2}(\mathfrak{g})\right] \subset U_{n_1+n_2-1}(\mathfrak{g}) .$$

For this reason as a basis in the vector space of the height weight vector we can use instead of (25.9) the set of elements

$$\mathbf{C}_2^{k_1}\mathbf{C}_4^{k_2}\mathbf{X}_1^{n_1}\mathbf{X}_2^{n_2}\mathbf{X}_3^{n_3}\mathbf{X}_4^{n_4}\mathbf{X}_5^{n_5} \in U(\mathfrak{g}),$$

where $k_1, k_2, n_r \geq 0$ for $r = 1, 2, 3, 4$ and $n_5 = 0, 1$.

25.4 Conclusion

We have studied the structure of enveloping algebra $U\big(\mathrm{so}(5)\big)$ as the vector space of adjoint representation of the Lie algebra $\mathrm{so}(5)$. We explicitly decomposed this representation into the irreducible components and found the highest weight vector in any such components.

Our result can be useful for further study of tensor products of the representations of the Lie algebras and for study of ideals in their enveloping algebras. The details of such study for $\mathrm{sl}(2)$ case see in [2].

The method described can be used for other simple Lie algebras. We really obtained some results for the Lie algebra $\mathrm{sl}(4)$, but the general solution for $\mathrm{sl}(n)$, $\mathrm{so}(n)$, etc., is still an open problem.

Acknowledgements The authors are grateful to M. Havlíček and S. Pošta for valuable and useful discussions. This work was Partially supported by GACR 201/05/0857.

References

1. Dixmier J.: Algebres enveloppantes. Gauthier-Vllars, Paris (1974)
2. Kirillov A.A.: Elementy teorii predstavlenij. Nauka, Moscow (1972), in russian
3. Flath E.: J. Math. Phys. **30**, 1076–1077 (1990)
4. Konstant B.: Am. J. Math. Phys. **85**, 327–404 (1963)

Index

Printing: Krips bv, Meppel, The Netherlands
Binding: Stürtz, Würzburg, Germany